Physical Principles of the Plant Biosystem

Physical Principles of the Plant Biosystem

George E. Merva
Michigan State University
East Lansing, Michigan

The cover art is an enhanced image of root hairs from a still-frame videotape provided by:

Professor Alvin J. M. Smucker
Crop and Soil Science
Michigan State University
East Lansing, Michigan

For information, contact:
ASAE
2950 Niles Road
St. Joseph, Michigan 49085-9659 USA
Phone: 616.429.0300 Fax: 616.429.3852 E-mail: hq@asae.org

ASAE Textbook Number 9

Pamela DeVore-Hansen, Editor
William Thompson, Cover Design
Information Publishing Group

LCCN 94-79226 ISBN 0-929355-57-1

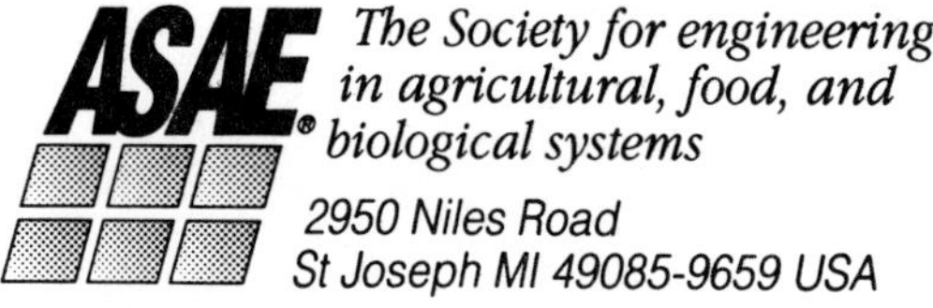

Contents

Preface

The biosystem is comprised of a complex interaction between the soil, the atmosphere, and the plants and animals which live in it. A chance alteration of one element may yield both desirable and undesirable consequences. Minimizing the undesirable while enhancing the desired end result is the principle aim of the biosystem engineer. This text is an attempt to bridge existing gap between traditional engineering sciences by tying together important aspects of energy transfers with related areas pertinent to the green plant environment. This material lays the engineering groundwork to tie physical and biological concepts together in the context of the biosystem. While the text is primarily intended for junior and senior engineering students, beginning graduate students may also find the information useful.

Only those elements of plant anatomy and physiology which are emendable to modification by engineering means have been emphasized. Modification of the green plant itself by genetic manipulation is not treated. This subject is biochemical and should be treated by experts in the area.

Much of the text first appeared published under a different title. When the original book went out of print, several institutions continued using portions of the material and I received a number of requests to republish. In doing so, I have included material on finite differences which I have found to be particularly suited to the analysis and solution of differential equations through the use of spreadsheets on personal computers. I have found that students view partial differential equations in a new light after using spreadsheet to solve problems because they are able to better grasp the importance of boundary and initial conditions to the solution of the problem. In addition, the material on chemical energy has been rewritten and has been separated from the properties of water which now occupies a separate chapter. Finally, the treatment of psychometrics has been broadened and emphasizes properties of water in the atmosphere between temperatures of –10° C to +50° C.

Most portions of the text assume familiarity with college level mathematics and physics. No attempt is made to review those aspects of the basic sciences that students beyond the second year of college should know.

The book is dedicated to my wife, Betty, and my son, Michael, who did not complain through the many evenings I spent at my computer.

Acknowledgments

The author gratefully acknowledges the suggestions of his colleagues Merrill Pack, Elmer Robinson, Malcolm Campbell, Walter Gardner, Gaylon Campbell, and Fred Koehler who reviewed individual portions of the manuscript as it was in preparation. Any errors or omissions which have found their way into the material are entirely the fault of the author.

Special thanks are also due to Glenn Schwab, Edward Hiler, Bahari Vasnavi, and John Blaisdell whose reviews of earlier drafts of the manuscript helped direct the course of the present work, and to Carl W. Hall who encouraged the author to become involved in this area of study.

Thanks to the many students who have used this text.

Chapter 1

Biosystems Engineering as an Engineering Science

A brief historical development of engineering is useful to help the student better understand bioengineering as it applies to the environment of the green plant. The reader will find the historical background helpful because it provides a foundation for this new and challenging topic as an engineering science.

Engineering is an ancient science, one of the oldest of all sciences. Engineering competes with medicine for this ranking. Engineering and medicine have much in common. Medicine arose out of humankind's attempt to solve problems related to illnesses. Engineering, grew from humankind's attempts to solve problems related to existence. Medicine developed largely by trial and error, and the body of knowledge to which it gave rise was tested, treasured, and preserved of its benefit to humans. Engineering developed because people needed practical solutions to problems that affected their ability to function as a society. While engineering did not immediately create written records as did medicine, engineers left physical evidence, much of which predates the beginning of written history.

The first engineer may have used a lever to move a weighty load, or placed a log over a stream to form a bridge. The engineer who had the most impact on society and commerce was undoubtedly the person or persons who discovered the wheel. This is engineering at its finest. The wheel arose out of a need to move a load while conserving physical effort. The discovery/invention may have been serendipitous, or it may have been the result of a complex thought process culminating in a workable solution. Regardless of the cause and however crude the first wheel, its importance was quickly grasped. In the hundreds of generations since its discovery, the shape has become the focal point of engineering. The wheel's shape has been modified by the addition of teeth to serve as an energy transmission device. With this engineering addition, it is the basis of most modern machinery.

Many civilizations developed independently during society's genesis. However, engineering was always a valuable part of society's base of knowledge. The Great Wall of China, a hydrological as well as an engineering feat, could not have been built without highly developed engineering. The Great Wall was constructed over mountain ridges along a hydrological divide. This eliminated the need to pierce the wall with culverts or bridges which could have been used by enemies to gain access. Thus, the Great Wall demonstrates the use of both military strategy and civil engineering strategy.

The pyramids of Egypt are another engineering marvel. Although the technique used to move the stones used to construct the pyramids has not been determined, significant engineering was an indispensable part of their construction.

Historical Development of Engineering

Since about the 1700s engineering has developed from both a disciplinary and an occupational approach. The disciplinary approach led to the application of knowledge derived from basic sciences and can be applied to problem solving regardless of the occupation in which the problem occurred. A good example of disciplinary development is mechanical or electrical engineering which is based on physics. In mechanical engineering, an understanding of basic physical phenomena is applied to solve problems. The codification and understanding of the phenomena are the domain of the physicist, while the application of the phenomena is done by the engineer. Similarly, chemical engineering developed because chemists applied their technical knowledge to solve everyday problems. In both cases, the application of knowledge differentiates the engineer from the theoretical scientist who is more inclined to study and explain a phenomena rather than apply it.

The occupational development of engineering arose because certain problems required the knowledge of skilled practitioners in specific occupations. Civil engineering developed from military engineering because roads and bridges which were needed for war were also needed for civilian commerce. Industrial engineering was developed to solve problems peculiar to manufacturing. Aeronautical engineering stemmed from problems encountered in air transportation and manufacturing. In like manner, agricultural engineering which stemmed from problems peculiar to the occupation of farming required expertise that could not be adequately addressed unless the engineering practitioner was intimately familiar with agriculture.

Subject Matter of Engineering

Certain subject matter is common to most engineering disciplines and is to be found in every course of study. These subjects include mathematics, physics, chemistry, communications, and broadening subjects such as the humanities. However, the occupational subject matter makes a particular engineering discipline unique. For instance, civil engineers, in addition to the common subjects, may study areas peculiar to a given segment of their occupation such as soils, hydrology, economics, traffic management, etc. Industrial engineers study statistics, psychology, metallurgy, and so forth.

The diversity of its occupational area makes agricultural engineering especially complex. Specialized subjects of agricultural engineering include areas found in no other engineering discipline. In particular, biology, animal and plant physiology, nutrition, and bacteriology as well as soil science, soil physics, hydrology, etc., are important for the agricultural engineer.

Development of Agricultural Engineering

Traditional Agricultural Engineering

The classical definition of agricultural engineering as given in the by-laws of the American Society of Agricultural Engineering is:

> *Agricultural Engineering is the application of principles of science and engineering to the production of feed and fiber for the benefit of mankind.*

Historically, agricultural engineering within the Society has been divided into four areas: Power and Machinery deals with machines unique to agriculture where the machine function affects or is affected by the biology of the crop; Structures and Environment treats the design of buildings and other structures and incorporates the specific requirements of animals and harvested crops; Electrical Power and Processing is devoted to the application of electrical power to agriculture and emphasizes the handling of biological products; and Soil and Water treats the management of water as related to the soil environmental needs of green plants and the prevention or alleviation of pollution.

For years, agricultural engineering students had rural backgrounds. In addition to the common areas of mathematics, physics, chemistry, etc., a familiarity with agriculture was supplemented with basics in biology design emphasizing agricultural problems. However, interest in preserving the environment and its natural resources has changed the classical picture. Today's agricultural engineer is as apt to have come from an urban background as from a rural one. The preservation of natural resources and the environment requires much of the same background as does traditional agricultural engineering.

Contemporary Agricultural Engineering

The decades from the 1950s through the 1980s saw significant changes in agriculture. Agriculture changed rapidly in the United States. While, in the past, one farmer might provide sufficient food for himself and perhaps 10 people, today one man feeds 80 people or more and the number continues to increase. This has resulted in altering engineering needs. Instead of traditional field machinery, today machines are needed to prepare, and process the product once it leaves the fields. With fewer individuals growing food for more people, methods of preparing, processing, and storing agricultural products becomes more critical. Environmental concerns require more engineers trained to recognize and implement strategies to protect both the population and the environment. Finally, the advent of genetic manipulation to alter crop structure and induce resistance to pathogens has introduced requirements into the educational track.

These changes, along with rising educational costs, have placed demands on training that are altering educational priorities. Increasing the number of courses which an engineering student must take has not proved to be a satisfactory solution. A five-year curricula may be the answer but few such curricula exist. Basic science requirements and mathematics have increased. Additionally, improved communications are necessary to speed technology transfer. These changes, along with fewer students with farming backgrounds, have given rise to a broadened definition of the role engineering plays in food production. Today it would be appropriate to call the agricultural engineer a biosystems engineer, a title defined more clearly as:

> *Biosystems engineering is the application of scientific principles of biology and engineering to the sensing, control, and modification of the environment of a biological system to complement its natural physiological, physical, chemical, and/or biochemical response and enhance its useful end products.*

The changes indicated above must be reflected in a realistic fashion in educating today's contemporary engineer. Prior biological education may have required only an agricultural background with the biological knowledge obtained by close association with the environment. The contemporary approach must stress basic facts presented in a predictive mode. The above definition implies the control of temperature, humidity, airflow, and atmospheric components to stabilize and complement the metabolic response of the biological product. In this approach, soil, plant, atmospheric, and water relations are controlled to enhance productivity and ensure increased useful output. This can be accomplished only by synthesizing and combining traditional areas with new tools presented in a predictive mode using an engineering science approach.

The material in this text is an attempt to contribute to this effort by combining and synthesizing information obtained from courses such as basic biology and mathematics, extracting elements that can be controlled by engineering means, and presenting them in such a fashion that they can be used analytically. In this mode, situations can be analyzed and designs created which are economically feasible and profitable, environmentally sound, sociologically practical, and agriculturally fulfilling.

Bioengineering Principles

The last sentence in the previous section describes the goal of *Physical Principles of the Plant Biosystems*. In this section, an overview of the subject matter and the intent of the book is presented. A cursory scan of the contents of this section will acquaint the reader with the aims and level of difficulty of the material. The subject matter assumes a grasp of basic biology as well as basic science common to the first two years of engineering and builds upon them. Mathematics through a first course in partial differential equations would be helpful although so a knowledge of ordinary differential equations is sufficient.

Course Content

Mathematics is a basic tool. Since most engineering problems deal with phenomena involving friction in some way or the other, many of the useful models are nonlinear in this text. Nonlinear phenomena are difficult to model in a straightforward closed form. For this reason, chapter 2 deals with mathematical modeling and introduces finite difference methods of analyzing differential equations. While the subject is not covered in as much detail as it would be in a complete text, there is sufficient information that students can grasp the fundamentals of finite difference modeling and apply the concept to some interesting and fundamental problems. The information gained in this chapter will enable the student to read, comprehend, and utilize a more detailed presentation such as might be obtained from a text or a formal course.

Chapter 3 examines water in the green plant environment in detail and introduces concept of the chemical potential of water. The thermodynamics of water in the environment are presented with many examples that will help the student grasp the various forms and energy transformations for which water is responsible. Chapter 4 continues the study of water and concentrates on the development of water potential. Chapter 4 is essential to the understanding of the movement and role of water within the green plant. It lays the foundation for an understanding of the state and movement of water in soil.

Chapter 5 is a review of basic biology as applied to green plants. It does not include taxonomy or detailed descriptions of items of little significance from a predictive engineering point of view. However, it does emphasize the anatomical parts of green plants that are important in the reproductive, growth, and productive aspects of agriculture. Also included in this chapter is an overview of the basic concepts of plant physiology along with a discussion of enzymatic reactions that are modifiable using environmental control techniques. Some ecological aspects of plants are discussed here also, specifically as they may be affected by engineering decisions.

Chapter 6 deals with the elements of green plant physiology. Detailed descriptions of phenomena such as the Kreb's cycle are not covered since there is little that the engineer can do to alter a single portion of the cycle. However, the distinction between photosynthesis and respiration, between digestion and assimilation, and an overall concept of the processes involved in metabolism are presented. This chapter should enable the engineer to discuss physiological aspects of the green plant with trained professionals and be adequately conversant with the jargon and nomenclature of the subject to determine what effects environmental modifications might have on the response of the plant.

Chapter 7 presents some concepts of soil science and soil physics that are important to the plant environment. Soil is not considered here as a construction material, but rather as to how its structure relates to the green plant. The effect that soil texture has on the retention of water, minerals, and organic matter, as well as how the presence of organic matter affects the above entities is considered. The earlier portions of chapter 7 contain material which might be found in a basic soil science course and may be skipped by students who have taken such a course. Latter sections of chapter 7 should be included in an engineering science course dealing with the plant biosystem.

Chapter 8 discusses radiant energy. Basic concepts of radiation are presented with particular attention to certain aspects such as ozone and chlorofluorocarbon molecules and their interaction with ultraviolet radiation. An introduction to the energy balance of the plant environment occurs in this chapter in sufficient detail to enable the student to appreciate the importance of radiation and the energy source for green plants.

Chapter 9 is devoted to the development of models to analyze thermal phenomena. Here, temperatures within the biosphere and in the soil are modeled and their effect on the phenomenon of green plant growth is considered. Temperatures in stored biological products and the effect of temperatures on enzymatic phenomena which occur in stored biological products are discussed as well. Finally, the application of temperature as a control to alter predictable enzymatic reactions and the effect of induced temperatures in stored biological products on enzymes and metabolism is considered as is the overall concern of temperature on plant growth.

Chapter 10 continues the study of water movement with an analysis of the saturated flow of water in a porous media such as soil. The discussion of saturated flow is applied to the removal of excess water from the root environment and this concept is extended to the point where drain placement for water removal can be analyzed. Also discussed in chapter 10 is the unsaturated flow in the soil toward plant roots and the effects of soil texture on the water holding and unsaturated transmission properties of soil water.

Student Expectations

Obviously, as in any branch of engineering, the application of scientific principles to the solution of practical problems cannot be done without practice. Practice in this case means applying scientific principles to situations in which there is a problem that can be solved using the principles. However, it is hoped that the student will learn to apply the principles in an organized manner.

The importance of using dimensions throughout the analysis and solution of problems is the first step in developing good engineering organization. A number is meaningless unless its application is specified. This requires that every number be accompanied by a dimension or some other identifying parameter. Throughout this text complete numerical communications will be used in the analytical process. The student is encouraged to learn completeness in numerical communications and use this technique to aid in the analytical process. This applies to homework turned in for grading, as well as to intermediate steps in the analytical process. There is no substitute for accuracy.

For practice, examples with solutions are sprinkled liberally throughout the text. These are accompanied by problems in the exercises at the end of each chapter. Often these are realistic problems requiring unique approaches for which no answers are given. These problems are more in the manner of a situation that the practicing engineer will encounter. The student is encouraged to use any approach possible to analyze these problems, drawing on information not only from the text, but from physics, chemistry, mathematics, computer technology, etc. A “no holds barred” approach is strongly encouraged.

Chapter 2

Mathematical Modeling in the Biosystem

In any engineering work the use of mathematical modeling is essential. Engineering is a predictive science, and modeling, particularly mathematical modeling, enables an engineer to predict the behavior of a design while keeping expense at a minimum.

Before examining the use of mathematical models, one must know and understand how to express the numerical results of a calculation. These attributes enable the engineer to determine when a computation is sufficiently precise. To interpret data and calculations, and avoid unattainable precision, only those digits that are "real" should be used. This is part of the art of engineering and is accomplished by using significant digits.

Significant digits are simply the number of digits in a numerical expression that indicate how precise the number is. The proper use of significant digits can speed up a computation by limiting the size of the numbers that enter the calculation. This does not alter the accuracy or precision of the answer. An example may serve to illustrate this point.

Example 2.1. Significant digit use
The gas tank on a car was filled and the car was driven 29 km. The tank required 3.7 L of gasoline to refill. What was the fuel efficiency expressed in L/100 km?

Solution

$$\frac{37\ \text{L}}{29\ \text{km} \times 100\ \text{km}} \times 100\ \text{km} = 12.758\,621\ \frac{\text{L}}{100\ \text{km}} \qquad 2.1(\text{a})$$

During the trip, 3.77 L of gasoline was used. Because of the slope of the pavement near the pump, the tank was not completely refilled at the end of the trip. Instead of 29 km, the actual distance was 28.98 km (a slight error resulting from worn tires on the car). Thus, the actual efficiency should have been:

$$\frac{3.77\ \text{L}}{28.98\ \text{km} \times 100\ \text{km}} \times 100\ \text{km} = 13.008\,972\ \frac{\text{L}}{100\ \text{km}} \qquad 2.1(\text{b})$$

Given the errors which were inherent in the observations leading to the calculation, it would have been more precise to say that the efficiency was about 13 L/100 km and round the number to 13 because the numbers used in the calculation had only two significant digits.

The uncertainty in the above calculations could have been avoided, along with the need to write a string of unnecessary digits, by paying attention to the rules of significant digits:

1. The number of significant digits in a number include all digits except those zeros at either end that serve only to fix the position of the decimal point.

1,203 has 4 significant digits
1,200 has 2 significant digits

2. In a decimal expression, zeros standing alone to the right of the decimal point should be considered as significant.

12 has 2 significant digits
12.500 has 5 significant digits
12.0 has 3 significant digits

3. In a numerical expression written with a power of 10, the zeros included in the number should be considered as significant.

16×10^{-3} has 2 significant digits
1.60×10^{-2} has 3 significant digits
1.6×10^{3} has 2 significant digits

4. In a two digit number that ends in a zero, there are two significant digits. If there are three digits and the number ends in a zero, assume that only two digits are significant unless it is obvious that more significant digits are present.

10 is assumed to have two significant digits.
100 is assumed to have only two significant digits.

The rules for manipulating significant digits are:

Rule 1. In simple multiplication and/or division, the answer can have no more significant digits than the least number of significant digits in the multiplier, multiplicand, divisor, or dividend.

Rule 2. In addition or subtraction, the answer can have no more significant digits, or decimal places, than are contained in the least precise number.

Rule 3. When performing a mathematical operation by hand calculation, numbers other than the least precise number may be rounded to one more significant digit than the least precise number. This eliminates the necessity of carrying numbers with many decimal places, since these will be lost in the final rounding process anyway.

Rule 4. When using a calculator or computer, retain as many significant digits as appear in the numbers entering into the calculation. Round only the final answer.

Rule 5. For transcendental functions, use the most precise argument to obtain a working value, do not round as indicated in rule 3 when using an exponential or other transcendentals.

Rule 6. When the number to be dropped is 5 or less, round as follows:

a. If the digit to the right of 5 is zero, or if there are no digits to the right of 5, do not round up.

To 2 digits, 1.550 rounds to 1.5
To 3 digits, 1.8354 rounds to 1.84
To 2 digits, 1.55 rounds to 1.5

b. If the digit to the right of 5 is greater than zero, round up.

To round 16.51 to 2 digits, round to 16
To round 16.557 to 3 digits, rounds to 16.6
To round 16.551 to 3 digits, round to 16.6

Example 2.2. Multiplying and dividing significant digits
Using the rules of operating with significant digits, calculate the fuel efficiency of example 2.1.

Solution
3.7 L of gas were used to travel 29 km. Since only 2 digits are significant:

$$\frac{3.7 \text{ L}}{100 \text{ km} \times 29 \text{ km}} \times 100 \text{ km} = 12.75 \frac{\text{L}}{100 \text{ km}} = 13 \frac{\text{L}}{100 \text{ km}}$$

The answer given in example 2.2 is as precise as can be obtained given the information available.

Example 2.3. Addition of numbers
Sum 1.023 + 1.603 + 0.005 66 + 18.0 + 491 = 511.6316 = 512

Solution
The least precise number in the above string has 3 significant digits. The answer should be rounded to 512.

Mathematical Modeling Techniques

The mathematical techniques used in the biosystem may not be the same as those used in the biological sciences. In a basic science such as biology, the analysis is often statistical. Conclusions may be based on the results of many observations, because biology is a deductive science. In engineering, the statistical approach is usually shunned, because engineering is a predictive science. An engineering activity usually produces a design or a numerical value. Designs are a final product rather than a means to an end. Statistically, complete engineering designs are not meant to fail. Failure is to be avoided at all cost and can be avoided by introducing a factor of safety into the calculation. The factor of safety places the probability of failure extremely low. Every attempt is made to ensure that the design is adequate. The factor of safety may not always be sufficient unless the designer is thoroughly familiar with the limits to which the design and materials being used will be stressed.

In engineering work, modeling is often mechanistic. The mathematical models used are linear, and energy degradation due to friction or heat is disregarded in the calculations. Engineering design performed in this manner is as much art as science. The agreement between predicted and observed values is a reflection of the skill of the engineer in interpreting the model used in the design process. An unskilled engineer can be led to serious error, as is demonstrated by the example below.

Example 2.4. Design without considering friction
An engineer has been assigned to design a nozzle to irrigate lawns. He plans to have the nozzle discharge 0.189 L/s so that the velocity of the water leaving the sprinkler at an angle of 30° with the ground surface is 23.9 m/s. What will be the wetted radius of the sprinkler if frictional losses of the

water droplets with the air are neglected? Assume a reference frame with x positive horizontally and y positive vertically so $g = -9.81$ m/s^2.

Solution

The vertical velocity of the water leaving the sprinkler is:

$$v_v = 23.9\ \frac{m}{s}\ \sin\left(\frac{\pi}{6}\right) = 11.9\ \frac{m}{s} \qquad 2.2(a)$$

The horizontal velocity is:

$$v_h = 23.9\ \frac{m}{s}\ \cos\left(\frac{\pi}{6}\right) = 20.7\ \frac{m}{s} \qquad 2.2(b)$$

The total time required for the vertical motion to rise and return to the ground can be computed from the equation describing the motion of a body under the influence of gravity. Assume v_v denotes the vertical velocity at $t = 0$.

$$m\left(\frac{d^2y}{dx^2}\right) = mg \qquad 2.2(c)$$

By integrating once, using the boundary conditions, and setting dy/dt equal to zero, one obtains the time required for the body to rise to maximum height and in the absence of friction the total time is twice the time to rise, or:

$$t = 2\ \frac{v_v}{g} \qquad 2.2(d)$$

and the horizontal travel will then be found by multiplying the horizontal velocity by the total time, or:

$$x = v_h \times t = v_h \times \left(\frac{2v_v}{g}\right) \qquad 2.2(e)$$

or

$$x = \frac{2 \times 11.9 \times \frac{m}{s} \times 20.7\ \frac{m}{s}}{9.812\ \frac{m}{s^2}} = 50.2\ m \qquad 2.2(f)$$

Note: When the sprinkler was tested, the actual wetted radius was 12.5 m. The observed difference was due to the drag caused by air friction acting on the water droplets. A correct analysis was conducted and yielded the results shown in figure 2.1 in which the path of a 3 mm water droplet acted on by air friction is compared to the path of the droplet described above without friction.

From the results of example 2.4, a mathematical model that more accurately described the phenomenon was needed. Such models are difficult to use because they are non-linear and are more complex. In addition they usually require careful selection of the model parameters. While the calculations involved are not straightforward, the use of a more accurate model does remove much of the guesswork from the design. Unfortunately, such models must be derived from basic principles that require a knowledge of the circumstances and conditions related to the situations for which the models are to be applied.

The development of models often requires the solution of differential equations. Often they will be ordinary differential equations in which the differential involves only one parameter, usually time. This is the case involving the motion of the water droplet depicted in figure 2.1. Often, however, differential equations involve two or more parameters. These require more care in their formulation and solution.

Regardless of the type of differential equation, a closed form or exact solution is not usually possible and other methods must be applied to determine the solution. Fortunately, in most engineering situations precise solutions to many significant digits are not always required. For most applications, if a solution within 5 to 10% of the correct value is obtained, a safety factor will ensure satisfactory performance of the design. Thus, an approximate solution of the equation is satisfactory. With the advent of computers and new solution methods, it is relatively easy to solve even complicated models.

We will use one such technique, the finite difference method, and demonstrate its application in this chapter. Other approximate techniques for the solution of differential equations, such as the finite element method and boundary integral methods, exist. The finite difference method is easy to visualize and can be very accurate depending only on the degree of precision desired and the computing time available. It forms a solid basis upon which other methods can be learned and utilized.

In the finite difference method, a mathematical expression modeling a phenomena is defined at specific nodes (points). The expression may be defined either on the solution line for an ordinary differential equation or on the solution plane or surface for a partial differential equation. The model obtained using the finite difference method is precise at the nodes, but is only approximate between nodes. The closer together the nodes, the more accurate the result. The computer time required to obtain an answer, however, increases exponentially increasing numbers of nodes.

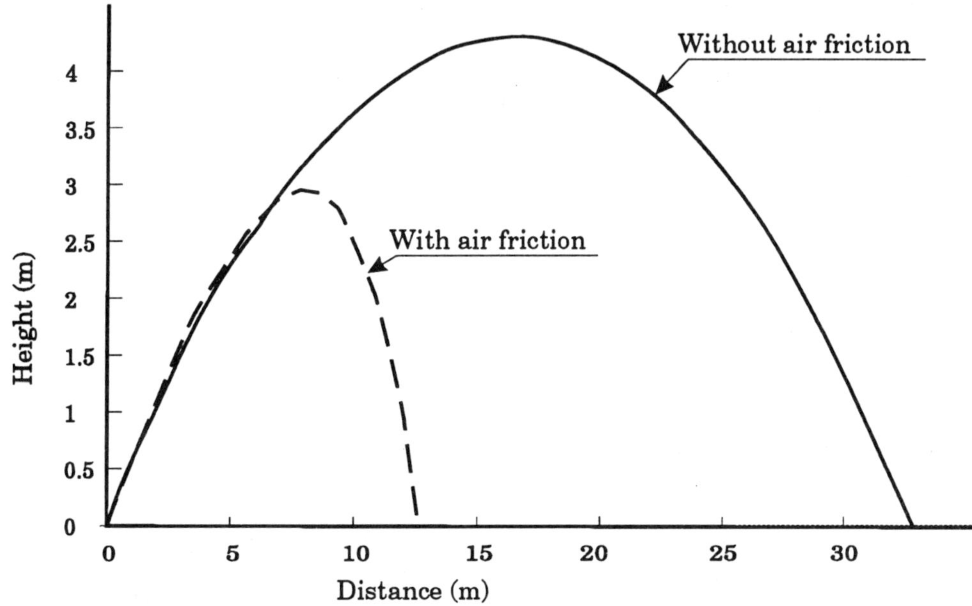

Figure 2.1. The path of a water droplet from a sprinkler irrigation nozzle must be calculated by including the effect of air friction, if the prediction is to be accurate.

Development of the Finite Difference Approach

The finite difference method, based on the Taylor Series Expansion, should be familiar to engineering students from basic calculus. Using the Taylor Series, a mathematical function is expanded at some point within the range of the independent variable. The result is a power series representation of the function that accurately represents the function in a region at and near the point of expansion. The size of the region depends on the degree of the expansion, i.e., the number of terms in the expanded expression. We will review the Taylor Series by first expanding a function. This will demonstrate the accuracy of the result near the point of expansion.

Series Expansion of a Function of One Variable

The Taylor Series applied to some function f(x) which is continuous and has continuous derivatives is:

$$f(x) = \left[f(x)\Big|_{x=a} \frac{(x-a)^0}{0!} \right] + \left[\frac{df(x)}{dx}\Big|_{x=a} \frac{(x-a)^1}{1!} \right] + \left[\frac{d^2 f(x)}{dx^2}\Big|_{x=0} \frac{(x-a)^2}{2!} \right] + \left[\frac{d^3 f(x)}{dx^3}\Big|_{x=a} \frac{(x-a)^3}{3!} \right] + \ldots + R_{err} \qquad 2.3$$

where the function f(x) is expanded at a point x = a in the range of the independent variable. The zero factorial, 0!, is evaluated as unity (1), as is anything raised to the *0th* power. On the right side of the equality, the function and each of its derivatives is evaluated at point a and multiplied by the appropriate power of the difference between x and point a. The result is then divided by the factorial of the power. We illustrate the application of equation 2.1 in example 2.5 by expanding a function about a point within the range of the function.

The remainder term in equation 2.1 is included to contain any errors that have accumulated because not all of the differentials were used. In example 2.5, this term has been omitted in the final expression and the equality has been replaced with an approximation. In engineering practice, often even the approximation is omitted. This is possible because the exponent of (x-a) and the factorial term in the denominator diminish the contribution of each additional term. For engineering purposes, if the series is being used to obtain a value very near the point where the expansion was made, the terms beyond some order of differentiation can be neglected. Usually, depending on the level of accuracy desired, the inclusion of the second derivative term is adequate.

Example 2.5. Taylor Series Expansion
Expand the function y = f(x) = exp(−x) in a Taylor Series about the point x = a = 0.

Solution
In accord with equation 2.1, it is necessary to obtain the derivatives of f(x) and evaluate them at the point a = 0. The appropriate values are given in the table below.

The expansion is:

$$f(x) = \left[e^{-0} \frac{(x-0)^0}{0!} \right] - \left[e^{-0} \frac{(x-0)^1}{1!} \right] + \left[e^{-0} \frac{(x-0)^2}{2!} \right] - \left[e^{-0} \frac{(x-0)^3}{3!} \right] + \left[e^{-0} \frac{(x-0)^4}{4!} \right] - \left[e^{-0} \frac{(x-0)^5}{5!} \right] + \left[e^{-0} \frac{(x-0)^6}{6!} \right] + \ldots + R_{err} \qquad 2.4(a)$$

which reduces to:

$$f(x) \approx 1 - \frac{x}{1!} + \frac{x^2}{2!} - \frac{x^3}{3!} + \frac{x^4}{4!} - \frac{x^5}{5!} + \frac{x^6}{6!} \qquad 2.4(b)$$

Rank	Derivative	Value at x = 0
0	e^{-x}	1
1	$-e^{-x}$	-1
2	e^{-x}	1
3	$-e^{-x}$	-1
4	e^{-x}	1
5	$-e^{-x}$	-1
6	e^{-x}	1

Example 2.6. Accuracy of a Taylor Series Expansion
Compare the actual value of $e^{-0.5}$ and e^{-1} with the value predicted by the Taylor Series representation of e^{-x} expanded about $x = 0$ using successively increasing orders of the expansion.

Solution
The correct value of e^{-x} to 9 significant digits is:

$$e^{-1} = 0.367\,879\,441$$

In the table following, the nearer the evaluation is to the point where it was expanded, the better the agreement between the expansion and the actual function. At $x = 0.5$, using only the second derivative, the expansion is valid to one significant digit. At $x = 1$, three derivatives are required for one significant digit accuracy. Note also that if the function is evaluated at zero, i.e., at the point of expansion, the agreement is perfect irrespective of the number of derivatives used. In engineering work, often only one or two derivatives are used to obtain an approximation.

Order of Derivative	Expansion at x = 0.5	Expansion at x = 1.0
0	1.000 000 000	1.000 000 000
1	5.000 000 000	0.000 000 000
2	0.625 000 000	0.500 000 000
3	0.604 166 667	0.333 333 333
4	0.606 510 417	0.375 000 000
5	0.606 532 118	0.366 666 667
6	0.605 630 660	0.368 055 556
7	0.606 530 660	0.367 879 441

The Taylor Series Expansion of a function applies to a small region around which the expansion occurred. The size of the region is determined by the behavior of the function. A function that changes rapidly with a small change in the independent variable will have a very small region of accuracy while one that changes only slightly will have a much larger region.

On the other hand, the adequacy of an expansion also will depend on the accuracy desired. If only two significant digit accuracy is necessary, then a second derivative expansion may be sufficient over a wide range near the point of expansion. On the other hand, if three or more significant digits are desired, the range of accuracy may be very small. This is left for the student to verify as an exercise.

Difference Representation of a Differential

The Taylor Series representation of a function given in equation 2.1 can be used to obtain algebraic approximations for the differentials in a differential equation. By using these approximations and expressing the differential equation at specific points of the independent variable separated by some finite distance, it is possible to obtain estimates of the solution of a differential equation that cannot be solved by existing closed-form techniques. This provides the engineer with a useful tool for modeling.

To develop the expressions, we first select an appropriate function for expansion. For ordinary differentials we will expand about x. Note that for convenience we assume that x is the independent variable and y is the dependent variable. These will change depending on the nature of the equation. The student should be prepared to substitute the notation appropriate to the independent variable.

Although using the variable x in this way may be awkward, it is just a value of the independent variable somewhere in the range of the function. The increment Δ is a distance separating points (nodes) in the range of the independent variable where the function has been approximated by the Taylor Series. At the nodes, the approximations are very accurate while at locations between the nodes the expansion is only approximate.

Expanding the function as indicated up to and including the second derivative yields the expression:

$$f\left(x + \Delta x\right) = f(x) \pm \frac{df(x)}{dx}\frac{\Delta x}{1!} + \frac{d^2 f(x)}{dx^2}\frac{\Delta x^2}{2!} \qquad 2.5$$

If only the positive Δx in equation 2.5 is used, we can write:

$$f\left(x + \Delta x\right) = f(x) + \frac{df(x)}{dx}\Delta x + \frac{d^2 f(x)}{dx^2}\frac{\Delta x^2}{2} \qquad 2.6(a)$$

while using the negative Δx yields:

$$f\left(x - \Delta x\right) = f(x) - \frac{df(x)}{dx} + \frac{d^2 f(x)}{dx^2}\frac{\Delta x^2}{2} \qquad 2.6(b)$$

Additional expressions can be obtained from equation 2.6(a) and 2.6(b). By adding the equations we obtain an approximation for the second derivative:

$$\frac{d^2 f(x)}{dx^2} = \frac{f\left(x + \Delta x\right) - 2f(x) + f\left(x - \Delta x\right)}{\Delta x^2} \qquad 2.7$$

and by subtracting equation 2.6(b) from 2.6(a), an approximation for the first derivative can be obtained.

$$\frac{df(x)}{dx} = \frac{f\left(x + \Delta x\right) - f\left(x - \Delta x\right)}{2\Delta x} \qquad 2.8$$

If the spacing Δx between points at which the function is assumed to be defined is kept small, we can assume the term involving the second derivative, multiplied by the second power of a small number, is negligible and can be ignored. We then obtain two additional expressions for the first derivative. From equation 2.6(a):

$$\frac{df(x)}{dx} = \frac{f(x + \Delta x) - f(x)}{\Delta x} \qquad 2.9$$

and similarly, using equation 2.6(b):

$$\frac{df(x)}{dx} = \frac{f(x) - f(x - \Delta x)}{\Delta x} \qquad 2.10$$

Equation 2.7 is the central difference approximation for the second derivative. Equation 2.8 is the central difference approximation for the first derivative. Equations 2.9 and 2.10 are, respectively, the forward difference and backward difference approximations for the first derivative. The expressions contain the functional representations. With knowledge of the initial and/or boundary conditions, the expressions transform complex differential equations into algebraic expressions from which models of physical phenomena can be obtained.

Obtaining a solution to a differential equation using the above expressions is accomplished by solving repetitively for values of the dependent variable at locations along the axis of the independent variable separated by some distance Δ. The solution "marches" along the independent variable axis and each new value of the function is obtained in terms of the previous known values. The procedure requires only that one use known initial conditions or boundary conditions to begin the solution process.

It is helpful to have a shorthand aid to write out the solution. Since the function is defined only at specific points of the independent variable separated by the distance Δ, it is convenient to rewrite the expressions given by equations 2.7, 2.8, 2.9, and 2.10 as follows.

Central difference form, second derivative:

$$\frac{d^2 f(x)}{dx^2} = y'' = \frac{y_{i+1} - 2y_i + y_{i-1}}{\Delta x^2} \qquad 2.11$$

Central difference form, first derivative:

$$\frac{df(x)}{dx} = y' = \frac{y_{i+1} - y_{i-1}}{2\Delta x} \qquad 2.12$$

Backward difference form, first derivative:

$$\frac{df(x)}{dx} = y' = \frac{y_i - y_{i-1}}{\Delta x} \qquad 2.13$$

Forward difference form, first derivative:

$$\frac{df(x)}{dx} = y' = \frac{y_{i+1} - y_i}{\Delta x} \qquad 2.14$$

We illustrate the application of the above expressions to the solution of an ordinary differential equation using the following example. The independent variable in this example is time rather than x.

Example 2.7. Ordinary differential equation procedure
The ordinary differential equation governing the fall of a sphere under the influence of air friction has a frictional retarding force which is a function of the square of the velocity. In this fashion, the frictional force always opposes the motion of the sphere, regardless of the sign of the velocity vector. We will use this equation to examine the fall of a 3-mm diameter drop of water under the influence of air friction.

A drop of water approximates a sphere even though some deformation occurs during the fall. In a reference frame oriented so the y-axis is positive upward, the differential equation governing the fall of the sphere under both friction and gravity is:

$$m \times y'' = m \times g^2 - c' \times y' \qquad 2.15(a)$$

where differentials are a function of t rather than x. The differential equation can be simplified by dividing by the mass and letting $c = c'/m$. Assume that the water drop falls from a height of 10 m beginning with an initial velocity of zero. Since the y-axis is positive vertically, the acceleration of gravity is –9.81 m/s². What is the speed with which the drop strikes the ground?

The differential equation becomes:

$$y'' = g - c(y')^2 \qquad 2.15(b)$$

From elementary fluid mechanics the friction coefficient acting on the water droplet can be written as:

$$c = C_D \times \rho\left(\frac{A}{2}\right) \qquad 2.15(c)$$

where ρ is the density of air (at 10° C this is about 1.2931×10^{-3} kg/m³), C_D is the coefficient of drag (about 0.0241 at a velocity of 6 to 10 m/s), and A is the projected area of the water drop. Dividing by the mass of the water droplet gives $c \approx -0.156/m$. The sign is negative because friction opposes the velocity of fall.

By applying equation 2.11 to the second derivative and using equation 2.13 for the first derivative, we obtain:

$$y_{i+1} - 2y_i + y_{i-1} + = g\Delta t^2 - c\left(\frac{y_i - y_{i-1}}{\Delta t}\right)^2$$

from which:

$$y_{i+1} = g\Delta t^2 + 2y_i - y_{i-1} - c(y_i - y_{i-1})^2 \qquad 2.15(d)$$

Before equation 2.5(d) can be solved, we must initialize the solution process. To do this, we apply the initial and boundary conditions. At t = 0, i = 0 and the initial condition yields $y_i = y_0 = 10$ m. Also, at t = 0 the velocity is 0 and from equation 2.13,

$$\frac{dx}{dt} = 0 = \frac{y_i - y_{i-1}}{\Delta t} \tag{2.15(e)}$$

so that:

$$y_{i-1} = y_i = 10 \text{ m} \tag{2.15(f)}$$

We now set i = 0 and choose a convenient value for Δt of 0.01 s. A larger value of Δt would mean fewer steps in the solution process, but it would require estimation at times t not equal to a multiple of Δt, and the accuracy at intermediate times would be compromised.

Using the expression derived above for y_{i+1} and noting that the acceleration of gravity is -9.81 m/s^2 we find, at i = 0:

$$\begin{aligned} y_i &= -9.81 \frac{\text{m}}{\text{s}^2} \times (0.01 \text{ s})^2 + 10 \text{ m} \\ &= 9.999 \text{ m} \end{aligned} \tag{2.15(g)}$$

and at time Δt = 0.01 s later, (i = 1):

$$\begin{aligned} y_2 &= \left(-9.81 \frac{\text{m}}{\text{s}^2}\right)(0.0001 \text{ s}) + 2 \times 9.999\text{m} - 10\text{m} \\ &\quad -\left(\frac{-0.156}{\text{m}}\right)(9.999\text{m} - 10\text{m})^2 \\ &= 9.997 \text{ m} \end{aligned} \tag{2.15(h)}$$

The equation can be solved repetitively using values obtained above for y_i and y_{i-1} to determine y_{i+1}. This is left as an exercise for the student. The solution shows that the velocity when the water drop strikes the ground is about -7.76 m/s with an acceleration of -0.43 m/s^2.

Two-dimensional Taylor Series

To derive the finite difference expressions for partial differentials, we must have a Taylor Series Expansion for a function of two variables. The two-dimensional expression is relatively easy to obtain. We do this by first working with a parametric form of the function of two independent variables, z = f(x,y) in which z is the dependent variable (the solution surface) and x and y are the independent variables which describe a position on the solution plane. We obtain the parametric form by letting:

$$x = x_i \pm \Delta x \times t \quad \text{so that } \frac{dx}{dt} = \Delta x \tag{2.16}$$

while:

$$y = y_j \pm \Delta y \times t \quad \text{so that } \frac{dy}{dt} = \Delta y \tag{2.17}$$

where x_i and y_j are points on the x and y axes, respectively, and Δx and Δy are small distances over which the function f(x,y) will be approximated in a two-dimensional Taylor Series. The parametric variable t is arbitrary and now we can write:

$$f(x,y) = f(x_1 \pm \Delta x \times t, y_1 \pm \Delta y \times t) = f(t) \qquad 2.18$$

and use the Taylor Series Expansion in one variable, equation 2.1, to obtain the expansion for the function f(t), about some arbitrary point. For convenience, we will choose this point to be 0:

$$f(x,y) = f(t)\big|_{t=0} = f(0) + f'(0)(t-0) + f''(0)\frac{(t-0)^2}{2!} + R_{err} \qquad 2.19$$

To obtain the proper form we obtain expressions for the total differentials f´(t) = d[f´(x,y)] and f´(t) = d{d[f(x,y)]} as follows:

$$f'(t) = df(x,y) = \frac{\partial f(x,y)}{\partial x}\frac{dx}{dt} + \frac{\partial f(x,y)}{\partial y}\frac{dy}{dt} \qquad 2.20$$

or in condensed form using operator notation with information from equation 2.14(a) and 2.14(b)

$$\left(\frac{\partial}{\partial x}\Delta x + \frac{\partial}{\partial y}\Delta y\right) f(x,y) \qquad 2.21$$

and similarly:

$$f''(x,y) = \frac{\partial f'(x,y)}{\partial x}\Delta x + \frac{\partial f'(x,y)}{\partial y}\Delta y \qquad 2.22$$

In operator notation, again using information from equations 2.14(a) and 2.14(b):

$$\left[\frac{\partial}{\partial x}\left(\frac{\partial}{\partial x}\Delta x + \frac{\partial}{\partial y}\Delta y\right)\Delta x + \frac{\partial}{\partial y}\left(\frac{\partial}{\partial x}\Delta x + \frac{\partial}{\partial y}\Delta y\right)\Delta y\right] f(x,y) \qquad 2.23(a)$$

or, after collecting terms:

$$\left[\frac{\partial^2}{\partial x^2}\Delta x^2 + \frac{\partial^2}{\partial x \partial y} 2\Delta x + \Delta y^2 \frac{\partial^2}{\partial y^2}\Delta y\right] f(x,y) \qquad 2.23(b)$$

which can be written in shorthand notation as:

$$\left(\frac{\partial}{\partial x}\Delta x + \frac{\partial}{\partial y}\Delta y\right)^2 f(x,y) \qquad 2.23(c)$$

We can now make a direct substitution back into equation 2.21, since, at t = 0:

$$f(0) = f\left(x_i, y_j\right) \tag{2.24}$$

and at t = 1:

$$f(t)\big|_{t=1} = f\left(x_i \pm \Delta x, y_j \pm \Delta y\right) \tag{2.25}$$

Using equation 2.19 with 2.23, 2.24, and 2.25, we can now expand the function f(t) at t = 1 about a point t = 0 to obtain:

$$f\left(x_i \pm \Delta x, y_j \pm \Delta y\right) = f\left(x_i, y_j\right) + \frac{\partial f\left(x_i, y_j\right)}{\partial x}(\pm \Delta x)\,\frac{\partial f\left(x_i, y_j\right)}{\partial y}(\pm \Delta y) + \frac{\partial^2 f\left(x_i, y_j\right)}{\partial x^2}\frac{\Delta x^2}{2!} + \frac{\partial^2 f\left(x_i, y_j\right)}{\partial y^2}\frac{\Delta y^2}{2!} + 2\,\frac{\partial^2 f\left(x_i, y_j\right)}{\partial x \partial y}\frac{(\pm \Delta x)(\pm \Delta y)}{2!} + \cdots + R_{err} \tag{2.26}$$

Customarily, partial differentials are written in a shorthand using a subscript to indicate the process of differentiation. Multiple subscripts are used to show second-order and higher differentiation. If this convention is adopted and the functional notation is shortened by using i and j to indicate position in the solution plane, equation 2.26 becomes:

$$f(i \pm 1, j \pm 1) = f(i,j) + f_z(i,j)(\pm \Delta x) + f_y(i,j)(\pm \Delta y) + f_{xx}(i,j)\frac{\Delta x}{2!} + f_{yy}(i,j)\frac{\Delta y}{2!} + 2\,f_{x,y}(i,j)\frac{(\pm \Delta x)(\pm \Delta y)}{2!} + \cdots + R_{err} \tag{2.27}$$

In partial differentiation, only one independent variable is incremented at a time and any other independent variable is held constant. Therefore, in writing the finite difference expressions to define the differentials, the terms involving the products of Δx and Δy vanish, because one or the other has the value zero. Using the same approach as was used for the Taylor Series in one independent variable, the following expressions are developed:

$$f(i-1,j) = f(i,j) - f_x(i,j)\frac{\Delta x}{1!} + f_{xx}(i,j)\frac{\Delta x^2}{2!} + \cdots + R_{err} \tag{2.28(a)}$$

$$f(i+1,j) = f(i,j) + f_x(i,j)\frac{\Delta x}{1!} + f_{xx}(i,j)\frac{\Delta x^2}{2!} + \cdots + R_{err} \tag{2.28(b)}$$

$$f(i,j-1) = f(i,j) - f_y(i,j)\frac{\Delta y}{1!} + f_{yy}(i,j)\frac{\Delta y^2}{2!} + \cdots + R_{err} \tag{2.28(c)}$$

$$f(i,j+1) = f(i,j) + f_y(i,j)\frac{\Delta y}{1!} + f_{yy}(i,j)\frac{\Delta y^2}{2!} + \cdots + R_{err} \tag{2.28(d)}$$

By neglecting terms involving the second-order and higher differentials, we obtain from equation 2.28, the backward first-order finite difference model for the first partial derivative in x:

$$f_x\,(i,j) = \frac{f(i,j) - f(i-1,j)}{\Delta x} \qquad 2.29(a)$$

Similarly, from equation 2.29, we obtain the forward finite difference model for the first partial derivative in x:

$$f_x\,(i,j) = \frac{f(i+1,j) - f(i,j)}{\Delta x} \qquad 2.29(b)$$

and for the partial derivatives in y:

$$f_y\,(i,j) = \frac{f(i,j) - f(i,j-1)}{\Delta y} \qquad 2.29(c)$$

and

$$f_y\,(i,j) = \frac{f(i,j+1) - f(i,j)}{\Delta y} \qquad 2.29(d)$$

By subtracting equation 2.28(a) from equation 2.28(b), the central difference form for the first-order partial differential with respect to x can be obtained:

$$f_x\,(i,j) = \frac{f(i+1,j) - f(i-1,j)}{2\Delta x} \qquad 2.30$$

Subtracting 2.28(c) from 2.28(d) the central difference form for the first-order differential with respect to the second independent variable we obtained:

$$f_y\,(i,j) = \frac{f(i,j+1) - f(i,j-1)}{2\Delta y} \qquad 2.31$$

The central difference forms for the second derivatives are obtained by adding equations 2.28(a) and 2.28(b) to obtain:

$$f_{xx}\,(i,j) = \frac{f(i+1,j) - 2f(i,j) + f(i-1,j)}{\Delta x^2} \qquad 2.32$$

and similarly, by adding equations 2.23(c) and 2.24:

$$f_{yy}\,(i,j) = \frac{f(i,j+1) - 2f(i,j) + f(i,j+1)}{\Delta y^2} \qquad 2.33$$

Finite Differences Applied to a Partial Differential Equation

There are three types of partial differential equations that are of primary interest in engineering work: parabolic, hyperbolic, and elliptic. The heat flow equation is an example of a parabolic partial differential equation. It has the form:

$$\frac{\partial^2 \phi}{\partial x^2} = \frac{\partial \phi}{\partial t} \qquad 2.34$$

Of the remaining types, hyperbolic partial differential equations have the form:

$$\frac{\partial^2 \phi}{\partial x^2} - \frac{\partial^2 \phi}{\partial t^2} \qquad 2.35$$

The elliptical partial differential equation is written:

$$\frac{\partial^2 \phi}{\partial x^2} + \frac{\partial^2 \phi}{\partial y^2} = 0 \qquad 2.36$$

We will examine the application of the finite difference method to the solution of the heat flow equation because it is typical of the solution method for partial differential equations. The solution yields a model of the heat flow in an insulated bar that is initially at some constant temperature T_1. At time $\tau = 0$, one end of the bar is placed in contact with a body at constant temperature T_2. The other end of the bar remains in contact with a body at temperature T_1. The bar then gains thermal energy with time. We must determine how the temperature throughout the rod is changing with time τ.

Heat Flow Equation

The differential equation governing heat flow can be written:

$$\frac{\partial^2 T(x,\tau)}{\partial x^2} = \frac{1}{\alpha}\frac{\partial T(x,\tau)}{\partial \tau} \qquad 2.37$$

where

$T(x,\tau)$ = temperature as a function of time and distance
x = distance
τ = time
α = thermal diffusivity

with boundary conditions:

$$T(0, \tau) = T_1 \qquad 2.38(a)$$

$$T(X, \tau) = T_2 \qquad 2.38(b)$$

$$T(x, 0) = T_2 \qquad 2.38(c)$$

The left side of the partial differential equation is a second partial derivative so a second partial finite difference expansion must be used. We will solve the model explicitly for values at time $\tau = \Delta\tau$ based on the temperature distribution along the bar at time τ. For the right side of the partial differential equation we will use the forward difference form of the first partial derivative with time. With these choices, the differential equation transforms into the algebraic expression:

$$\frac{T(i+1, j) - 2T(i, j) + T(i-1, j)}{\Delta x^2} = \frac{1}{\alpha}\frac{T(i, j+1) - T(i, j)}{\Delta\tau} \qquad 2.39$$

In equation 2.39, at the initial instant $\tau = 0$, all the temperatures along the bar are known (see the initial and boundary conditions). The only unknowns are Δx, $\Delta\tau$, and α. These parameters can be grouped into a ratio λ where:

$$\lambda = \frac{\alpha\Delta\tau}{\Delta x^2} \qquad 2.40$$

and using equations 2.40 and 2.39 we can obtain:

$$T(i, j+1) = \lambda T(i+1, j) + (1 - 2\lambda)T(i, j) + \lambda T(i-1, j) \qquad 2.41$$

When using this method to solve the parabolic differential equation, there is a limitation on the selection of Δx and $\Delta\tau$. Unless $0 < \lambda \leq 1/2$, the solution will not converge, i.e., no solution can be found. Thus, the dimensions of the grid in the solution plane, i.e., the x,τ plane, are not independent but are restricted by the inequality above. The magnitude of λ must be such that:

$$0 < \frac{\alpha\Delta\tau}{(\Delta x)^2} \leq \frac{1}{2} \qquad 2.42$$

For convenience, we can choose $\lambda = 1/4$ so that equation 2.41 becomes:

$$T(i, j+1) = \frac{T(i-1, j) + 2\,T(i, j) + T(i+1, j)}{4} \qquad 2.43$$

Equation 2.43 can be programmed easily into a personal computer spreadsheet to solve. Solving the partial differential equation governing heat flow is demonstrated by example 2.8.

Example 2.8. Solution of the heat flow equation
Solve the partial differential equation of heat flow in an aluminum bar insulated so no heat is lost by convection along the bar length. The bar is 200 mL long and is initially at 10° C. It is placed between two masses that have constant temperatures of 10° C and 100° C. The thermal diffusivity of aluminum is 81.95 mm^2/s. Use $\lambda = 1/4$.

Solution
If we choose Δx to be 20 mL, then, for λ to be 1/4:

$$\Delta t = \frac{(\Delta x)^2}{4\alpha} = \frac{(20\text{mm})^2}{4 \times 81.95\,\frac{\text{mm}^2}{\text{s}}}\ 1.22\text{s} \qquad 2.44$$

Using the boundary and initial condition given below:

$$T(0, \tau) = 100° C$$

$$T(x, 0) = 10° C$$

$$T(X, \tau) = 10° C$$

The subscript i relates to Δx and increases at 20 mm/distance increment, while the subscript j increases time $\Delta\tau$ by 1.22 s/time increment. The solution is accomplished on a personal computer using the spreadsheet form as given in table 2.1.

In table 2.1, the letters across the top of the table represent letters that identify each column on a typical microcomputer spreadsheet. Row 1 identifies the values of i (x increments) and column A identifies the values of row number, while column B identifies the time steps (time increments). Row 2 contains the actual distances to the nodes of the bar, T(x,0), and column C contains the values of time for which the computations are made. Column D contains the temperature of the source at x = 0, i.e., T(0,τ). Column N contains the constant temperature of the low temperature end of the bar, T(X,τ). In this form, the distance increment, i, is traced across the page while time, τ, is incremented down the page. The values of temperature at time at some interior node, i are determined by a weighted average of the values of temperature at the previous time step as given in equation 2.43 at nodes i-1, i, and i+1. For example, the spreadsheet formula for cell E5 is:

$$E5 = \frac{D4 + (2 \times E4) + F4}{4} \qquad 2.45$$

whose contents are:

$$T(1, 2) = \frac{T(0, 1) + \left(2 \times T(1, 1)\right) + T(2, 1)}{4}$$

$$= \frac{100 + (2 \times 32.5) + 10}{4}$$

$$= 43.75 = 43.8$$

and this expression is repeated, with the appropriate adjustment of cells, at every cell of the spreadsheet encompassed by the cells C3:N3 and extending down the spreadsheet to the desired row (time). The temperature at time τ = 1.22 s at a distance of 20 mm (1 node) from the heated end of the block is given by equation 2.44.

Laplace's Equation

The use of finite differences as applied to an elliptical differential equation can be demonstrated as applied to commonly encountered problems involving potential flow. A typical problem deals with flows of electricity, heat or fluids. The equation is:

$$\frac{\partial^2\phi}{\partial x^2} + \frac{\partial^2\phi}{\partial y^2} = 0 \qquad 2.46(a)$$

The boundary conditions for this equation usually take one or more of the following forms:

$$\phi(X, y) = \Phi_1 \qquad 2.46(b)$$

$$\phi(x, Y) = \Phi_2 \qquad 2.46(c)$$

with possibly:

$$\frac{\partial\phi(X, y)}{\partial y} = 0 \qquad 2.46(d)$$

$$\frac{\partial\phi(X, y)}{\partial x} = 0 \qquad 2.46(e)$$

The first two conditions of equation 2.46 state that the values of the potential are known and fixed at the boundaries denoted by X = x and Y = y. The last two conditions indicate that the potential is reflective across the boundary at right angles to the direction in which the differential is taken. The term *reflective* is used because $\partial\phi(X,y)/\partial x = 0$ implies that $\phi(X+\Delta x,y) = \phi(X-\Delta x,y)$. If the boundary passes through the point (X,y) where X lies on the boundary, then the point across the boundary is a mirror image of the point immediately within the boundary.

The procedure for finding the remainder of the temperatures involves replicating the formula at each cell of the spreadsheet. The calculation mode of the spreadsheet operation is then used. The temperatures at the forward time are expressed in terms of the temperatures at the preceding time and surrounding nodes. Figure 2.2 shows the temperatures along the bar at increments of $\Delta\tau$.

Since only second partial derivatives occur in equation 2.37, it is necessary to use the second partial derivative finite difference form to transform the partial differential equation into a finite difference algebraic equation. The result is:

$$\frac{\phi(i+1, j) - 2\phi(i, j) + \phi(i-1, j)}{(\Delta x)^2} + \frac{\phi(i, j+1) - 2\phi(i, j) + \phi(i, j-1)}{(\Delta y)^2} = 0 \qquad 2.47$$

Table 2.1. Spreadsheet solution of the heat flow in an aluminum bar

A	B	C	D	E	F	G	H	I	J	K	L	M	N
1		i=>	0	1	2	3	4	5	6	7	8	9	10
2	j	τ\x	0	20	40	60	80	100	120	140	160	180	200
3	0	0.00	100	10.0	10.0	10.0	10.0	10.0	10.0	10.0	10.0	10.0	10.0
4	1	1.22	100	32.5	10.0	10.0	10.0	10.0	10.0	10.0	10.0	10.0	10.0
5	2	2.44	100	43.8	15.6	10.0	10.0	10.0	10.0	10.0	10.0	10.0	10.0
6	3	3.66	100	50.8	21.3	11.4	10.0	10.0	10.0	10.0	10.0	10.0	10.0
7	4	4.88	100	55.7	26.2	13.5	10.4	10.0	10.0	10.0	10.0	10.0	10.0
8	5	6.10	100	59.4	30.4	15.9	11.1	10.1		—	—	—	10.0
9	6	7.32	100	62.3	34.0	18.3	12.0	—	—	—	—	—	10.0
10	7	8.54	100	64.7	37.2	20.7	—	—	—	—	—	—	10.0
11	8	9.76	100	66.6	39.9	—	—	—	—	—	—	—	10.0
12	9	11.0	100	68.3	—	—	—	—	—	—	—	—	10.0
13	10	12.2	100	—	—	—	—	—	—	—	—	—	10.0
14	11	13.4	100	—	—	—	—	—	—	—	—	—	10.0
15	12	14.6	100	—	—	—	—	—	—	—	—	—	10.0
16	13	15.8	100	—	—	—	—	—	—	—	—	—	10.0
17	14	17.1	100	—	—	—	—	—	—	—	—	—	10.0

where i and j denote the location of nodes on the solution plane, i advancing in the x direction and j advancing in the y direction. Equation 2.39 can be simplified for solution if Δx is equal to some multiple of Δy. For convenience, we can choose the nodal grid such that Δx is equal to Δy in which case equation 2.46 becomes:

$$\frac{\phi(i+1,j)+\phi(i,j+1)+\phi(i-1,j)+\phi(i,j-1)}{4}=\phi(i,j) \qquad 2.48$$

an expression in which the value of the potential at a node is expressed in terms of the potentials immediately surrounding the node.

A relaxation technique now enables a solution surface $\phi(x,y)$ to be approximated. In the relaxation technique, the (x,y) plane, the solution plane over which the function $\phi(x,y)$ is to be determined, is overlain with a grid and the boundary conditions are defined at the boundaries. When $\phi(X,Y)$ = a constant (where X and Y are values of the independent variables), the known values are assigned to the nodes. When the first partial of the function has a known value, the first-order central difference form for the first partial derivative is used. This is usually applied as in the second pair of equation 2.46(d,e) where the first partials are equal to zero, i.e., there is no gradient of potential across the boundary. The reflective boundary condition guarantees there is no movement of the entity governed by the potential gradient across the boundary. Any motion of the entity (heat, water, electricity, etc.) driven by the potential gradient must be in the direction perpendicular to the first partial derivative which specifies the boundary condition.

An impermeable boundary in the solution plane must be treated as a reflective boundary. This follows from the argument given above. If the fluid cannot flow across the boundary it must flow along the boundary. This makes the boundary a flow line. The potentials on opposite sides of a reflective boundary have the same value.

In effecting a solution using the relaxation technique, an iterative process is required. In this process, the unknown values of the potential are calculated successively at each node within the solution plane. The process is repeated until the values calculated at a node do not differ significantly from those determined during the previous iteration. Using this method, a value $\phi(i,j)$, bordered by one or more known values $\phi(X,Y)$, will be computed using equation 2.47. This would be repeated for each of the unknown values of the potential throughout the solution plane.

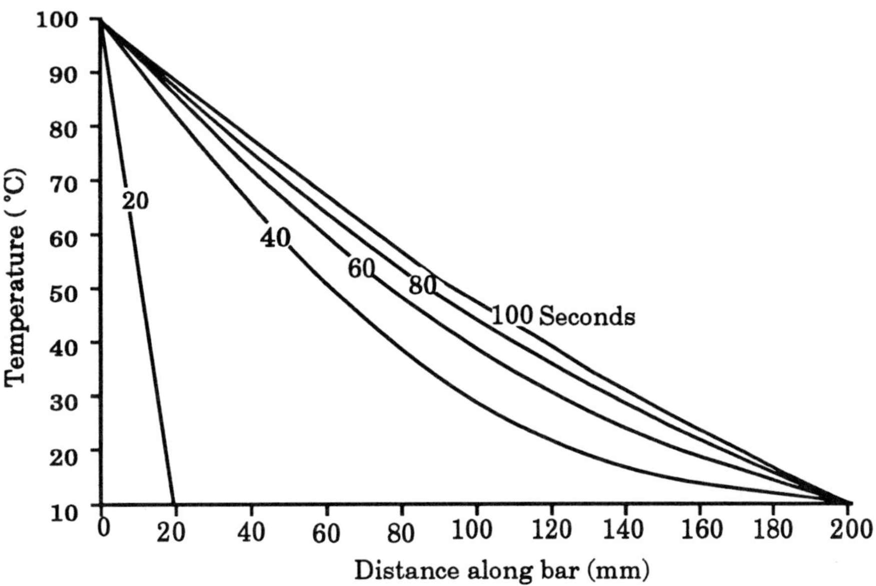

Figure 2.2. The temperature at different times is given as a function of distance along the aluminum bar of example 2.8 heated to 100° C at one end.

At boundaries where the gradient condition $\partial\phi(i,j)/\partial x = 0$ (or $\partial\phi(i,j)/\partial y = 0$) holds, since this condition implies there is no gradient across the boundary, the potential value $\phi(i+1,j) = \phi(i-1,j)$ (or $\phi(i,j+1) = \phi(i,j-1)$). Thus, along a boundary parallel to the x axis and along the farthest y border, one can write:

$$\frac{\phi(i-1, j) + \phi(i+1, j) + 2\phi(i, j-1)}{4} = \phi(i,j) \qquad 2.49$$

At a corner where the gradient boundary conditions holds along both boundaries, the reflective condition yields:

$$\frac{2\phi(i-1, j) + 2\phi(i, j-1)}{4} = \phi(i, j) \qquad 2.50(a)$$

or

$$\frac{\phi(i-1, j) + \phi(i, j-1)}{2} = \phi(i, j) \qquad 2.50(b)$$

The solution approach may be best illustrated by an example.

Example 2.9. Relaxation solution technique
A solution plane is shown in figure 2.3. The potentials at the nodes along the A-B, and B-C boundaries are defined as given. Along the A-D and C-D boundaries the gradient conditions:

$$\frac{\partial\phi(i, j)}{\partial y} = 0$$

and

$$\frac{\partial\phi(i, j)}{\partial x} = 0$$

apply.

At point E, a first estimate of the potential could be written:

$$\frac{9 + \phi_3 + 12 + \phi_4}{4} = \phi_E$$

A value for either ϕ_4 or ϕ_3 could now be estimated using the value found for ϕ_E.

At node G, the expression would be:

$$\frac{7 + \phi_H + 2 \times \phi_5}{4} = \phi_G$$

B					C
10+	+9	+8	+7	+7	+7
12+	+E	+ϕ_4	+	+ϕ_5	+G
12+	+ϕ_3	+	+	+	+H
13+	+	+	+	+	+ϕ_8
14+$_A$	+	+F	+K	+ϕ_7	+D

Figure 2.3. A solution plane for a relaxation solution to Laplace's Equation.

The point ϕ_H can be assumed to be zero or some other appropriate number for the first pass through the solution plane. A similar equation holds for point K.

At point d, since both immediately adjacent points are on reflective boundaries, the correct equation to estimate the value at point D would be:

$$\frac{(\phi_8 + \phi_7)}{2} = \phi_D$$

The above process using the appropriate equations for the point is performed for each node on each pass through the solution plane. This process is repeated until the value of ϕ computed for a node does not differ significantly from the previously computed value. When this condition is attained, the solution is said to have converged.

The question of significance is critical here. The use of a computer is essential for the computations may be repeated several hundred times before convergence is reached. At least one more significant digit than the number of significant digits that are desired in the answer should be used in determining when convergence has occurred. If the answer is desired to two significant digits, the newly calculated value should agree to at least three significant digits.

Within the boundaries of the solution plane at a point or points where the potential is constant, the constant value must be maintained for all passes. The surrounding nodal values must relax to the known nodal value. In a computer solution where an iterative procedure is used to traverse the nodes, this is accomplished by equating the node to its constant value immediately after every recalculation of the node. A decision statement to omit the particular node from the calculation procedure is another alternative to be considered. If a spreadsheet is used, the appropriate calculation formula can be omitted at the node with a constant value.

In the solution plane of this example, an answer to two significant digits is given below. The locations on the number within the row and column correspond to a node on the solution surface, and the number itself is the distance of the solution surface from the solution plane.

+10.00	+9.00	+8.00	+7.00	+7.00	+7.00
+12.00	+10.3	+9.1	+8.3	+8.0	+7.9
+12.00	+11.1	+9.9	+9.2	+8.8	+8.6
+13.00	+11.6	+10.5	+9.7	+9.2	+9.1
+14.0	+12.0	+10.7	+9.9	+9.4	+9.2

Figure 2.4. Solution surface for a relaxation solution of Laplace's Equation. The numerical values indicate positions of the solution surface above the solution plane.

Exercises

2.1. Verify the final expression in example 2.6 by performing the expansion and grouping terms.

2.2. By expanding about zero, show that the number "e" can be found without the use of a calculator. If eight terms are used, to how many significant digits is your answer accurate?

2.3. To what number of significant digits would the expansion of e^{-x} be accurate if only the second derivative were used and the expression was applied to calculate $e^{-0.1}$?

2.4. Calculate exp(−1/2) by expanding the function about zero using fifth-order differentials. To how many significant digits is your answer accurate?

2.5. Estimate $2^{1/2}$ by expanding $x^{1/2}$ about zero using as many orders of differentiation as necessary to obtain two significant digit accuracy.

2.6. Write a short computer program to solve the falling water droplet under the influence of friction using the variables given in example 2.3.

2.7. Using the program written for exercise 4 above, assume there is no frictional resistance (i.e., c = 0) and find the time required for the droplet to strike the ground. Verify your answer by solving the problem using a closed form solution of the differential equation. [1.41 s]

2.8. Compare the energy of the falling water droplet with and without friction. How many times more energetic is the non-friction solution? What does this indicate regarding the influence of friction on the potential for soil erosion?

2.9. By expanding the terms on the extreme left, show the correctness of equation 2.17.

2.10. Derive equation 2.27 and 2.28 from equation 2.22, 2.23(c), 2.24, 2.25.

2.11. What would the time increment be in example 2.5 if λ is set equal to 2? (2.44 s)

2.12. Using a spreadsheet, solve equation 2.37 if $\lambda = 1/2$. What advantage would there be in using the larger value of λ? What disadvantages might there be?

2.13. Using a spreadsheet, solve the heat flow equation for an aluminum rod 20 cm long if the heat source is sinusoidal with a period of 3 s. The initial temperature of the rod is 10° C and the temperature at the distal end of the rod is constant at 10° C. Carry the solution out for a sufficiently long time so that one complete pulse of heat has had time to be noticed at a node 2 cm from the distal end of the rod.

2.14. Using a computer program that you write yourself, solve example 2.9 if the potential along the edge A-D is constant at 14.00.

Chapter 3

Properties of Water in the Green Plant Biosystem

Enormous quantities of water are needed to sustain life. Each individual needs about 250 L/day for everyday activities such as bathing, washing dishes, cooking, etc. More water per capita is needed for industrial activities but the per capita use for agriculture and food production far exceeds the industrial usage. Maxwell (1965) estimates from 60 000 to 76 000 L/person per day are needed for food production, most of it to satisfy the evapotranspiration demand of growing crops.

Besides the obvious uses, water is critical to the processes that nurture and sustain life. It serves as the substrate for all biological and biochemical processes of growth and development. It is the transport medium to move energy and nutrients through the body, and it is essential to maintain the homeothermic balance of animal life. Evaporation helps remove excess heat and maintain a temperature range that is conducive to health. Finally, water is instrumental in the waste removal processes.

Water also plays a crucial role in weather. In conjunction with energy from the sun, water is the central element in the hydrologic cycle. Phase changes of water significantly modify the temperature of the biosphere, that thin layer which supports and makes life on the surface of the earth. In addition to thermal effects, the mechanical effects of freezing water, along with the recrystallization of chemical compounds as water evaporates from a solution help in the natural geologic weathering processes which continually produce soil.

In addition to the myriad of uses mentioned above, water is broken down during the process of photosynthesis and contributes both hydrogen and oxygen to the production of sugars in the green plant as well as releasing oxygen to the atmosphere. Obviously, the importance of water requires that its chemical and physical properties be investigated to appreciate its behavior and affects on and in the green plant environment. The object of this chapter is to review the physical and chemical concepts associated with water to develop an understanding of the behavior of water in the biosphere. This will develop a firm foundation upon which predictive models can be constructed.

Physical, Chemical, and Thermodynamic Aspects of Water

Potential energy is the energy an object has by virtue of its position in a gravitational field. The quantity of potential energy a mass possesses depends on the datum to which it is related. With respect to the datum, it also may represent the work that was necessary to raise the mass to the position in a gravitational field. Tremendous quantities of energy are stored in the upper atmosphere by evaporating water. Water condensing and falling as rain or snow returns some of the potential energy that it obtained from the action of the sun during evaporation.

Example 3.1. Potential energy of rainfall
Compute the energy per unit volume of water that is potentially available in the anvil cloud of a thunderstorm at 12 km above the earth's surface.

Solution
The density of water is $1000\ \frac{\text{kg}}{\text{m}^3}$. Therefore, potential energy per cubic meter of water is:

$$\frac{1000\ \text{kg}_{\text{H}_2\text{O}}}{\text{m}^3} \times \frac{9.81\text{m}}{\text{s}^2} \times \frac{12\ 000\ \text{m}}{1} = 1.18 \times 10^8 \text{N}$$

The kinetic energy of a mass is the energy it possesses by virtue of its velocity. Strictly speaking, while potential energy depends on a gravitational field, kinetic energy does not. It is a function only of the inertial property of the mass. A moving mass must do work to lose kinetic energy. In the case of water, raindrops striking the ground often perform work by loosening soil particles. In addition, water moving over the ground surface also may perform work by detaching soil particles and/or transporting them from one point to another in the process of accelerated erosion. While it is possible for water droplets to strike and tear plant leaves, air friction acting on the falling droplets absorb sufficient droplet energy that damage to the plant does not occur.

Example 3.2. Kinetic energy of water falling as rain
If the energy per unit volume of water in example 3.1 were to be completely converted into kinetic energy, what would the velocity of a 5-mm diameter raindrop be upon striking the earth's surface?

Solution
The volume of a 5-mm diameter raindrop is:

$$V = \frac{\left(\dfrac{5\text{mm}}{1000\,\dfrac{\text{mm}}{\text{m}}}\right)^3 \times \pi}{6} = 6.5 \times 10^{-8}\,\text{m}^3$$

The potential velocity, assuming all potential energy is converted into kinetic energy, is:

$$\text{velocity} = \sqrt{\left(\frac{2 \times 1.18 \times 10^{-8}\,\dfrac{\text{kg m}}{\text{s}^2}}{6.5 \times 10^{-8}\,\text{m}^3 \times 1000\,\dfrac{\text{kg}}{\text{m}^3}}\right)} = 3.59 \times 10^{12}\,\frac{\text{m}}{\text{s}}$$

Obviously, rain falling at this speed would be disastrous to the surface of the earth. Friction is man's friend in this case, slowing the droplet to about 11 m/s so that erosion of the surface due to rainfall energy could be controlled.

Elastic or strain energy is energy manifested by the tendency of a stretched object to return to its original position. In the case of a solid, this is most clearly seen when a spring returns to its original position after being displaced. However, water displays a similar behavior in the form of surface tension. Surface tension, the manifestation of strain energy exists in water because of the polar property of the water molecule. In the water molecule, two hydrogen atoms separated by an angle of 104.5° are attached to an oxygen atom as shown in figure 3.1. Since the hydrogen atoms carry a positive charge while the oxygen atom is negatively charged, the water molecule is slightly polarized. Within liquid water, the attraction of individual water molecules for one another cancels out any excess force. However, at the water's surface, molecules are attracted only to the interior of the liquid and therefore form a layer of molecules that possess an attraction to each other, creating the surface tension effect as depicted in figure 3.2. Surface tension is measured as a force per unit width and, for water in the green plant environment, nominally has a value of 0.072 to 0.074 N/m.

The internal energy of water manifests itself on the atomic level in terms of the motion of the atoms making up the molecule. Normally, the water molecule exhibits three degrees of vibrational freedom, in addition to two degrees of rotational freedom and three degrees of translational freedom. Energy can be stored in any combination of the various motions. As the energy level rises, the motions becomes more intense. In particular, at the boiling point, the vapor pressure of the liquid which is a function of the momentum being carried by the individual molecules through a unit of area perpendicular to the surface of the fluid, equal to the atmospheric pressure and the molecules break free of the surface, carrying away energy equal to the latent heat of vaporization.

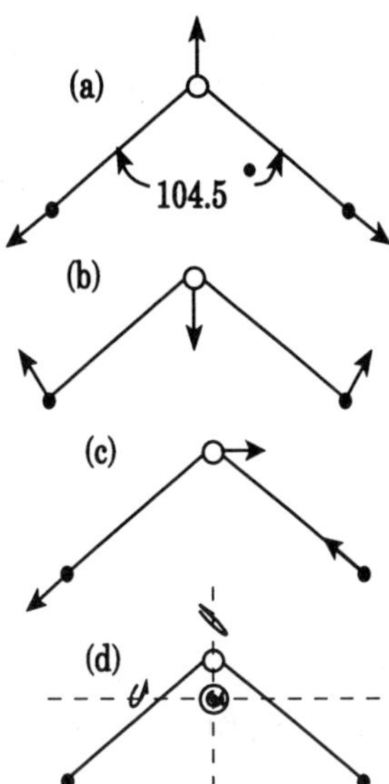

Figure 3.1. The structure of a water molecule showing the position of the two hydrogen atoms with relation to the oxygen atom. The three modes of vibration are designated by the arrows in (a), (b), and (c). Figure 3.1(d) shows the three rotational degrees of freedom which can be assumed by the water molecule.

In addition to the energy of motion of the individual molecules, a substantial amount of energy is stored by the attraction of the atoms within the molecule. This is the energy required to break down the atomic structure into atomic hydrogen and oxygen.

Energy Transfer

Energy is transferred from one location to another primarily by conduction, convection, and radiation. *Conduction* occurs through excitation of the molecular structure by contact with an adjacent molecule. It is the primary mode of energy transfer in a solid and also acts in a liquid and a gas, although to a lesser degree. In the green plant environment conduction warms the soil. Some conduction occurs through plant parts, however, it is of little significance compared to other modes of transfer.

Convection is energy transfer due to mass transfer by fluid motion. When vaporized water is moved by the action of air currents, vast quantities of latent energy are transferred. When air movement occurs within a room, the effect is termed convection. Generally, when air moves over the surface of planet earth, it is termed *advection*. Advection is responsible for global energy circulation.

Radiation is the primary means by which energy reaches the earth's surface. Radiation is an electromagnetic phenomenon in which individual quanta or bits of energy are transferred through electrical and magnetic waves.

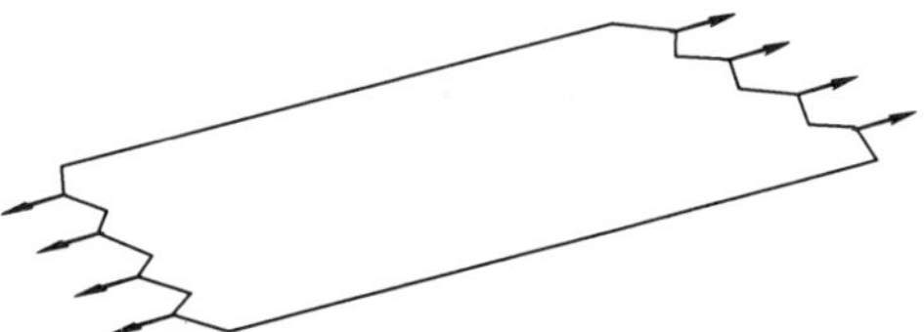

Figure 3.2. The attraction of water molecules to each other when there are no molecules above the surface is responsible for the phenomena of surface tension. The parameter quantifying surface tension has dimensions of force per unit width of the surface.

Because it is electromagnetic, radiant energy reaches the earth through empty space. A specific transfer medium is unnecessary. While the equations governing conduction and convection are basically parabolic differential equations, the equation modeling transfer of energy by radiation is a direct function of temperature to the fourth power. The model is known as the Stefan-Boltzman Law after it's discoverers and has the form:

$$R_{total} = \varepsilon \sigma T^4 \qquad 3.1$$

where

R_{total} = energy flux $J/m^3 \cdot s$
ε = emissivity or absorptivity of the surface
σ = Stefan-Boltzman constant, $5.670\,32 \times 10^{-8}\ J/m^2 \cdot s \cdot K^4$
T = temperature K

The sun, radiating at a temperature of about 6,000 K is Earth's energy source. It radiates at all frequencies, but has it's peak radiation in the band of wavelength frequencies clustered about the visible band. Also in this band is the wavelength of energy which photosynthesis must have to initiate the production of sugars, the primary food for the green plant. Radiation is examined in greater detail in chapter 8.

As mentioned above, latent energy transfer, in conjunction with advection, is responsible for transporting immense quantities of energy over the earth's surface. At sea level, water boils at 100° C and absorbs 2264 kJ for every kilogram of water vaporized. At lower temperatures, more energy is absorbed by vaporizing water while at higher elevations, i.e., lower atmospheric pressures, less energy is needed. The latent heat of vaporization is:

$$l_v = 2503 - 2.39T \qquad 3.2$$

where

l = latent heat of vaporization (kJ/kg)
T = temperature (C)

Equation 3.2 allows one to calculate the latent energy in kilojoules required to evaporate a kilogram of water as a function of temperature. Since this amount of energy is released when the water condenses, it is obvious that vast quantities of energy are released during a rainstorm.

Example 3.3. Latent energy transfer
Virtually all water which falls as precipitation on land originates as evaporation from the ocean where it picks up latent heat of vaporization. How much heat is released into the atmosphere by 1000 mm (1 m) of rain falling on 1 km² of land? Assume that the condensation occurs at the dew point temperature of 20° C. This is the normal annual rainfall over much of the North Central U.S.

Solution
The total amount of latent energy released would be:

$$1\,m_{rain} \times \left(10^3 \frac{m}{km}\right)^2 \times \left(\frac{1000\ kg}{m^3}\right) \times \left(2490 - 2.13 \times 20°\ C\right) \frac{kJ}{kg_{water}} = 2.46 \times 10^{12}\ kJ \qquad 3.3$$

The conversion of liquid water to ice releases about 333 kJ of energy for every kilogram of water frozen at 0° C. This value varies somewhat at colder temperatures. However, since we are seldom concerned with phase changes from liquid to solid in the green plant environment, this text will use the value 333 kJ/kg.

Example 3.4. Latent heat of fusion
In some areas the temperature is low enough to promote winter sports, but nature does not cooperate by producing snow. Snow-making is used to supplement nature. Compute the energy released when enough snow is produced to cover 30 ha of ski slope 25 cm deep. The specific gravity of the snow (ratio of snow density to water density) is 1/6.

Solution
The mass of water required to freeze 25 cm of snow on 30 ha is:

$$30\text{ ha} \times 10\,000\,\frac{\text{m}^2_{\text{snow}}}{\text{ha}} \times 0.25\,\text{m}_{\text{snow}} \times \frac{1\ \text{m}^3_{\text{water}}}{6\ \text{m}^3_{\text{snow}}} = 12\,500\ \text{m}^3_{\text{water}} \qquad 3.4$$

The energy released during freezing is:

$$12\,500\ \text{m}^3_{\text{water}} \times 1000\left(\frac{\text{kg}}{\text{m}^3}\right) \times 335\,\frac{\text{kJ}}{\text{kg}_{\text{water}}} = 4.16 \times 10^9\,\text{kJ} \qquad 3.5$$

In addition to latent energy due to change of phase, considerable energy in the green plant environment is stored or released due to the interaction between the change in temperature and the specific heats of air, water, and water vapor. While there is a difference in the specific heat depending on whether the action takes place at constant pressure or constant volume, in the green plant environment, all actions occur at atmosphere pressure, thus, only the constant pressure values are of concern. Table 3.1 presents values of the latent energies and specific heats at constant pressure for air, water, and water vapor.

Example 3.5. Influence of the specific heat of air with the latent heat of fusion
What temperature increase would occur if a 10 km depth of atmosphere were required to absorb the energy produced by 1000 mm (1 m) of rain falling on 1 km^2 of land? Assume the energy release occurred at 20° C and the density of air is 1.278 kg/m^3.

Solution
One meter of rain on 1 km^2 surface is 1000 m^6 and has a mass of 1×10^9 kg. At 2455 kJ/kg, this releases 2.46×10^{12} kJ. The quantity of air is:

$$1\ \text{km}^3 \times 10^9\,\frac{\text{m}^3}{\text{km}^3} \times 1.278\,\frac{\text{kg}}{\text{m}^3} = 1.278 \times 10^9\,\text{kg}_{\text{air}} \qquad 3.6$$

The temperature rise would be:

$$\frac{2.46 \times 10^{12}\,\text{kJ}}{1.278 \times 10^9\,\text{kg}} \times \frac{1}{1\,\dfrac{\text{kJ}}{\text{kg}\cdot{}^\circ\text{C}}} = 1925^\circ\ \text{C} \qquad 3.7$$

Since this happens over the course of a year, the amount of energy released is not readily appreciated.

Table 3.1. Properties of water, water vapor, and air

Substance	Latent heat: Fusion	Latent heat: Vaporization	Specific Heat at Constant Pressure
Ice	333 kJ/kg	2836-0.213T kJ/kg	2.03 kJ/(kg·K*)
Liquid water		2503-2.39T kJ/kg	4.186 kJ/(kg·K)
Water vapor			1.88 kJ/(kg·K)
Air			1.01 kJ/(kg·K)

* 1 K = 1° C

Water Vapor and the Perfect Gas Law

Air and water vapor, at the temperatures and pressures encountered in the green plant environment, both behave as perfect gasses. Because of this, many of the models that are needed to understand the behavior of water and energy in the plant environment can be derived from the perfect gas law.

Historically, the perfect gas law had its beginnings in the work of Robert Boyle (1627-1691). In his experiments, Boyle showed that the pressure of a gas varied inversely with volume if the temperature was held constant. About a century later, Jacques A. Charles (1764-1823) and Joseph L. Gay-Lussac (1788-1850) showed that the volume of a gas varied directly with temperature if the pressure was held constant. During the same period, Amadeo Avogadro (1776-1856) and his colleagues defined the standard conditions (STP) to which gasses should be reduced for comparison purposes and discovered that, at STP, all gasses had the same numbers of molecules. This value became known as Avogadro's Number, 6.023×10^{23} molecules for any gas at a temperature 0° C, a pressure of one atmosphere pressure and, a volume of 22.414 L. The value 6.023×10^{23} was likewise found to be the number of molecules in a mass of a substance in grams which was numerically equal to the atomic weights of the atoms comprising a molecule of the substance.

It remained to develop the absolute scale of temperature before all the previous developments could be melded into a single comprehensive mathematical description of the behavior of a gas. The absolute scale of temperature resulted from the behavior of a gas under constant pressure and/or constant volume with change in temperature. The linear nature of the change in pressure and/or volume seemed to indicate that there should be a temperature at which either the pressure and/or volume of a gas should become zero. This temperature was determined to be the absolute zero. The temperature scale was named after it's developer, Lord Kelvin.

Example 3.6. Estimation of absolute zero
Using a temperature scale with 100 divisions between the freezing and boiling points of water and an arbitrary quantity of gas held at a constant pressure, the following measurements were obtained:

$$T_1 = 0° \qquad V_1 = 0.2271\ \text{L}$$
$$T_2 = 37° \qquad V_2 = 0.2579\ \text{L}$$

The slope of the volume/temperature relationship was found to be:

$$T_1 - T_0 = 0.2271\ \text{L} \times \frac{(0-37)}{(0.2271 - 0.2579)}\frac{\text{deg}}{\text{L}} = 272.8°\ \text{absolute}$$

and by the technique described above.

The temperature of melting ice appears to be 272.8° absolute. More precise measurements using more significant digits reveal that the Kelvin absolute temperature of melting ice is 273.15 K. For most engineering work, the value 273 will suffice.

$$T_1 - T_0 = 0.2271 \text{ L} \times \frac{(0-37)}{(0.2271 - 0.2579)} \frac{\text{deg}}{\text{L}} = 272.8° \text{ absolute}$$

With the introduction of the Kelvin absolute temperature scale, it was possible to relate the pressure and volume of one mole of a gas to its temperature. This was done by determining the proportionality constant as shown in example 3.7.

Example 3.7. Calculation of the perfect gas constant
Using Avogadro's principle that all gases at 101.325 (atmospheric pressure at sea level) and 273.15 K, the freezing point of water, possess 6.022×10^{23} (one mole) of molecules and occupies 22.414 L, calculate the value of PV/T.

Solution

$$\frac{101.325 \text{ kPa} \times 22.414 \frac{\text{L}}{\text{mol}}}{0 + 273.156 \text{ K}} = 8.3143 \frac{\text{kPa} \cdot \text{L}}{\text{mol} \cdot \text{K}} \qquad 3.8$$

The value obtained, 8.314 kPa·L/mol·K, is the gas constant, the proportionality constant for the perfect gas law.

The statement of the perfect gas law is then:

$$PV = RnT \qquad 3.9$$

where

P = pressure
V = volume (L)
T = temperature (K)
n = number of moles of gas present (mol)
n = m/M where m is the mass of gas, and M is the gram molecular weight
R = gas constant per mole [8.314 kPa·L/mol·K] (or equivalently kPa·m^3/kmol·K)

John Dalton (1766-1844) discovered the law of partial pressures which states that the total pressure exerted by a gas is the sum of its component pressures. This law enabled researchers to determine that water vapor was a component of the total atmospheric pressure. In their experimentation it was discovered that the atmosphere had a finite capacity for water, determined somewhat by the total atmospheric pressure, but primarily by the temperature of the air. The finite capacity was termed the saturation vapor point: the point beyond which the addition of vapor to the atmosphere would cause condensation of the excess. The partial pressure of water vapor at which the air becomes saturated is the saturation vapor pressure, denoted generally by the symbol $e°$. Very accurate values of the saturation vapor pressure are obtained from tables; however, in the range of temperatures and pressures encountered in the green plant environment, calculation of the saturation vapor pressure can be performed using the Clausius-Clapeyron equation. Because the properties of water change as the temperature drops through the freezing point, it is necessary to change the parameters of the equation. Values of the parameters for the Clausius-Clapeyron equation are given in table 3.1 for temperatures from –20° C to 50° C. The Clausius-Clapeyron model is:

$$e^{\circ} = \exp\left(a + \frac{b}{T}\right) \qquad 3.10$$

where a,b are parameters from table 3.2, and exp is the exponential function. Example 3.8 demonstrates use of the Clausius-Clapeyron equation.

Example 3.8. Calculation of saturation vapor pressure
Determine the saturation vapor pressure at 25° C (298 K).

Solution
Application of equation 3.4 using the appropriate values from table 3.1 yields:

$$\begin{aligned} e^{\circ} &= \exp\left(19.0177 - \frac{5327}{298}\right) = 3.132\ 593 \text{ kPa} \\ &= 3.13 \text{ kPa} \end{aligned} \qquad 3.11$$

a value accurate to the second significant digit within ± 1.25%. This is sufficient for engineering calculations in most cases.

Table 3.2. Coefficients for psychrometric models

a_1 =	19.0177 T ≥ 0° C
b_2 =	–5327 T ≥ 0° C
a_1 =	22.0233 T ≤ 0° C
b_2 =	–6148 T ≤ 0° C

One of the useful properties of water vapor in the air is the water vapor density, also known as the absolute humidity. The water vapor density can be obtained from the perfect gas law, by virtue of Dalton's Law, by using the vapor pressure as the pressure term and solving for the mass density. The water vapor density is:

$$\rho_{vapor} = \frac{m_{H_2O}}{V} = \frac{eM_{H_2O}}{RT} \qquad 3.12$$

where

ρ_{vapor} = absolute humidity or water vapor density (kg_{H_2O}/m^3)
e = partial pressure of water vapor
M_{H_2O} = gram molecular weight of water (kg_{H_2O}/kmol)
R = the perfect gas constant (8.314 kPa·m³/kmol·K)
T = absolute temperature (K)

Example 3.9. Calculating water vapor density
Calculate the water vapor density in 20° C saturated air.

Solution
It is necessary to determine the pressure of the water vapor in the air. Since the air is saturated with water vapor, we can use the Clausius-Clapeyron equation for saturated air:

$$e^\circ = \exp\left(19.0177 - \frac{5327}{20 + 273}\right) = 2.309 \text{ kPa} = 2.31 \text{ kPa} \qquad 3.13$$

By equation 3.5, the water vapor density is:

$$\rho_v = \frac{2.309 \text{ kPa} \times 18 \frac{\text{g}}{\text{mol}}}{\frac{8.314 \text{ kPa} \cdot \text{L}}{\text{K} \cdot \text{mol}} \times (20 + 273\text{K})} = 0.017\,06 \frac{\text{g}}{\text{L}} = 0.017\,06 \frac{\text{kg}}{\text{m}^3}$$
$$= 0.0171 \frac{\text{kg}}{\text{m}^3} \qquad 3.14$$

to two significant digits.

Another parameter of water vapor is the specific humidity, so named because it is specific to the mass of air. The specific humidity is the ratio of the water vapor density to the density of moist air or:

$$q = \frac{\rho_{vapor}}{\rho_{moist\ air}} = \frac{e \times M_{H_2O}}{P \times M_{moist\ air}} \qquad 3.15$$

where

q	= specific humidity ($kg_{H_2O}/kg_{moist\ air}$)
e	= partial pressure of water vapor
P	= atmospheric pressure
M_{H_2O}	= gram molecular weight of water ($kg_{H_2O}/kmol$)
$M_{moist\ air}$	= gram molecular weight of moist air ($kg_{moist\ air}/kmol$ the ratio $M_{H_2O}/M_{moist\ air}$ is, to engineering accuracy, 0.622)

A more useful parameter of water vapor is the mixing ratio. This parameter is based on the density of dry air, i.e., air without water vapor. The most significant reason for using the mixing ratio in calculations is that it forms an unchanging basis. Furthermore, if air is to be moved mechanically, dry air has the higher density, therefore, calculations based on dry air will err, if at all, on the conservative side. The mixing ratio can be obtained by subtracting the partial pressure of water vapor from the atmospheric pressure when the air density is being calculated. The mixing ratio can be calculated using:

$$\chi = \frac{e \times M_{H_2O}}{(P - e) \times M_{dry\ air}}$$
$$= \frac{e}{(P - e)} 0.622 \qquad 3.16$$

where the ratio 0.622 is commonly used as being sufficient for engineering accuracy, especially in the range of temperatures and vapor pressures common to the green plant environment.

Example 3.10. Specific humidity and mixing ratio calculations
Compare the specific humidity and mixing ratio calculations for saturated air at 20° C.

Solution
From example 3.9, the saturated vapor pressure is 2.31 kPa. Therefore, specific humidity is:

$$q = \frac{2.31 \text{ kPa}}{100 \text{ kPa}} 0.622 = 0.014368\,2 \frac{kg_{H_2O}}{kg_{moist\ air}} \quad 0.0144 \frac{kg_{H_2O}}{kg_{moist\ air}}$$

The mixing ratio is:

$$\chi = \frac{2.31 \text{ kPa}}{100 - 2.31} 0.622 = 0.147\,079 \frac{kg_{H_2O}}{kg_{dry\ air}} = 0.147 \frac{kg_{H_2O}}{kg_{dry\ air}}$$

The relative humidity (rh) is an important parameter. Relative humidity is the ratio of the actual amount of water in the air to the amount of water the air could hold if it were completely saturated. The actual amount of water in the air does not change with a change in temperature unless the temperature drops below the saturation point and condensation occurs. However, the saturation vapor pressure corresponding to the air temperature does change with temperature, therefore, the relative humidity, for a given amount of water vapor in the air, will vary with temperature. The actual vapor pressure can be calculated if the relative humidity is known and vice-versa.

$$rh = \frac{e}{e^o} \qquad 3.17$$

Psychrometry

A very important aspect of modeling water and especially water vapor in the air is the science of psychrometry. Psychrometry deals with determining the status, movement, and control of water vapor. To accomplish this, it is necessary to use the energy relations introduced earlier in this chapter, and to introduce several new terms common to water vapor in the atmosphere.

Dew Point Temperature

The dew point temperature is the temperature at which the air becomes saturated with water vapor. Generally, the amount of water vapor in the atmosphere, a direct function of the vapor pressure, does not change during the day. However, as the temperature changes, the relative humidity, i.e., the amount of water present in the air in proportion to what the air could hold if it were saturated, does change. This can be illustrated through an example.

Example 3.11. Effect of air temperature on relative humidity
A parcel of air at 25° C has a relative humidity of 30%. What is the relative humidity if the air cools to 20° C?

Solution
The saturation vapor pressure at 25° C is:

$$e^o = \exp\left(19.0177 - \frac{5327}{(25 + 273)}\right) = 3.132\,592\,9 \text{ kPa} \qquad 3.18$$

The vapor pressure is $e = 0.3 \times 3.132\,592\,9 = 0.939\,777\,8$ kPa.

At 20° C the saturation vapor pressure is:

$$e^{\circ}_{20} = \exp\left(19.0177 - \frac{5327}{293}\right) = 2.308\ 995\ 6\ \text{kPa.}$$

Since the actual amount of water present does not change, the actual vapor pressure also does not change, therefore, the relative humidity at 20° C is:

$$\text{rh} = \frac{2.308\ 995\ 6}{0.939\ 777\ 8} = 0.407\% = 41\%$$

As the parcel of air in example 3.11 continues to cool, the relative humidity continues to increase, eventually reaching the point at which the saturation vapor pressure is equal to the actual vapor pressure. The temperature at which this occurs is the dew point temperature.

Example 3.12. Calculation of the dew point temperature
At what temperature will the vapor pressure of the water vapor in the air in example 3.11 become equal to the saturation vapor pressure?

Solution
The saturation vapor pressure is related to the temperature by equation 3.4. Therefore, the absolute temperature at which the vapor pressure is reached is:

$$e^{\circ}_{20} = 0.939\ 78 = \exp\left(19.0177 - \frac{5327}{T}\right) \qquad 3.19$$

from which T is:

$$T = \left(\frac{5327}{19.0177 - \ln(0.939\ 777\ 8)}\right) = 279.19\ \text{K} = 6.2°\ \text{C} \qquad 3.20$$

The dew point temperature for the parcel of air at 25° C and 30% rh.

The dew point temperature for any parcel of air can be calculated providing the temperature and relative humidity are known. The appropriate equation is:

$$T_{\text{dew point}} = \frac{C_2}{(C_1 - \ln(e))} \qquad 3.21$$

Enthalpy

An important parameter related to water vapor in the air is the heat content or enthalpy. This parameter is used when it is necessary to determine the energy needed to raise the temperature of a parcel of air. It is used to determine how much energy must be removed from a parcel of air to lower its temperature. The enthalpy of the air changes as the temperature or water content of the air changes. It is determined by summing up the changes in internal energy of the mixture of dry air and water vapor.

The changes in internal energy due to a temperature or water vapor change are measured from an arbitrary datum. This is proper since only changes in enthalpy are significant for calculation purposes. Under conditions normal to the green plant atmosphere, the enthalpy is calculated from 0° C and includes the internal energy of the dry air, the internal energy due to the specific heat of water from 0° C to the dew point temperature, heat of vaporization of water at the dew point temperature, and the internal energy of the water vapor from the dew point temperature to the temperature under consideration. All water vapor calculations in the SI system of units use the mass of water vapor per kilogram of dry air.

Using a base of 0° C, the enthalpy change of air between 0° C and 55° C can be obtained from:

$$H = H_{dry\ air} + H_{vaporization} + H_{water\ vapor} \quad 3.22$$

The enthalpy of dry air at temperature T is:

$$\begin{aligned} H_{dry\ air} &= \left(m_{dry\ air}\, c_{p_{dry\ air}} [T - 0]\right) \\ &= m_{dry\ air}\, c_{p_{dry\ air}} T \end{aligned} \quad 3.23$$

or, since calculations are based on a kilogram of dry air:

$$H_{dry\ air} = c_{p_{dry\ air}} T \quad 3.24$$

The water vapor term requires additional calculations since the enthalpy of the liquid water must be considered as the temperature is increased to the dew point. This can be calculated as:

$$H_{liquid} = \chi \times c_{p_{liquid}} (T_d - 0) \quad 3.25$$

As water condenses or vaporizes at the dew point, the internal energy change due to change of phase must be considered. For vaporization:

$$H_{vaporization} = \chi \times l_v \quad 3.26$$

where l_v is the latent heat of vaporization. From table 3.1, using the more accurate form:

$$\begin{aligned} H_{vaporization} &= \chi \times l_v\,(T) \\ &= \chi \times \left(2503 - 2.39\ T_{dew}\right) \end{aligned} \quad 3.27$$

The enthalpy of water vapor is calculated based on the difference in temperature between the dew point temperature and the temperature of the parcel of air.

$$H_{vapor} = \chi \times c_{vapor} (T_d - T) \quad 3.28$$

If only the difference in enthalpy is needed, equation 3.24 to 3.28 can be modified. If the parcel of air is initially at a temperature exceeding the dew point, the enthalpy difference can be calculated from:

$$H = c_{p_{dry\ air}} \Delta T + \chi \left[\left(c_{p_{vapor}} + 2.39 \right) \Delta T \right] \qquad 3.29$$

If the initial temperature of the air parcel is less than the dew point, or if air is being cooled to below the dew point, it is preferable to use equation 3.23 through 3.27 to determine the enthalpy change.

Example 3.13. Using enthalpy to calculate energy requirements
Air at 17° C and 50% rh is heated to 30° C. (a) What is the increase in enthalpy of the air?
(b) How much energy was required?

Solution
The vapor pressure of the air being heated is:

$$\begin{aligned} e^\circ &= \exp\left(19.0177 - \frac{5327}{290}\right) = 1.9131\ \text{kPa} \\ &= 0.5 \times e^\circ = 0.956\ 56\ \text{kPa} \end{aligned} \qquad 3.30$$

and the dew point is:

$$T_d = \frac{5327}{19.0177 - \ln(0.573\ 96)} \ 279\text{K} = 6.45^\circ\ \text{C} \qquad 3.31$$

The mixing ratio is then:

$$\chi_{initial} = \frac{0.956\ 56\ \text{kPa}_{H_2O}}{(100 - 0.956\ 56)\ \text{kPa}_{dry\ air}} \times 0.622 = 0.006\ 01 \ \frac{\text{kg}_{H_2O}}{\text{kg}_{dry\ air}} \qquad 3.32$$

Initially the enthalpy of the dry air is:

$$\begin{aligned} H_{dry\ air} &= \left(1 \times \frac{1\ \text{kJ}}{\text{kg}_{dry\ air}\ \text{C}} \times 17^\circ\ \text{C}\right) \\ &= 17 \ \frac{\text{kJ}}{\text{kg}_{dry\ air}} \end{aligned} \qquad 3.33$$

The enthalpy of the moist air at 17° C is the enthalpy of water from 0° C to the dew point:

$$H_{liquid} = 0.006\ 01 \ \frac{\text{kg}_{water}}{\text{kg}_{dry\ air}} \times 4.186 \ \frac{\text{kJ}}{\text{kg}_{water}} \times (6.45^\circ\ \text{C}) = 0.1624 \ \frac{\text{kJ}}{\text{kg}_{dry\ air}} \qquad 3.34$$

The enthalpy due to vaporization is:

$$H_{vaporization} = 0.006\ 01 \ \frac{\text{kJ}}{\text{kg}_{dry\ air}} \times \left((2503 - 2.39)\ 6.45^\circ\ \text{C}\right) = 14.95 \frac{\text{kJ}}{\text{kg}_{dry\ air}} \qquad 3.35$$

The enthalpy of the water vapor from the dew point to the temperature being considered is:

$$H_{vapor} = 0.006\,01 \frac{kJ}{kg_{dry\ air}} \times 1.88 \frac{kJ}{kg_{dry\ air}} \times (17°\ C - 6.45°\ C) \tag{3.36}$$
$$= 0.1192 \frac{kJ}{kg_{dry\ air}}$$

So that the total enthalpy of the air at 17° C is:

$$H_{17} = 17 \frac{kJ}{kg_{dry\ air}} + 0.1624 \frac{kJ}{kg_{dry\ air}} + 14.95 \frac{kJ}{kg_{dry\ air}} + 0.1192 \frac{kJ}{kg_{dry\ air}} \tag{3.37}$$
$$= 32.2 \frac{kJ}{kg_{dry\ air}}$$

At 30° C, the enthalpy of the dry air is:

$$H_{dry\ air} = \left(1 \times \frac{1\ kJ}{kg_{dry\ air}\ C} \times 30°\ C\right) \tag{3.38}$$
$$= 30 \frac{kJ}{kg_{dry\ air}}$$

The values of enthalpy for water to the dew point and vaporization will remain the same while the enthalpy of the water vapor from dew point to 30° C will be:

$$H_{vapor} = 0.006\,01 \frac{kJ}{kg_{dry\ air}} \times 1.88 \frac{kJ}{kg_{dry\ air}} \times (30°\ C - 6.45°\ C) \tag{3.39}$$
$$= 0.2660 \frac{kJ}{kg_{dry\ air}}$$

And the enthalpy of the air at 30° C is:

$$30 \frac{kJ}{kg_{dry\ air}} + 0.1624 \frac{kJ}{kg_{dry\ air}} + 14.95 \frac{kJ}{kg_{dry\ air}} + 0.2660 \frac{kJ}{kg_{dry\ air}} \tag{3.40}$$
$$= 45.4 \frac{kJ}{kg_{dry\ air}}$$

The increase in enthalpy is:

$$H = 45.4 \frac{kJ}{kg_{dry\ air}} - 32.2 \frac{kJ}{kg_{dry\ air}} = 13.2 \frac{kJ}{kg_{dry\ air}} \tag{3.41}$$

The previous calculations can be streamlined if the temperature of the initial condition of the air parcel exceeds the dew point. For this condition, the calculations involving the liquid water cancel out. Furthermore, only the change in the latent heat of vaporization must be considered. Under these conditions, the change in enthalpy can be calculated from the change in temperature using equation 3.29 providing the mixing ratio is known. The equation is:

$$H = c_{p_{dry\ air}} \Delta T + \chi \left[\left(c_{p_{vapor}} + 2.39 \right) \Delta T \right] \qquad 3.29$$

Example 3.14. Enthalpy change
For the conditions of example 3.13, compute the change in enthalpy using equation 3.41.

Solution
The enthalpy change will be:

$$H = \left[1 \frac{kJ}{kg_{dry\ air} \times K} \times + 0.006\ 01 \frac{kg_{H_2O}}{kg_{dry\ air}} \left(1.88 \frac{kJ}{kg_{dry\ air} \times K} + 2.39 \frac{kJ}{K} \right) \right] \times (30 - 17) = 13.3 \frac{kJ}{kg_{dry\ air}} \qquad 3.42$$

While equation 3.41 shortens the calculation, the user must be aware that it is not universal. A second caution is also in order. If calculations involve temperatures below 0° C, then it is necessary to use the latent heat of fusion. Since enthalpies are measured from 0° C, values for ice, water, or water vapor at temperatures less than 0° C will be negative.

Wet Bulb Temperature

We have discussed relative humidity without showing how it is measured in practice. Obviously, one way to determine the relative humidity would be to cool a parcel of air until it reached the dew point temperature. At the dew point temperature, the air becomes saturated with water vapor and condensation occurs. The vapor pressure at saturation is the actual vapor pressure of the air when the air was at its initial temperature. The change of temperature from its initial value to the dew point at which saturation does not alter the vapor pressure. The temperature change only alters the relative humidity. If the initial temperature and the temperature of the surface upon which condensation is occurring is known, the Clausius-Clapeyron equation can be used to determine the two vapor pressures from which the relative humidity ratio is determined.

The more common method of determining the relative humidity involves the use of wet and dry bulb thermometers along with a psychrometric chart. We will first discuss the theory and use of the wet and dry bulb thermometer, then examine the psychrometric chart.

If the bulb of a thermometer is covered with a wetted fabric and exposed to an air stream ($v \geq 4.6$ m/s), evaporation of the water will occur, carry away the latent heat of vaporization and cool the bulb. The rate at which evaporation is occurring, i.e., the rate at which latent heat is removed, is a function of the amount of moisture present in the air. The cooling will occur at a rate related to the vapor pressure of the air. The temperature reached by the wet bulb thermometer is the wet bulb temperature.

The wet bulb temperature is greater than the dew point temperature and is only equal to the dew point temperature if the air is completely saturated with water, i.e., the air is already at dew point. For any other condition, a relationship between the rate of evaporation, i.e., the rate of heat removal from the wetted bulb and the relative amount of water the atmosphere can hold at the temperature of a dry bulb, determines the wet bulb temperature. An expression relating the vapor pressure in the air and the wet bulb temperature can be developed as follows.

The rate of heat removal from the wetted bulb as a function of the difference in moisture content of the atmospheric air is:

$$E = l_v \left(\frac{e^\circ_{wet\ bulb} M_{water}}{P\ M_{air}} - \frac{e_{air} M_{water}}{P\ M_{air}} \right)$$

$$= \left(\frac{l_v M_{water}}{P\ M_{air}} \right) \left(e^\circ_{wet\ bulb} - e_{air} \right) \quad 3.43$$

where

E	=	energy removed (kJ/kg_{air})
l	=	heat of vaporization of water [(2590 – 2.39)kJ/kg]
$e^\circ_{wet\ bulb}$	=	vapor pressure of the saturated air surrounding the wetted bulb
M_{water}	=	gram molecular weight of water (18 kJ/kg_{water})
e_{air}	=	vapor pressure in the atmosphere (kPa)
M_{air}	=	gram molecular weight of air (28.964 kg/kmol)

The rate at which the air can absorb energy is a function of the specific heat of the air and the difference in temperature between the air and the wet bulb:

$$E = c_{p,\ air} \left(T_{air} - T_{wet} \right) \quad 3.44$$

where

E	=	energy absorbed in the moving air (kJ/kg_{air})
$c_{p,air}$	=	specific heat of the air (1 $kJ/kg_{air} \times {}^\circ C$)
T_{air}	=	air temperature (° C)
T_{wet}	=	wet bulb temperature (° C)

By equating equation 3.26 and 3.27 and solving for e_{air}, one obtains:

$$e_{air} = e^\circ_{wet} - \frac{c_p\ P\ M_{air}}{l_v\ M_{water}} \left(T_{air} - T_{wet} \right) \quad 3.45$$

The equation shows the basis upon which the actual air temperature is obtained from the dew point temperature where:

$$e^\circ_{wet\ bulb} = \exp \left(19.0177 - \frac{5327}{\left(T_{wet} + 273 \right)} \right) \quad 3.46$$

and all other symbols are as defined above. To calculate the relative humidity, it is necessary to determine the saturation vapor pressure at the dry bulb temperature and form the ratio e_{air}/e°_{air}.

The reader is cautioned that equation 3.45 is approximate; it is necessary to know the actual atmospheric pressure at the time the determination is made. Further, too slow an air speed around the wick, a dirty wick, or water containing impurities used to wet the wick may alter the outcome. More careful calculations may also require knowledge of the Prandt and Schmit dimensionless numbers to determine the exact rate of heat removal. However, for most engineering applications, equation 3.45 will provide a satisfactory answer. If the air temperature is below freezing, however, the equation becomes invalid because of the change in the slope of the saturation vapor pressure-temperature line.

Example 3.15. Use of the wet bulb thermometer
A sling psychrometer displays a dry bulb temperature of 22° C and a wet bulb temperature of 18° C. What is the relative humidity of the air?

Solution
The saturation vapor pressure at the temperature of the bulb is:

$$e^{o} = \exp\left(19.0177 - \frac{5327}{295}\right)$$
$$= 2.611\ 886\ \text{kPa} \qquad 3.47$$

The latent heat of vaporization at the wet bulb temperature is:

$$l_v = 2503 - 2.3918\ \frac{\text{kJ}}{\text{kg}_{H_2O}}$$
$$= 2459.98\ \frac{\text{kJ}}{\text{kg}_{H_2O}} \qquad 3.48$$

The amount to be subtracted is:

$$\frac{c_p\ P\ M_{air}}{l_v\ M_{water}}\left(T_{air} - T_{wet}\right)$$
$$\frac{\dfrac{1\ \text{kJ}}{\text{kg}_{\text{dry air}}\ \text{K}}\ 100\ \text{kPa} \times 28.964\ \dfrac{\text{kg}_{\text{dry air}}}{\text{kmol}}}{2459.98\ \dfrac{\text{kJ}}{\text{kg}_{H_2O}} \times 18\ \dfrac{\text{kg}_{H_2O}}{\text{kmol}}}\left(22 - 18\right)\text{K} = 0.261\ 646\ \text{kPa} \qquad 3.49$$

And the actual vapor pressure at the wet bulb temperature is:

$$e = \exp\left(19.0177 - \frac{5327}{291}\right) - 0.261\ 646\ \text{kPa}$$
$$= +\ 2.037\ 774\ \text{kPa} - 261\ 646\ \text{kPa} = 1.776\ 13\ \text{kPa} \qquad 3.50$$

$$\text{rh} = \frac{1.776\ 13}{2.611\ 09} = 0.68 = 68\% \qquad 3.51$$

Psychrometric Chart

The psychrometric chart is a graphical formulation of the properties of water vapor in the atmosphere. The information required to enter the psychrometric chart generally includes the dry bulb air temperature and either the wet bulb temperature or the relative humidity. Given this basic information, it is possible to determine the dew point temperature, the enthalpy, and the mixing or humidity ratio as well as the specific volume (the volume per kilometer of dry air).

Figure 3.3 is the psychrometric chart for temperatures from 0° C to 50° C, the normal range of temperatures of interest in the green plant environment. On the psychrometric chart, the dry bulb or ambient temperatures are along the bottom of the chart. Wet bulb temperatures are indicated by the dotted lines slanting from upper left to lower right on the chart. Relative humidities are indicated by the solid lines curving from lower left to upper right, while enthalpies can be obtained from the solid lines which almost parallel the wet bulb temperature lines. The solid lines which slant steeply from upper left to lower right delineate the specific volume of dry air while the horizontal lines parallel to the dry bulb temperature axis enable one to read the mixing or humidity ratio.

Example 3.16. Effect of temperature on relative humidity
Use the psychrometric chart to determine the relative humidity of a parcel of air at 25° C which initially had a relative humidity of 30%, but cooled to 20° C.

Solution
Using figure 3.3, enter the psychrometric chart on the horizontal axis at a dry bulb temperature of 25° C, and proceed vertically upward until the 30% rh line is encountered. Since the vapor pressure exerted by the water vapor present in the parcel of air will not change as the parcel cools, neither will the mixing or humidity ratio. Therefore, one can proceed horizontally in the direction of cooler dry bulb temperatures until the vertical 20° C is encountered (at this point just above the 40% rh line). The relative humidity at 20° C is then 41% which agrees with the value calculated in example 3.10.

Example 3.17. Calculation of dew point temperature using the psychrometric chart
As the parcel of air in example 3.10 cools, at what temperature is the dew point reached?

Solution
By continuing to the left in the direction of cooling temperature from the intersection of the 25° C and 30% rh line, the 100% rh line is encountered just to the right of the 6° C vertical line. Thus, it appears that the dew point temperature is about 6.2° C, which agrees with example 3.12.

The enthalpy and changes in enthalpy can be quickly calculated from the psychrometric chart. The values on the psychrometric chart are calculated from extensive computer programs. Thus, while chart values may not agree immediately with hand calculated values, differences will be completely adequate for engineering purposes.

Example 3.18. Increases in enthalpy from the psychrometric chart.
Perform example 3.13 using the psychrometric chart.

Solution
Enter the psychrometric chart at the vertical 17° line and proceed upward until the 50% rh line is intersected. Then, using a triangle, align the triangle parallel to a solid slanting line which intersects the enthalpy axis. Keeping it parallel to this line, move the triangle until it intersects the point defined by 17° C vertical line and 50% rh line. Read and note the value of enthalpy for this condition numerically as 32.2. Since the mixing or humidity ratio remains constant as the air is heated, proceed horizontally to the right until the 30° C line is intersected. From this point, keeping the triangle parallel to the solid enthalpy line, proceed upward to the left and read the enthalpy numerically as 45.3. The difference between the two enthalpy values gives the increase in enthalpy or:

$$H = 45.3 \frac{kJ}{kg_{dry\ air}} - 32.2 \frac{kJ}{kg_{dry\ air}} = 13.1 \frac{kJ}{kg_{dry\ air}} \qquad 3.52$$

This value agrees to the third significant digit with the value calculated in example 3.13. Since the increase in enthalpy must be due to the energy added, the stated value also represents the energy which must be added to the air.

One of the most important uses of the psychrometric chart is in conjunction with a sling psychrometer. This instrument consists of two thermometers mounted horizontally to rotate about a vertical axis. The bulbs of the thermometers are farthest from the axis and one bulb is covered with a cloth which is wetted with distilled water. When the two thermometers are rotated by swinging them, air rushing past the wet cloth evaporates the water on the cloth, cooling the wet bulb thermometer. This yields the wet bulb temperature. The relative humidity can then be determined from a psychrometric chart.

Example 3.19. Using a sling psychrometer to determine properties of moist air
What are the relative humidity, mixing ratio, dry air density, and enthalpy for the air of example 3.15 ($T_{dry\ bulb} = 22°$ C, $T_{wet\ bulb} = 18°$ C).

From the psychrometric chart, at the intersection of the vertical 22° C line and the slanting, dashed 18° C wet bulb line, the following values can be read:

relative humidity = 68%
mixing ratio = 11.35 $kg_{H_2O}/kg_{dry\ air}$
specific volume = 0.851 $m^3/m_{dry\ air}$ so the dry air density is 1.175 $kg_{dry\ air}/m^3$
enthalpy = 50.7 $kJ/kg_{dry\ air}$

All values that could be calculated using the equations presented in this chapter, but, within engineering accuracy, can be obtained much faster using the psychrometric chart.

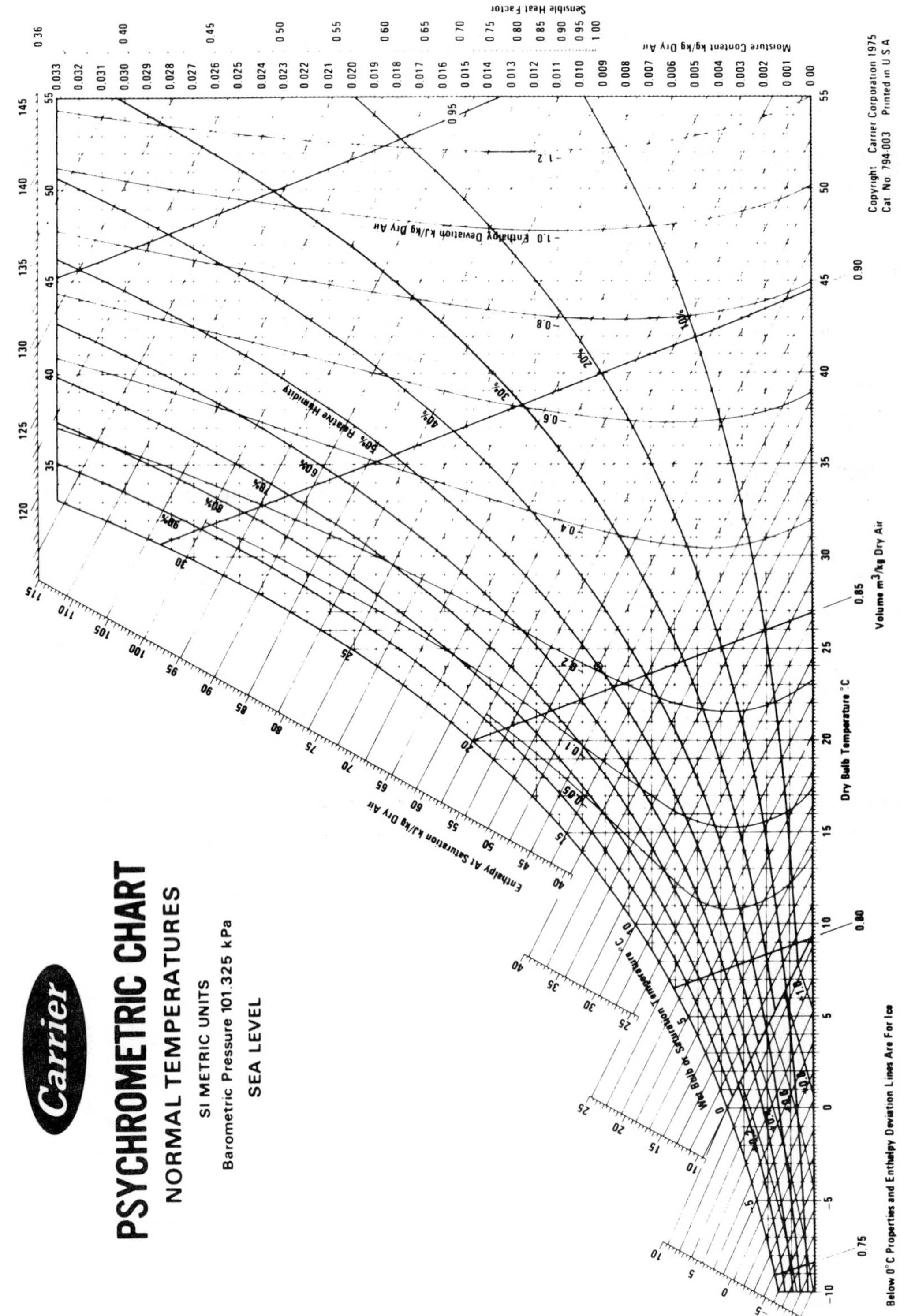

10 ASAE STANDARDS 1990

Figure 3.3. The psychrometric chart describing the properties of water vapor at standard pressure (101.325 kPa) between 0° C and 40° C.

Exercises

3.1. An inventive farmer asks you to design a snow-making machine which he will use to generate snow to cover his blueberry crop to protect it from winter damage. He has 8 ha of blueberries. Answer the questions below to decide whether or not the idea is practical:

(a) If water is available at 20 L/min at 0° C and ambient air at –5° C will be used to carry away the latent heat of fusion, at what rate (m^3/s) must air be supplied to prevent the air temperature from rising above 0° C?

(b) Your initial idea is to spray the water into an airstream by centering the nozzle in a 0.4 m diameter orifice. At what velocity (m/s) must the air exit the orifice?

(c) If the snow which could be produced has a density 1/6 of that of the water, how long would it take (days) to cover the 8 ha with an average 1 cm depth of snow?

3.2. On a particular date the time temperature was 15° C with 70% rh. Assuming the actual vapor pressure of the air did not change, and that latent heat of vaporization released during dew formation is sufficient to prevent the nighttime temperature from falling further, what temperature do you expect to be reached at night?

3.3. In terms of percent by volume, normal dry air has constituents:
78.09 N_2 , 20.95 O_2 , 0.93 Ar, 0.03 CO_2.

While moist air at 30° C and 80% rh has constituents:
74.94 N_2 , 20.10 O_2 , 0.89 Ar , 0.03 CO_2 , 4.04 H_2O.

(a) Compute the apparent gram molecular weight of dry air at 30° C and moist air at 30° C and 80% rh. Use six significant digits in computing gram molecular weights.

(b) Compute the density of the dry and moist air using the apparent gram molecular weights calculated in 3.3(a).

(c) Would you expect to find the higher vapor pressure near the top of the plant canopy or within the plant canopy near the soil surface?

3.4. A 1.4 kW heating element is adding energy to water. At the time the water begins to boil, the room is at 25° C and 75% rh. The dimensions of the room are 4 m × 4 m × 2.13 m.

(a) If no air exchange takes place, how long will it take for the air in the room to reach 100% rh?

(b) Air is to be removed from the room by a fan placed above the boiling water. If the air entering the 76.2 mm diameter fan duct is at 30° C and 80% rh, at what rate should the fan remove air (m_3/min)? Makeup air enters the room at 25° C and 75% rh.

(c)What should the fan watt rating be if it is assumed to operate at 90% electrical efficiency and 40% fan efficiency?

(d)If the fan is only 50% efficient in removing the water vapor and the temperature of the inside surface of a window in the room is 20° C, how long will it take before the window begins to sweat. Assuming water vapor is distributed uniformly.

Chapter 4

Water Movement in the Soil-Plant-Atmosphere Continuum

Water is a vital element in every phase of the soil-plant-atmosphere continuum. It is important to know as much as possible about its movement, both in the green plant as well as in the biosphere, the medium in which the plant grows. Water is the solvent and the medium for the biochemical processes that control plant growth. Water transports nutrients through the soil to plant roots. It serves as the medium to transport food through the green plant. It is active in both food storage and as well as food metabolism which supplies the energy for development and growth. It is a necessary ingredient in photosynthesis, furnishing the hydrogen and oxygen atoms which combines with carbon from carbon dioxide to form the sugars which are the foodstuff of the growing plant. Oxygen from water and/or carbon dioxide is released to the atmosphere as a result of photosynthesis, thus providing the essential gas which supports animal life on the earth.

Transpiration, the evaporation of water from the intercellular spaces cools the plant. The water vapor passes through the stomata, the openings in the leaf's surfaces, thereby taking with it the radiant energy through the latent heat of vaporization. In this way, water maintains leaf temperatures at the level necessary for biochemical activity.

Example 4.1. Transpiration and water use
The leaf area index, LAI, is the ratio of actual leaf area to ground surface area occupied by the canopy. Measurements of transpiration from a crop with 3.1 LAI were found to remove 1.8 $kg_{water}/m^2leaf \times day$ during the period from 9:30 A.M. to 3:30 P.M., when 60% of the water use of the crop occurs. Estimate the water use in mm_{water}/day for the crop. Estimate also the rate of energy removal by an individual leaf in $J/h \times m^2$.

Solution
The amount of water evaporated from the ground surface per day is:

$$\frac{1.8\ kg_{water}}{m^2_{leaf} \times d} \times \frac{1}{0.6} \times \frac{3.1\ m^2_{leaf}}{m^2_{soil}} \times \frac{m^3_{water}}{1000\ kg_{water}} \times \frac{1000\ mm}{m} = 9.3\ \frac{mm_{water}}{m^2_{soil} \times d} \qquad 4.1$$

The average energy removal for an individual leaf is:

$$\frac{1.8\ kg_{water}}{m^2_{leaf} \times d} \times \frac{d}{6\ h} \times \frac{1}{0.6} \times \frac{2470\ kJ}{kg_{water}} = \frac{1240\ kJ}{m^2_{leaf} \times h} = 1.24\ \frac{MJ}{m^2_{leaf} \times h} \qquad 4.2$$

Direct measurements on transpiring and non-transpiring leaves have shown that leaf temperatures as much as 2° C below ambient can be maintained by transpiration.

Turgor Pressure and the Green Plant

The young green plant owes its form to the pressure developed within individual green plant cells (see chapter 5). Pressure is important for germination as it expands the cells of the developing embryo to force it to emerge from the seed. Pressure in root cells causes the roots to penetrate the soil pores. Cellular pressure is active in emergence forcing the shoot to break through the ground surface and exposing the leafy portion of the plant to the sunlight, initiating photosynthesis.

Pressure within individual cells is maintained using water absorbed from the soil by the plant roots. Through thermodynamic processes, water is transported through the plant by the vascular system to the cell walls. Osmosis moves water through the semipermeable plasma membrane surrounding the protoplasm, the living portion of the green plant cells. Turgor pressure is created by the semipermeable membrane which is confined by the microfibrilic structure of the cell wall. Through the action of thousands of cells expanding against one another, structure is given to the various portions of the young green plant.

Water vapor which exits from the leaves due to transpiration reduces the density of the air surrounding the plant. Lowered air density promotes air movement to create circulation patterns which remove the oxygen-laden moist air and replace it with drier, carbon dioxide-laden air, supporting photosynthesis and enhancing growth.

Water Deficiency and Stress

A deficiency of water stresses the green plant and slows or even halts the biochemical processes of photosynthesis and metabolism. Excessive water in the soil causes similar stress conditions. When the soil is saturated there are no pathways for oxygen to diffuse through the soil. There is no support for the respiratory processes of metabolism essential for root growth. Products of metabolism such as carbon dioxide are prevented from diffusing away from the metabolizing sites. This results in carbon dioxide saturation and eventually halts all metabolic processes. Metabolic processes in the roots appear to play a role in the absorption of water by the root. Plants stressed by excessive soil water lose turgor pressure and wilt. Thus, excessive water may prevent the root from withdrawing water from the soil to replenish water lost by transpiration. Unless the excessive soil water is removed, death of the plant soon follows.

Forces and Potentials Causing Water Movement

Since water is so important to the growth of the green plant, it is important to understand the forces which move water through the biosphere. There is no single force responsible for water movement. Instead, a contingent of forces acting in concert is responsible. Gravitational forces are responsible for mass movement of liquid water when the soil is at or near saturation. Electrical, physical, and chemical forces bind water molecules to each other, to the soil particles, and to the chemical constituents resulting from growth processes.

Electrical forces bind polar water molecules to charges on soil particles. Electrical transfers are also involved in photosynthesis. Physical forces such as those due to surface tension cause water to move against gravity, as when a column of soil is seen to wet upward from a source of water at a lower elevation. Physical forces also cause vapor-saturated air to become more buoyant than dry air, promoting circulation of air within the green plant canopy. Chemical forces bond water to plant membranes. Chemical forces result in osmosis, the movement of water through semipermeable membranes. Finally, thermodynamic transfers of energy are involved in the change of phase from liquid to vapor which occurs during transpiration.

Modeling water movement through the soil-plant-atmosphere-continuum is a function of potential energy differences. As the potential energy of position increases when a mass is elevated some distance above a datum, a potential for water movement represents a measure of the energy which is available to move water.

The potential energy of a mass, m, is extensive, i.e., the amount of energy required depends on the size or amount of the mass. To model water movement, an intensive potential is useful. An intensive potential does not depend on size or amount but is defined at a point and is independent of mass or volume. What is desired is a single parameter which incorporates all of the forces involved in moving water. These various forces arise from different phenomena, i.e., electrical, mechanical, chemical, etc. The additive nature of intensive parameters will prove to be useful. This will allow for pointwise definition of the potentials. Once the forces acting on a water particle are defined at some point, the intensity of the potential at that point will be known.

An ideal choice is to approach the potential-causing water movement in the soil-plant-atmosphere continuum from a chemical viewpoint. Water is involved in processes in the green plant and much of the activity is due either to a change of phase or results from the presence or addition of solutes. The basis of the sought-after potential is Gibbs free energy, the energy which is available to move water. While a knowledge of the Gibbs free energy does not allow one to determine the *amount* of water movement that may occur, it does enable one to determine in which *direction* the movement will take place.

Chemical Potential

Water is unique because it exists both as a liquid and a gas at the temperatures of the green plant environment. When water is a vapor it behaves as a perfect gas. This property makes it easier to describe the properties of water.

For a gas, the state variables, i.e., those parameters needed to describe the beginning and ending states of a gas, are manipulated to determine what can be accomplished. We will briefly review the state variables that are important to better understand the ramifications of their application.

In a gas, the variables needed to determine the state of the system are the pressure, P, volume, V, and temperature, T, of the gas along with the total mass of the gas. In addition, it is necessary to know the mathematical operations that are permissible between the state variables. These include a knowledge of the actions that change the state parameters such as adding heat to the system and partitioning the heat between the components of the enthalpy, H. Partitioning can increase the internal energy, U, and modify the volume, V, or the pressure, P, to obtain work from the system.

Once heat has been added to a system, if no exchange of matter occurs between the system and it's environment, work can be obtained from the system. However, only part of the energy can become available for useful output, the remainder remains unavailable. Therefore, it is necessary to introduce another parameter to account for the unavailable energy before the system can be thermodynamically modeled. This parameter is the entropy, S. The unavailable energy, in a cyclical system is discarded to some lower temperature, is denoted by the entropy of the system multiplied by the temperature at which the unavailable energy is discarded. An example may help explain the concept.

Consider a small amount of water isolated in a piston to which heat may be added as in figure 4.1. The piston is surrounded by infinite volume of air at atmospheric pressure. To the water a small amount of heat energy, dQ obtained by radiation is added causing some of the water to change phase. The water that evaporates does so at constant temperature, since only enough heat was added to cause evaporation, so the temperature of the water did not change. The water that evaporates expands, pushing the piston outward against atmospheric pressure to do work, PdV. If the process is done very slowly with very small amounts of heat it fulfills the requirement of reversibility. In a reversible system, the transformation can be modeled as a change in internal energy equal to an increment of heat minus an increment of work. Consider the heat added positive and the work performed negative. Both actions, adding heat and obtaining work, alter the internal energy of the system, therefore:

$$dU = dQ - dW \qquad 4.3$$

where

dU = change in internal energy
dQ = amount of heat added to the system
dW = work obtained from the system

The change in internal energy is the difference between the heat added and the work obtained. Since the piston expands against a constant atmospheric pressure, it performs work. From the principles of basic physics:

$$dW = PdV \qquad 4.4$$

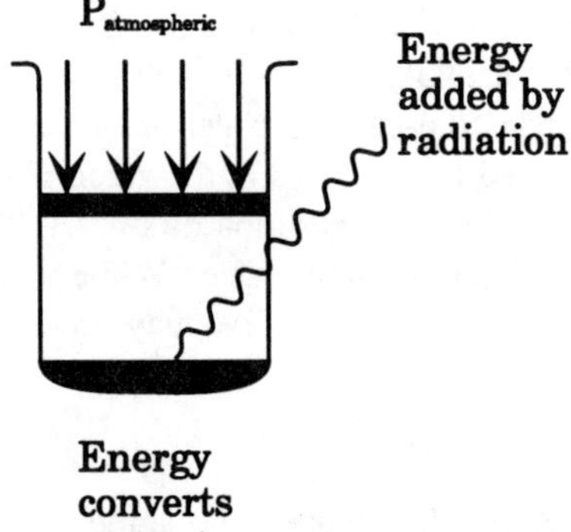

Energy converts water to vapor causing gas to expand

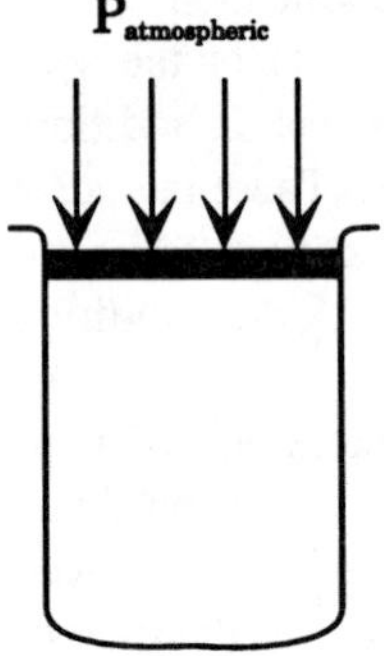

Figure 4.1. Work is done by the water vapor as it expands against atmospheric pressure.

Since the heat was added reversibly:

$$dQ = TdS \tag{4.5}$$

where

T = temperature that heat was added

S = entropy transferred to the system

By using equations 4.4 and 4.5, we can obtain:

$$dU = TdS - PdV \tag{4.6}$$

In equation 4.5, dU represents the change in internal energy which should be available to perform work, dW, while TdS represents that energy which is unavailable to accomplish work.

Example 4.2. Heat energy and work
A cylinder piston arrangement contains 100 g of water at 100° C. Sufficient heat is added to the water to cause it to evaporate so that the water vapor expands the piston against atmospheric pressure performing work. How much heat must be added to the water assuming the addition takes place reversibly and no heat is lost? How much work is performed during the expansion?

Solution
From table 3.1, the energy required to evaporate the water is:

$$0.1\ \text{kg} \times \left(2503\ \text{kJ} - 2.39\ \frac{\text{kJ}}{°\text{C}}\,(100°\ \text{C})\right)\frac{\text{kJ}}{\text{kg}} = 226.4\ \text{kJ} \qquad 4.7$$

The final volume of the gas is:

$$\frac{100\ \text{g}}{18\ \frac{\text{g}}{\text{mol}}} \times \frac{8.314\ \text{kPa}\cdot\text{L}}{\text{K}\cdot\text{mol}} \times \frac{373\ \text{K}}{100\ \text{kPa}} \times \frac{\text{m}^3}{1000\ \text{L}} = 0.172\ \text{m}^3 \qquad 4.8$$

Since:

$$100\ \text{kPa} = 100\ \frac{\text{N}}{\text{m}^2} = 100\ \frac{\text{N}\cdot\text{m}}{\text{m}^3} = 100\ \frac{\text{kJ}}{\text{m}^3} \qquad 4.9$$

The work performed by the gas against atmospheric pressure is:

$$\left(0.172\ \text{m}^3 \times 100\ \frac{\text{kPa}}{\text{m}^3}\right) = 17\ \text{kJ} \qquad 4.10$$

While 247 kJ of heat energy was added to the water, only 17 kJ of work was performed. The excess energy was transferred into rotational and vibrational energy of the water vapor molecules to increase the enthalpy of the water vapor.

A convenient expression of the available and unavailable energy was formulated by Willard Gibbs (1839-1903) using state variables of a gas in terms of its free or available energy. The Gibbs free energy function is:

free energy = total energy – unavailable energy

or

$$G = U + PV - TS \qquad 4.11$$

where
- G = free or available energy
- U = internal energy of the liquid/vapor system
- PV = heat energy due to molecular motion and available to do work
- TS = energy unavailable for work

In the form depicted above, the Gibbs free energy applies to a closed system, i.e., one not exchanging mass or energy with the outside environment. If the system is open, as occurs in a leaf where water is evaporating from the intercellular surfaces through the stoma of the leaf, it is necessary to add a term to the function to account for exchanges of mass. In an open system which we will use in the green plant environment, the expression is:

$$G = U + PV - TS + \Sigma_i n_i \qquad 4.12$$

where n_i represents number of moles of the *ith* component of the system and all other parameters are as previously defined.

In modeling a thermodynamic process, only the *changes* in the free energy are of interest. The *actual magnitude* of the internal energy of a system is unknown. To determine the changes in the system, since the free energy function is composed of state variables, we are allowed to differentiate:

$$dG = dU + PdV + VdP - TdS - SdT + \sum_i \left.\frac{\partial G}{\partial n_i}\right|_{i=j} dn_i \qquad 4.13$$

Making use of equation 4.6 allows equation 4.12 to be written:

$$dG = VdP - SdT + \sum_i \left.\mu_i\right|_{j \neq i} dn_i \qquad 4.14$$

The term $\partial G_i/\partial n_i$ is the partial molar Gibbs free energy with respect to the moles of the *ith* component. It is the chemical potential of the *ith* component and is denoted by the symbol μ_i. The chemical potential is the internal energy which is added to the system due to an increase in the amount of *ith* component.

The concept of chemical potential is difficult to visualize. If potential energy of position were being modeled and additional mass were added to that already above the datum, the potential energy of the total would be increased. Similarly, if additional mass is added to that already present, the internal energy of the overall system would increase. Since internal energy is potentially available to perform work, the potential of the system has increased.

The chemical potential is an intensive property of the system. While the Gibbs free energy depends on the whole of the system, the chemical potential, being the rate of change of the Gibbs function with respect to a component is intensive. This is important because the chemical potential gives information about the energy status at a point, independent of the size of the entire system.

Gibbs free energy relates to the energy which is available to perform useful work. However, it is extensive. It is more valuable to be able to define system parameters intensively so that a knowledge of the extent of the system is not necessary. This can be done as follows. From equation 4.4 above at n_i and T constant:

$$dG = VdP$$

and, after performing ∂n_i on both sides and inverting the order of differentiation:

$$\frac{\partial}{\partial P}\left.\frac{\partial G}{\partial n_i}\right|_{T,\,n_i} = \left.\frac{\partial \overline{G}}{\partial P}\right|_{T,\,n_i} = \left.\frac{\partial V}{\partial n_i}\right|_{T,\,n_i} = \overline{V}_i \qquad 4.15$$

Similarly, at n_i and P constant:

$$\frac{\partial}{\partial T}\frac{\partial G}{\partial n_i}\bigg|_{p,\,n_i} = \frac{\partial \overline{G}}{\partial T}\bigg|_{p,\,n_i} = \frac{\partial S}{\partial n_i}\bigg|_{p,\,n_i} = \overline{S}_i \qquad 4.16$$

The terms $\overline{V}_i$ and $\overline{S}_i$ are the partial molar volume and entropy with respect to the *ith* component of the system. They are both intensive because they involve a differentiation with respect only to the *ith* component. Using this approach, equation 4.14 has been converted into an expression whose right side is composed only of intensive parameters.

$$dG = \overline{V}_i\, dP - \overline{S}_i\, dT + \sum_i \mu_i\Big|_{j=i} dn_i \qquad 4.17$$

One additional term is encountered when one is dealing with a system composed of several components. The number of moles of a component is extensive, however, the mole fraction is intensive. The mole fraction is the number of moles of that component divided by the total number of all moles present in the system. Denoting the fraction of the *ith* component by N_i, we write the mole fraction as:

$$N_i = \frac{n_i}{\sum_i n_i} \qquad 4.18$$

The basis of the chemical contribution of water to the potential of a solution is the partial molar Gibbs free energy with respect to the number of moles of water present. The chemical potential of water is expressed as μ_w.

The actual potential depends on the datum from which the potential value is measured. However, the actual value of the potential is not as important as the difference in potential between two points. The difference in chemical potential determines the direction in which a process will proceed. In the case of water in the soil-plant-atmosphere continuum, it is sufficient to be concerned only with differences in chemical potential between points in the water pathway through the biosphere. This is accomplished by defining an expression for $d\mu_w$, the differential in the chemical potential of water between two points. We proceed as follows.

To obtain an expression for $d\mu_i$ it is necessary to express the chemical potential as a function of the parameters P, T, and the number of moles of the *ith* component. This yields:

$$\mu_i = \mu_i\,(P,\, T,\, n_i)\; i = 1, 2, \cdots k$$

Now differentiate to obtain the total differential of μ_i:

$$d\mu_i = \frac{\partial \mu_i}{\partial P}\bigg]_{T,\,n_i} dP + \frac{\partial \mu_i}{\partial T}\bigg]_{P,\,n_i} dT + \sum_i \frac{\partial \mu_i}{\partial n_i}\bigg]_{P,\,T,\,n_j} dn_i \qquad 4.19$$

From equations 4.14 and 4.15, equation 4.19 may be written:

$$d\mu_i = \overline{V}_i\, dP\Big]_{T,\,n_i} + \overline{S}_i\, dT\Big]_{P,\,n_i} + \sum_i \frac{\partial \mu_i}{\partial n_i}\bigg]_{T,\,P,\,n_j} dn_i \qquad 4.20$$

Since we are interested in the chemical potential of water, the equation can be rewritten:

$$d\mu_w = \overline{V}_w\, dP\Big]_{T, n_w} + \overline{S}_w\, dT\Big]_{P, n_w} + \frac{\partial \mu_w}{\partial n_w}\Big]_{T, P, n_s} dn_w + \frac{\partial \mu_s}{\partial n_s}\Big]_{T, P, nw} dn_s \qquad 4.21$$

It is convenient to express equation 4.21 in terms of mole fractions rather than numbers of moles present. When this is done, equation 4.21 becomes:

$$d\mu_w = \overline{V}_w\, dP\Big]_{T, n_w} + \overline{S}_w\, dT\Big]_{P, n_w} + \frac{\partial \mu_w}{\partial N_w}\Big]_{T, P, n_s} dn_w + \frac{\partial \mu_s}{\partial N_s}\Big]_{T, P, n_w} dN_s \qquad 4.22$$

Equation 4.22 is an expression for the chemical potential of water as a function of the state variables P, T, N_j. In equation 4.22 when the partial derivative to form $\overline{V}_i$ is taken, temperature and composition over all components are held constant. Similarly pressure and all components are constant for $\overline{S}_i$. In addition to holding T and P constant, all components for the last partials except the particular one for which the partial is taken, are assumed to be held constant. Thus, for $\partial\mu_w/\partial N_w$, only the water component is allowed to vary. Likewise, for $\partial\mu_s/\partial N_s$, the solute content is assumed to vary while the water content is held constant.

Example 4.3. Differential of chemical potential
Assume a system with two constituents contains n_1 moles of water and n_2 moles of solute. At constant atmospheric pressure and ambient temperature, write:

(a) The differential of the chemical potential of water with a change in the number of moles of water.

(b) The differential of the chemical potential of water when the number of moles of solute changes.

(c) The mole fractions of water and solute.

(d) The differential of the chemical potential of water in terms of mole fractions assuming the number of moles of solute remains constant.

Solution
(a) To obtain the differential of the chemical potential with a change in the number of moles of water, noting that only the water content changes from equation 4.19, we can write:

$$d\mu_w = \sum_i \frac{\partial \mu_i}{\partial n_i}\Big]_{P, T, n_j} dn_i$$

$$d\mu_w = \frac{\partial \mu_w}{\partial n_w} dn_w + \frac{\partial \mu_s}{\partial n_s} \times 0 \qquad 4.23$$

$$d\mu_w = \frac{\partial \mu_w}{\partial n_w} dn_w$$

(b) The differential of the chemical potential of water assuming only the number of moles of solute change is:

$$d\mu_w = \frac{\partial \mu_s}{\partial n_s} dn_s \qquad 4.24$$

The mole fractions of water and solute are:

$$N_w = \frac{n_w}{n_w + n_s}$$
$$N_s = \frac{n_s}{n_w + n_s} \qquad 4.25$$

The differential of chemical potential of water in terms of the mole fractions, however, is:

$$d\mu_w = \frac{\partial \mu_w}{\partial N_w} dN_w + \frac{\partial \mu_s}{\partial N_s} dN_s \qquad 4.26$$

because the mole fraction of the solute changes with a change in the number of moles of water, as can be seen from equation 4.2(c).

Potential of Water as a Vapor

An important chemical phenomenon occurring in the soil-plant-atmosphere continuum involves the change of phase of water from liquid to vapor. In figure 4.2, the bell jar, and the container of water can be considered a closed system. If the air above the liquid under the bell jar is not saturated, water will continue to evaporate until saturation is achieved. When saturation occurs, the vapor and the liquid are in equilibrium.

Since movement of water occurs from the container to the air, the air above the liquid is at a lower water potential than the liquid. The system in figure 4.2 has a constant temperature and a constant overall pressure, but the vapor pressure of the air is less than the vapor pressure of the liquid, Based on these conditions, we can develop a model for the chemical potential of water vapor. From equation 4.19, with no solute present in the liquid and considering the entire system to be closed, the chemical potential difference between the liquid and the vapor is:

$$d\mu_w = \overline{V}_w \, de \qquad 4.27$$

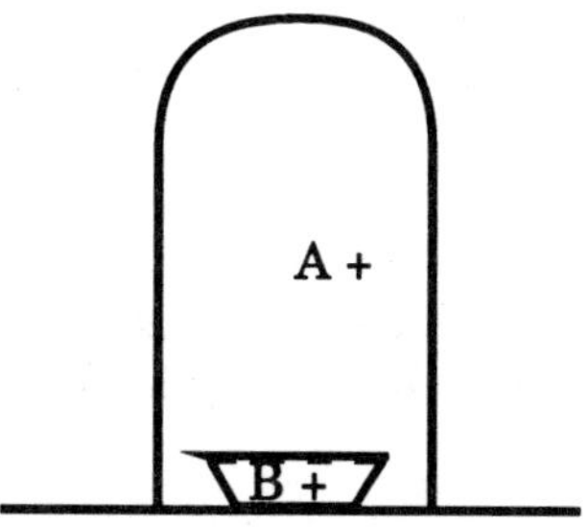

Figure 4.2. In an isolated system, the chemical potential of water at point A will be the same as that of the liquid water at point B when equilibrium is reached.

Here the vapor pressure of water is denoted by the letter e. Equation 4.32 can now be integrated between the chemical potential of the vapor, μ_w, occurring at the actual vapor pressure less than saturation, e, and the chemical potential, $\mu°$, which exists when the air is saturated and consequently, has vapor pressure e°. The integration becomes:

$$\int_{\mu°}^{\mu} d\xi = \int_{e} V_w \, d\eta \qquad 4.28$$

From the perfect gas law, the partial molal volume of water can be expressed in terms of vapor pressure:

$$V_w = \frac{n_w RT}{e} \qquad 4.29$$

from which:

$$\frac{\partial V_w}{\partial n_w} = \frac{RT}{e} = \overline{V}_w \qquad 4.30$$

and the integral becomes:

$$\int_{\mu°}^{\mu} d\xi = RT \int_{e°}^{e} \frac{d\eta}{\eta} \qquad 4.31$$

where dummy variables ξ and η have been used to represent μ and e under the integral sign.

Equation 4.31 integrates to:

$$\mu - \mu° = RT\ln\left(\frac{e}{e°}\right) \qquad 4.32$$

The left side of equation 4.32 is the chemical potential of water vapor in the atmosphere expressed in Joules per mole of water vapor. The datum for the chemical potential is the saturation vapor pressure where the air cannot hold additional water at the given temperature and pressure. In general, wherever the chemical potential for water is used, the datum will be based on the chemical potential of pure water calculated at the identical temperature, pressure, and elevation.

Example 4.4. The chemical potential of water vapor
What is the difference in the chemical potential of air at 27° C at 95% rh as compared to 93% rh? Which way would the vapor be expected to migrate if a barrier permeable to water vapor separates the two conditions?

Solution
From the expression for the chemical potential of water vapor (equation 4.32) we can write:

$$\mu_w - \mu_w{}^\circ = 8.314 \frac{kPa \cdot L}{K \cdot mol} \times (273 + 27)\ K \times \ln\left(\frac{e}{e^\circ}\right)$$
$$= 2494.2 \times \ln\left(\frac{e}{e^\circ}\right) \qquad 4.33$$

which yields:

$$(\mu_w - \mu_w{}^\circ)_{95} = 2494.2 \frac{kPa \cdot L}{mol} \times \ln(0.95)$$
$$= -\ 127.935 \frac{J}{mol} = -130 \frac{J}{mol} \qquad 4.34$$

and

$$(\mu_w - \mu_w{}^\circ)_{95} = 2494.2 \frac{kPa \cdot m^3}{mol} \times \ln(0.93)$$
$$= -\ 181.006 \frac{J}{mol} = -180 \frac{J}{mol} \qquad 4.35$$

The vapor will move from 95% rh toward 93% rh because the potential is lowest for 93% rh and movement is always toward the lower potential.

Potential of a Solution

A substance or substances dissolved in water alters the chemical potential of the solution. Water within the young green plant cell is always present as a solution. Often, especially under irrigated conditions, salts in the soil dissolve forming a soil-water solution. A model for the potential of a solution is therefore useful for analyzing the movement of water in the soil-plant-atmosphere continuum. A simple way of approaching the development of a model for the chemical potential of a solution is Raoult's Law. This law states that for a dilute solution, the presence of a solute lowers the vapor pressure of the solution. It also lowers the freezing temperature and raises the boiling temperature. A model for the chemical potential of a solution can be developed as follows:

By Raoult's Law:

$$e^\circ \times N_w = e$$

from which:

$$\frac{e}{e^\circ} = N_w \qquad 4.36$$

Equation 4.36 states that the air over a dilute solution will be at equilibrium with air at a lower relative humidity than that over pure water. Using equation 4.36 in equation 4.32 yields the chemical potential of a dilute solution of a non-ionic solute.

$$\mu - \mu^\circ = RT \times \ln(N_w) \qquad 4.37$$

Raoult's Law applies only to dilute solutions. If a particular solute is ionic, it will separate into pairs of ions, one with a positive charge and one with a negative charge. The effect is the same as if two molecules were present in the solution rather than one. Thus, a single molecule of sodium chloride (NaCl) in a dilute solution disassociates into an ion of sodium and an ion of chlorine. A dilute ionic solution can be approximated by assuming that one mole of solute creates two moles of ions. However, as the solute content increases, the disassociation will vary.

For non-dilute solutions, the effect of the solute is better described by the activity of the solution. Reference books such as a chemistry handbook can be consulted to determine the activity of solutions of various concentrations. For the effect of a solution that requires the use of the activity, the appropriate formulation is:

$$\mu_i - \mu_i^\circ = RT \times n(a_i) = RT \times \ln(c_i\, N_i) \qquad 4.38$$

where a_i is the activity (or c_i is the activity coefficient) of the *ith* component. Fortunately, in many cases in the green plant environment, the mole fraction is sufficiently close to unity that activities or activity coefficients are not needed.

Pressure Potential of Confined Water

Equation 4.37 (or 4.38) along with equation 4.22 under the conditions pictured in figure 4.3 allows the development of a model to explain the pressure within a green plant cell. Figure 4.3 shows an osmometer, a device in which a solution is contained within a semipermeable membrane exposed to a solution, in this case pure water. Both the solution and the water are at the same temperature. In figure 4.3, the osmometer cell is surrounded by a cage which supports the semipermeable membrane containing the solution. The arrangement is then immersed in pure water and allowed to come to equilibrium. We assume for convenience that the osmometer cell contains a transducer that provides a readout of the pressure within the cell.

When the osmometer is lowered into the distilled water, because the chemical potential of the solution within the osmometer is less than that of distilled water, water migrates into the osmometer. Since the semipermeable membrane is confined, the pressure increases until equilibrium is established. From equation 4.22, assuming that the differential in potential within the osmometer has disappeared, noting that the temperature is constant and there is no solute, we can write:

$$\frac{\partial \mu_w}{\partial N_w} dN_w = d\mu_w = 0 = \overline{V}_w dP + \frac{\partial \mu_w}{\partial N_w} dN_w \qquad 4.39$$

From equation 4.37 an expression for $\partial\mu_w / \partial N_w$ can be obtained:

$$\frac{\partial \mu_w}{\partial N_w} dN_w = d\mu_w = RT \times d[\ln(N_w)] \qquad 4.40$$

Equation 4.40 can be integrated from P_{atm}, the initial pressure where the potential is μ_w due to the mole fraction of water N_w, to the final pressure, P, where the potential difference is 0 and equilibrium is established. This yields:

$$\int_P^{P^\circ} \overline{V}_w\, d\xi = -RT \int_0^{\ln(N_w)} d\eta \qquad 4.41$$

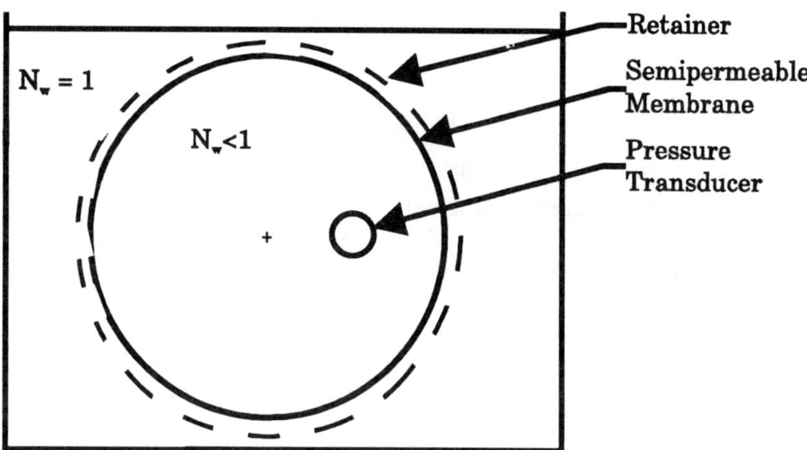

Figure 4.3. In the osmometer, the difference in the chemical potential of water between the inside and outside of the semipermeable membrane will cause water to migrate into the enclosure. Pressure built up within the enclosure, however, adds to the total potential bringing the entire system into equilibrium.

from which, by rearranging we obtain:

$$P - P^{\circ} = \pi = -\frac{RT\ln(N_w)}{\overline{V}_w} \qquad 4.42$$

where ψ_p is the positive pressure (turgor pressure) built up within the osmometer that results from the confinement of the semipermeable membrane by the cage.

The right side of equation 4.42 is divided by the partial molal volume of water to obtain the dimension of pressure. The pressure produced as a result of confinement of the solution within the osmometer is equal but it is opposite in sign to the chemical potential of the solution divided by the partial molal volume of pure water.

It is convenient to express the chemical potential in terms of energy per unit volume, i.e., pressure, rather than energy per mole. This form can be used whenever the effect of potential on water movement is discussed. The relative incompressibility of water makes it possible to divide the chemical by the partial molal volume of water, $\overline{V}w = 0.018$ kmol/m^3. We will express the result in multiples of atmospheres (100 kPa ≈ 1 atm) or some other convenient dimension such as meters of water (10.33 m_{water} ≈ 1 atm). This convention will be followed throughout the text.

Following this convention, equation 4.32 becomes:

$$\psi_v = \frac{\mu - \mu^{\circ}}{\overline{V}_w} = \frac{RT\ln\left(\frac{e}{e^{\circ}}\right)}{\overline{V}_w} \qquad 4.43$$

where potential is denoted by ψ and the subscript *v* indicates the potential of water vapor. Likewise, the water potential of a solution is:

$$\psi_s = \frac{\mu - \mu^o}{\overline{V}_w} = \frac{RT \times \ln(N_w)}{\overline{V}_w} \quad 4.44$$

Example 4.5. Pressure within a green plant cell
Assume the solution within a green plant cell is sucrose at a concentration equivalent to 0.5 $mol_{sucrose}/L_{water}$. If the activity coefficient of sucrose is very nearly unity, i.e., that the solution is dilute, calculate the osmotic pressure within a green plant cell in equilibrium with pure water at 25° C.

Solution
At a concentration of 0.5 mol/L, the mole fraction of water in the solution is:

$$N_w = \frac{n_w}{n_w + n_s} = \frac{\dfrac{1000\ g}{18\ \dfrac{g}{mol}}}{\dfrac{1000\ g}{18\ \dfrac{g}{mol}} + 0.5\ mol} = 0.991\ 08 \quad 4.45$$

Based on N_w as calculated, the water potential is:

$$\psi_s = \frac{RT \times \ln(N_w)}{\overline{V}_w}$$

$$\psi_s = \frac{\dfrac{8.314\ kPa \cdot m^3}{kmol \cdot K} \times 298}{\dfrac{0.018\ m^3}{kmol} \times \dfrac{100\ kPa}{atm}} \ln(0.991\ 08) \quad 4.46$$

$$= 1376\ atm \times \ln(0.991\ 08) = -12.3\ atm$$

The activity coefficient of sucrose at a concentration of 0.5 $mol_{sucrose}/L$ of water is $c_w = 0.9994$. Therefore, actual water potential would be:

$$\psi_s = 1376\ atm \times \ln(0.991\ 08 \times 0.9994) = -13.2\ atm$$

The answer assumes the solution behaves as a dilute solution and is within 6.8% of the actual value. In most cases, this would be sufficiently accurate for engineering purposes. If more accuracy is necessary, careful analysis of the solution to determine both the solute and its concentration would be needed.

The development of the water potential model for the osmometer has demonstrated that the potential is the work performed when energy is stored. In the osmometer, the stored energy results from the elasticity of the confining enclosure used to overcome the pressure. The pressure reduces the difference in potential to zero by transferring it to stored energy. In cases where the capacity to store energy is not as obvious, the concept is still valid. In the case of evaporating water, stored energy is in the form of latent energy of evaporation.

Example 4.6. Work and energy relationships in evaporating water
Compute the work performed on a unit volume basis by water evaporating into a constant volume container.

Solution
If the volume is constant, the vapor pressure of the water increases as evaporation occurs. Expressing vapor pressure as e:

$$dW = Vde \qquad 4.47(a)$$

From the perfect gas law, per mole of water:

$$V = \frac{RT}{e} \qquad 4.47(b)$$

we can write:

$$dW = RT\,\frac{de}{e} \qquad 4.47(c)$$

which, when integrated from the initial vapor pressure to the saturation vapor pressure yields:

$$\int_0^{\Delta W} d\xi = Rt \int_e^{e^\circ} \frac{d\eta}{\eta}$$

$$\Delta W = RT \ln\left(\frac{e^\circ}{e}\right) \qquad 4.47(d)$$

$$= -RT \ln\left(\frac{e}{e^\circ}\right)$$

Note that work done to store energy is the negative of the chemical potential.

Other Contributions to Water Potential

Capillarity is the movement of water due to the effect of surface tension. It is an important contribution to the potential affecting water causing water to rise in a small diameter vertical tube when one end is placed into a liquid. In soil, capillarity causes water to move horizontally, wetting a relatively large diameter from a water source which drips slowly onto the center of the moistened circle. Capillarity is responsible for wetting a soil which lies above a stationary water table. The potential due to the surface tension effect can be visualized as shown in the following development.

The energy stored in surface tension is similar to that stored by a stretched spring. However, the spring constant has dimensions of force per linear distance while the dimension of surface tension is force per width of surface. Figure 4.4 shows a capillary tube with one end immersed in pure water. Water has risen in the tube, indicating that a potential has been satisfied by storing energy. The force which has acted to lift the water is equal to the surface tension of water multiplied by the length of film acting, i.e., the circumference of the capillary, or:

$$F_{up} = 2\left(\pi r \Lambda\right) \qquad 4.48$$

where

r = radius of the capillary
Λ = surface tension (0.072 N/m at 22° C)

The weight of water supported by the energy stored in the surface film is:

$$F_{down} = \rho g \pi r^2 h \qquad 4.49$$

where

ρ = density of water (1000 kg/m³)
g = acceleration of gravity (9.81 m/s²)
h = height of the water column above the free water surface

Equating equations 4.48 and 4.49 and solving for the potential energy $\rho g h$ yields:

$$\rho g h = \frac{2\Lambda}{r} \qquad 4.50$$

Since the water potential is the negative of the energy stored, the capillary component of the water potential is:

$$\psi_\tau = -\kappa \rho g h = -\frac{2\kappa\Lambda}{r} \qquad 4.51$$

where ψ_τ is the component of water potential resulting from the effect of surface tension acting on a curved air-water interface and κ is a conversion factor that allows ψ_τ to be expressed in appropriate dimensions. The remaining illustrations in figure 4.4 show that it is not necessary for one end of the tube containing the curved air-

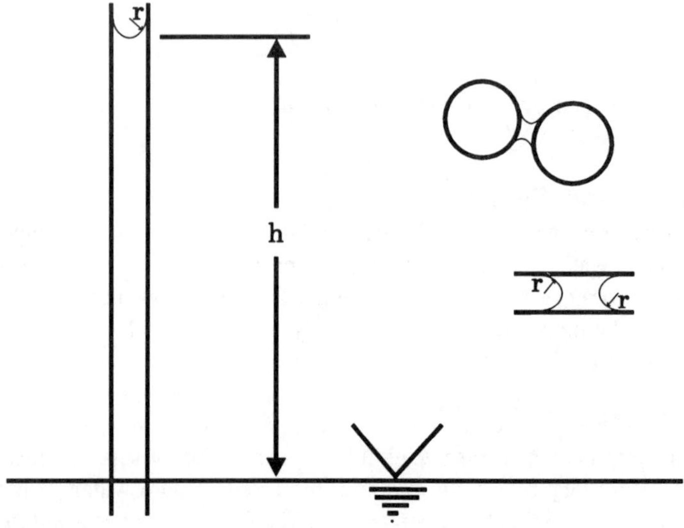

Figure 4.4. Surface tension lowers the chemical potential of water beneath the meniscus, causing water to move upward until the potential energy of the column of water is equal to the chemical potential of the water at every point within the column. In an isolated capillary, the potential is balanced by the meniscus at either end.

water interface to be immersed in water. As long as a curved air-water interface exists, the water lying under the curved air-water interface is at the same water potential as is predicted by equation 4.51.

Example 4.7. Water potential on a curved air-water interface
What is the water potential of a clean capillary 2μ in diameter?

Solution
A capillary 2μ in diameter has a radius of 1μ. Applying equation 4.37 yields:

$$\psi_\tau = -\frac{2\Lambda\kappa}{r_w}$$

$$\psi_\tau = -\frac{2 \times 0.072\ \frac{N}{m}}{1 \times 10^{-6} m} \times \frac{1\ kPa}{1000\ \frac{N}{m^2}} = -144\ kPa = -1.4\ atm \qquad 4.52$$

Capillaries this size are common within the green plant vascular system. They are small diameter holes, or pits, that penetrate the walls of certain cells within the vascular system of the plant.

The neck of water connecting two soil particles may have a curved air-water interface approaching a radius of 1μ. However, within the soil it is unlikely that values of ψ_τ less than -0.8 atm exist because of impurities in the soil-water solution and foreign particles within the soil which interfere with the surface tension film.

Example 4.8. Water potential due to capillarity
A hygienically clean tube is held vertically and one end is dipped into water, causing water to rise into the tube. The exposed end of the tube is then covered, creating a vacuum as the tube is removed. It is lain horizontally and the length of water is measured and found to be 50 mm.

(a) What is the diameter of the tube?

(b) What is the water potential of the water trapped within the tube?

(c) What was the water potential within the tube at the level of the free water surface when the tube was still inserted in the water?

Solution
(a) Assuming no water escaped from the tube when it was removed from the water, the diameter of the tube can be found from equation 4.37 since 100 kPa pressure will support a column of water 10.33 m high:

$$0.05\ m \times \frac{100\ kpa}{10.33\ m} \times \frac{1000 N\cdot m^2}{kPa} = \frac{2 \times 0.072 N\cdot m}{r}$$

$$r = 2.975 \times 10^{-4} \qquad m = 0.2975\ mm \qquad 4.53$$

The tube diameter is then twice the radius or 0.595 mm.

(b)The water potential within the tube is determined from the fact that a curved air-water interface still exists at each end of the water column. Therefore, the water potential within the tube is:

$$\psi_{\tau} = \frac{100 \text{ kPa}}{10.33 \text{ m}} \times 0.05 \text{ m} = 0.48 \text{ kPa} \qquad 4.54$$

Example 4.8 demonstrates that water potential due to capillarity does not require a vertical water column for its existence, only that a curved air-water interface exists.

(c) While the tube was still vertical, because the hydrostatic pressure at a given level within a fluid constant, the water potential within the tube at the level of the free water surface must be the same as that outside the tube, i.e., zero. Actually, while the tube is vertical, the potential within the tube but above the water at any level is diminished by ρgd, where d is the vertical distance from the curved air-water interface to the level at which the water potential is computed. In the horizontal tube removed from the water, the water potential is determined by the curved air-water interface, which was responsible for the water being raised up in the tube initially.

Matric Components

There are other forces that give rise to potentials affecting the movement of water. These occur both in the soil and within the green plant. Most of these components of potential are subtle and difficult to model. They include the attractive forces of electrical charge acting on the polar water molecules and adsorptive and chemical bonding forces. The combination of these components of the water potential is unique for each soil and green plant biochemical process. These components of the water potential function, along with the capillary potential described above, constitute the matric potential component of the total water potential function.

Within the green plant, very little is known or understood regarding the effects and consequences of the matric potential. Fortunately, as long as sufficient water from the soil reservoir is available to the roots of the green plant, potential differences along the water pathway within the green plant keep the cells well supplied with water. Transpiration keeps the leaves at a temperature sufficiently low to maintain the biochemical processes essential to plant growth.

The matric potential becomes most significant to water movement through the soil. Within the soil, minerals, particularly the clay colloid fraction, have the most important effect on the movement and availability of water. The wide variety of clay sizes, along with the interactions between electrical and other forces acting on water in the soil, makes it difficult to model the total matric component of water potential. In general, models of soil matric potential result from experimental measurement of the energy holding water within a particular soil. The result can be pictured graphically with the energy at which water is held expressed as a function of soil water content. However, a particular mathematical formulation which models the potential within the soil can only be approximated. Figure 4.5 shows the total soil water potential as a function of soil water content for Yolo light clay loam soils.

The matric component of the soil water potential involves the capillary component which acts near saturation (0 atm, 0 kPa) to about –0.8 atm (–80 kPa). Beyond –80 kPa, the electrical, chemical, and adsorptive effects which act on water molecules become significant. Thus, from 0 to –80 kPa, equation 4.51 holds, while below –80 kPa, a graphical form or a mathematical approximation of the matric potential must be used.

Example 4.9. Soil water potentials

The free water surface in a Yolo clay loam soil is 10 m below the ground surface. Estimate the soil water potential in the soil:

(a) 5 m below the ground surface.

(b) 500 mm below the ground surface where the volumetric soil water content is 22%.

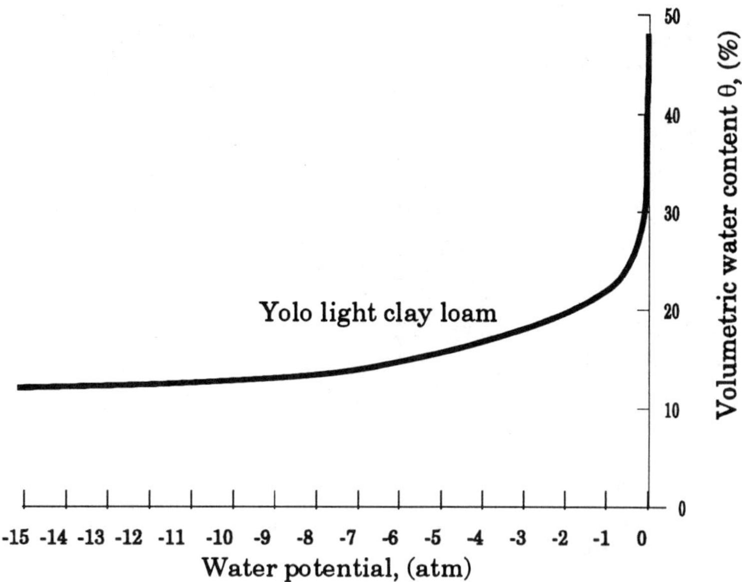

Figure 4.5. The soil characteristic curve relates the water potential of unsaturated soil water to the volumetric water content.

Solution

(a) At 5 m below the ground surface, assuming that the free water surface has remained stationary for a sufficiently long period of time, capillary rise has had sufficient time to act so that the soil water potential is in equilibrium with the free water surface. Therefore, the soil water potential will be:

$$\psi_\tau = \kappa \times 5\text{m} = \frac{100\text{ kPa}}{10.33\text{ m}} \times 5\text{m} = 48.4\text{ kPa} \qquad 4.55$$

where κ was expressed in a form which yielded an answer in kilopascals.

(b) The free water surface is so far removed from the point under consideration that the soil water potential must be ψ_m, and can be determined only from a graph of the water potential-water content relationship. From figure 4.5, since the soil is a Yolo clay loam, the water potential is estimated to be −1 atm.

The capillary component of the matric potential is responsible for the phenomenon of hysteresis. Hysteresis causes the soil to have a different water potential at the same water content depending on whether the soil is drying down or wetting up. The hysteresis effect is most pronounced in the range of water potentials from 0 to −80 kPa. Hysteresis can result as a consequence of a difference in pore sizes in a soil. As illustrated in figure 4.6, soil voids with a small radius pores at the bottom are being wet due to a rising water table. The small radius pores prevent the void volume from filling until sufficient force develops to overcome the surface tension due to the small pores. The water content of soil containing such voids would be less if the pore were full and losing water.

Under water loss or drying conditions as on the right of figure 4.6, however, the smaller diameter pores exert sufficient force to retain the water stored in the larger diameter pores above through the action of capillarity. The water potential at which the water is admitted or retained is determined by the pore radii. The volumetric water content of the soil varies depending upon whether the soil sample is wetting or drying. Hysteresis is most predominant at water contents below but very near saturation.

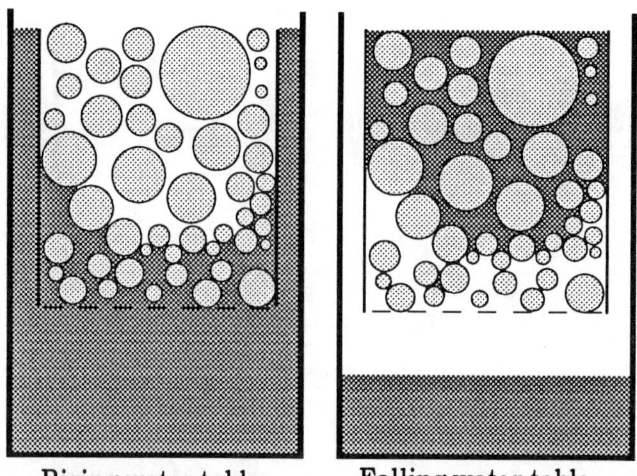

Figure 4.6. Surface tension is the cause of hysteresis in the soil water characteristic curve.

Example 4.10. Effect of ψ_τ in a rising water table
A capillary pore 0.05 mm in diameter opens upward into a larger diameter soil void. If there is no other entry into the macropore except at the top, where will the free water surface be in relation to the small capillary pore when the macropore begins to fill?

Solution
If the entire void were of the same diameter as the opening at the bottom, using equation 4.51 water would rise into the pore to a height of:

$$h = \frac{2\Lambda}{\rho g r} = \frac{2 \times 0.072 \frac{N}{m}}{1000 \frac{kg}{m^3} \times 9.81 \frac{m}{s^2} \times 0.025 \times 10^{-3} m} = 0.587 \text{ m} \qquad 4.56(a)$$

However, because the small diameter opening exists only at the bottom of the macropore, surface tension will act to *prevent* water from entering the macropore until the water outside exerts sufficient pressure to overcome ψ_τ. This will occur when the free water surface is 0.59 m above the point where the small diameter capillary empties into the macropore.

Example 4.11. Effect of ψ_τ in a falling water table
If the opening at the top of the macropore has a diameter of 0.04 mm, where will the free water surface be in relation to the top of the macropore when the macropore begins to drain?

Solution
Again, using equation 4.30:

$$h = \frac{2\Lambda}{\rho g r} = \frac{2 \times 0.072 \frac{N}{m}}{1000 \frac{kg}{m^3} \times 9.81 \frac{m}{s^2} \times 0.02 \times 10^{-3} m} = 0.734 \text{ m} \qquad 4.56(b)$$

Water will be retained in the macropore by the small diameter capillary at the top until the free water surface has fallen 0.73 m below the level of the 0.04 mm diameter capillary. It is obvious from examples 4.10 and 4.11 that the water content of the block of soil containing the macropore will be less under a rising water table than it will under a falling water table. It is left as an exercise to consider what conditions will exist if the soil is wetting from above as it would from a saturating rainfall.

Water Potential in Saturated Soil

In saturated soils a different phenomenon is responsible for the potential driving the movement of water. In an unsaturated situation, energy must be expended to cause the movement of water. For evaporation, heat energy must be added to the water to satisfy the latent heat requirement of vaporization. For the solute component, pressure must be applied to a semipermeable membrane container to raise the water potential from some negative value to a value nearer zero. Pressure can be used to drive water by capillarity from a pore in which it is held. In all of these situations, pressure, or the energy per unit volume in one form or another can be used to determine the value of the potential.

In the case of a fully saturated soil, however, the opposite is true. From a lower datum, the pressure of the water due to its position could be used to obtain energy. While all potentials studied up to this point are negative and require energy to attain a zero value, in saturated soil the potentials are positive and can be used to obtain energy. The concept of water potential for saturated conditions then becomes much the same as that of positional potential energy, which is familiar from physics. Along with position is hydrostatic pressure, also familiar from physics. The water potential function for saturated soil can be developed as follows.

Figure 4.7 illustrates a soil completely saturated with water above some datum. An elemental volume of water is located at point A at an elevation z_2 above the datum but some distance z_1 below the free water surface. The elemental volume of water at point A is seen to have positional potential energy with respect to the datum of:

$$PE_p = \rho g z_2 \tag{4.57}$$

At point A, the elemental volume of water is subjected to hydrostatic pressure of:

$$PE_h = \rho g z_1 \tag{4.58}$$

Thus, the total amount of energy stored in the elemental volume at point A is the sum of the positional and hydrostatic components, or:

$$\begin{aligned} \psi_g &= \rho g z_1 + \rho g z_2 \\ &= \rho g (z_1 + z_2) \\ &= \rho g z \end{aligned} \tag{4.59}$$

Equation 4.59 indicates that in a saturated soil the water potential is a function of the distance from the datum to the free water surface. Both positional and hydrostatic components are combined into a single gravitational potential. However, it is necessary to analyze the potential at a point very carefully because moving water dissipates energy due to friction and turbulence. Because of this, the actual distance from the datum to the free water surface will vary from point to point throughout the soil mass if the water is moving through the soil. Only when no movement is occurring will the water potential be determined everywhere by the distance from the datum to the free water surface.

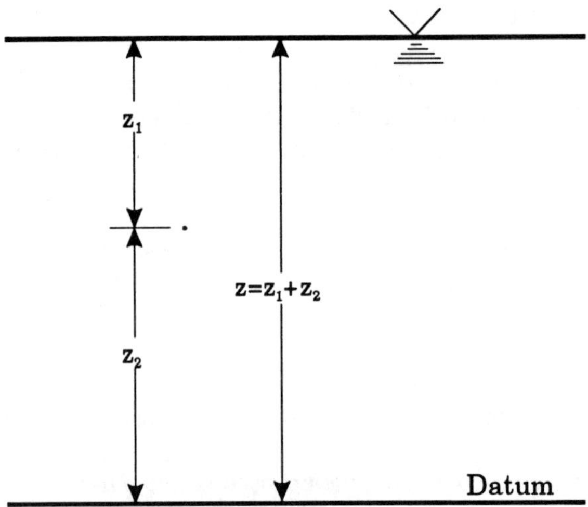

Figure 4.7. At the point, both potential energy of position and potential energy of hydrostatic pressure contribute to the water potential of a saturated medium.

Total Water Potential

The total water potential acting at any point in the soil-plant-atmosphere continuum is a summation of the contribution from any or all of the components of water potential derived above. The actual contribution of any component of the potential will require a careful analysis of the situation existing at the point where the potential is being defined. Obviously, the vapor component of the potential will contribute only if there is an air-water interface across which evaporation can occur. Similarly, for the solute component to contribute, there must be an impermeable membrane with a solution of low water potential on one side and water at a higher potential on the other. The matric potential will be most significant at very low values of water potential, while the capillary component will be effective only if there is a curved air-water interface present. Within the green plant, or where surfaces are absolutely hygienically clean, it may be possible to have contributions from the matric potential to values of −300 kPa or more.

Usually in a saturated soil, one in which the water potential is greater than zero, only the gravitational component of the water potential applies. However, there are exceptions especially if salts are present in irrigated arid regions. In this case, one must be careful to use the actual soil water potential, since the plant cell is an osmometer and effects of the solute potential may be experienced.

The total soil water potential is:

$$\psi_{tot} = \psi_v + \psi_s + \psi_p + \psi_m + \psi_g \tag{4.60}$$

with individual components:

$$\psi_v = \frac{\mu - \mu^o}{\overline{V}_w} = \frac{RT\ln\left(\frac{e}{e^o}\right)}{\overline{V}_w} \tag{4.61}$$

$$\psi_s = \frac{\mu - \mu^o}{\overline{V}_w} = \frac{RT \times \ln\left(N_w\right)}{\overline{V}_w} \tag{4.62}$$

and:

$$\psi_\tau = -\kappa\rho g h = -\frac{2\kappa\Lambda}{r} \quad 4.63$$

which operates at values of ψ_m generally greater than –80 kPa, and:

$$\psi_g = \kappa\rho g(z_1 + z_2) = \kappa\rho g z \quad 4.64$$

which is significant generally only when the soil is saturated; ψ_p is the component due to the buildup of confined pressure in a cell.

Water Potentials in the Soil-Plant-Atmosphere Continuum

Water movement in the soil-plant-atmosphere continuum is generally passive. From the soil to the roots, through the plant, and into the atmosphere, it occurs down the potential gradient from higher potential values to lower potential values since this applies for most crops except for rice, the soil is unsaturated. All values of water potential are less than zero. Exceptions to passive transfer may occur at locations within the green plant where metabolic processes cause active transfer. It is generally agreed that withdrawal of water from the soil by plant roots ceases at soil water potentials less than 12 to –18 atm. Again, some exceptions exist; for instance, wheat has been known to grow in soils where the water potential is less than –20 atm.

Green plant growth occurs most successfully at soil water potentials greater than –5 atm and probably is most significant at potentials greater than –1.5 atm. This is because the movement of water in the soil slows drastically at potentials much below about –1 atm. Roots can still extract water from the soil but the flow occurs so slowly that, during periods of high evapotranspiration demand, there is insufficient water available to the plant.

Figure 4.8 shows the decrease in water potential in pepper plant leaves during the removal of water from the soil. Up to a water potential of about –2 atm, very little change in leaf water potential occurs, except for normal daily fluctuations. However, once soil water potential drops below about –2 atm, a significant change in the measured water potential of the leaf was noted.

Figure 4.9 shows the effect of water potential on growth of cotton leaves. Note that as the plant is stressed due to lower water potential in the soil, growth (as measured by the extension of the leaves) slowed drastically and virtually ceased at a potential of –14 atm.

Since the flow of water through the soil-plant-atmosphere continuum occurs primarily through passive movement, the water potential within the roots of the green plant must be less than that in the soil. This decrease in water potential continues throughout the root system, up the stem, into the leaf, and through the stoma out into the atmosphere surrounding the plant canopy. Differences in water potential in the various portions of a green plant must continue to exist over a period of days as the soil water is depleted. Differences are necessary for water movement to occur. With decreasing soil water potentials, the differences increase.

While water potential differences are in decreasing order within the vessels of the plant through which water movement occurs, a different effect may occur within individual cells. For example, figure 4.9, assume that the water potential in the soil is approximately –1 atm, a water potential of as low as –6 atm or less may exist within the leaf. To prevent turgor pressure from being lost in the leaf cells, the potential within the cells must be of equal or lesser magnitude. The sum of the solute contribution to the water potential, ψ_s, and the turgor pressure within the cell, ψ_p, must be less than or equal to the water potential within the leaf on the cell wall. Since the turgor pressure, ψ_p, is created by the solute potential of the solution and the water potential of the leaf cell wall, the mole fractions of water and solutes in the plant cell control the turgor. As shown in example 4.13, the maintenance of turgor, and therefore the green plant structure, depends on the production of solutes by photosynthesis of the green plant. Should photosynthesis halt for some reason during the period when leaf water potential is decreasing, a rapid loss of turgor in the green plant would occur with accompanying severe wilting of the plant.

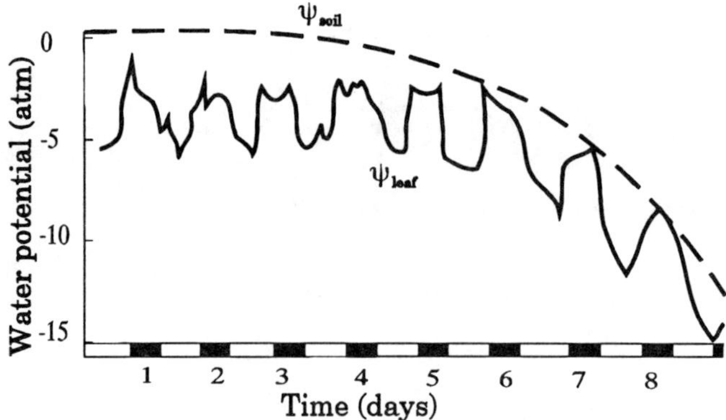

Figure 4.8. A reduction in water potential in the soil can significantly influence the water potential of the crop growing in the soil. The cotton leaves show a significant drop in water potential when the soil water potential drops below 12 atm. (Adapted from Slayter, 1967 after Gardner and Nieman, 1964)

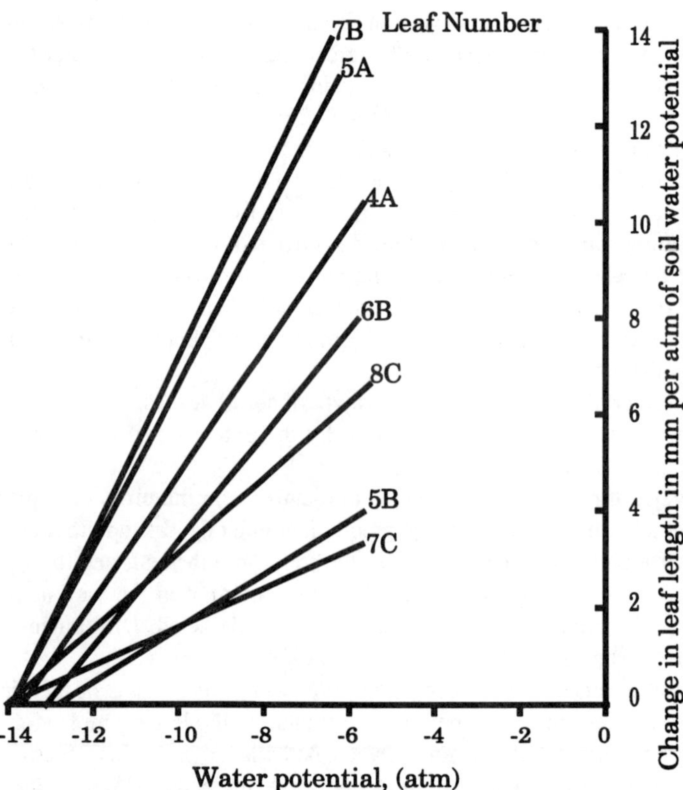

Figure 4.9. Cotton leaves grow more slowly when the soil water potential drops. (Adapted from Slayter, 1967 after Wadleigh and Gauch, 1948)

Example 4.12. Leaf cell solute potential
A cell within a leaf at 27° C has a turgor pressure of 5 atm. If the water potential of the cell walls of the leaf is –3 atm, what is the solute potential of the cell contents at equilibrium? What is the mole fraction of water in the cell assuming the activity coefficient is unity?

Solution
Since $\psi_{\text{cell wall}} = \psi_{\text{leaf}} = \psi_s + \psi_p$:

$$\begin{aligned}\psi_s &= \psi_{\text{leaf}} - \psi_p \\ &= -3 \text{ atm} - 5 \text{ atm} \\ &= -8 \text{ atm} \times 100 \frac{\text{kPa}}{\text{atm}} \\ &= -800 \text{ atm}\end{aligned} \tag{4.65}$$

the mole fraction of water in the cell is then:

$$\begin{aligned}N_w &= \exp\left(\frac{\psi_s \overline{V}_w}{RT}\right) = \exp\left(\frac{-800 \text{ kPa} \times 0.018 \dfrac{\text{m}^3}{\text{kmol}}}{8.314 \dfrac{\text{kPa}\cdot\text{m}^3}{\text{K}\cdot\text{kmol}} \times 300 \text{ K}}\right) \\ &= 0.992\ 43\end{aligned} \tag{4.66}$$

Example 4.13. Turgor and solute potential interaction
If the leaf in example 4.12 is photosynthetically active and produces sufficient solute that the mole fraction of water within the cell drops to 0.992, to what value can the leaf water potential drop before the leaf loses turgor?

Solution
If the mole fraction of water is 0.992, the solute potential is:

$$\begin{aligned}\psi_s &= \frac{R\ T \ln(0.992)}{\overline{V}_w} \\ &= \frac{\dfrac{8.314 \text{ kPa}\cdot\text{L}}{\text{K}\cdot\text{mol}} 300 \text{ K} \times \ln(0.992)}{0.018 \dfrac{\text{L}}{\text{mol}} \times \dfrac{100 \text{ kPa}}{\text{atm}}} \\ &= -11 \text{ atm}\end{aligned} \tag{4.67}$$

Since:

$$\psi_{cell} = \psi_s + 0$$

$$\psi_{cell} = \psi_s \qquad 4.68$$

$$= -11 \text{ atm}$$

The cell will loose turgor when ψ_{leaf} equals ψ_s which will occur at –11 atm of water potential.

The stomata controls the communication from the interior of the green plant leaf to the atmosphere surrounding the plant canopy. Since turgor controls the stomatal opening, it also controls the carbon dioxide entering the plant, and the oxygen and water vapor exiting the plant leaf. Thus, anything that causes the plant to lose turgor affects the photosynthetic process that produces the plant food and helps the green plant maintain turgor.

Exercises

4.1. How much energy in Joules is required to evaporate 6 mm of water from 1 ha of ground? If the evaporation occurs over a period of 6 h, what is the power expended?

4.2. What is N_w if 50 g of sugar ($C_{12}H_{22}O_{11}$) is dissolved in 500 g of water at 27° C?

4.3. What is the solute water potential of the solution in exercise 2?

4.4. What is the mole fraction and water potential of a solution of 50 g of salt (NaCl) dissolved in 500 g of water? (Assume complete disassociation.)

4.5. Use the definition of mole fraction and show $N_w = 1 - N_s$ where N_s designates the collective mole fraction of solutes.

4.6. Expand $\ln(N_w)$ in a Taylor Series about 1 and show that $\ln(N_w) \approx -(1 - N_w)$ or $\ln(N_w) = -N_s$.

4.7. Show that $N_s/\bar{V}_w = c_s$, the molar concentration in volume/(mol_{solute}). (Hint: Note that n_s is much less than n_w. Also, apply the rule regarding the use of significant digits relating to the operation of addition.)

4.8. A solution with c_s = 0.000 802 mol_{solute}/cm^3 is placed in an osmometer and allowed to equilibrate at constant temperature. After equilibration, is c_s the same, greater, or less? What is the pressure if the equilibration is with distilled water? Why?

4.9. A cell whose molar concentration is 0.0008 moles/cm^3 is in equilibrium in a solution with c_s of 0.0003 mol/cm^3 and T = 300 K. What is the osmotic pressure, ψ_p, within the cell?

4.10. The cell in example 9 is located in a horizontal capillary completely filled with distilled water. The ends of the capillary narrow so that the curved air water interfaces at each end have a radius of 1 μ. What is the osmotic pressure within the cell? (You may assume the c_s at equilibrium is 0.0008 mol_{solute}/cm^3.)

4.11. Assume that sufficient time has elapsed following the cessation of photosynthesis to allow the water potential of a cell in a developing fruit to approach the soil water potential. If ψ_{soil} is −10 atm and the cell sap of the fruit has c_s = 0.1 mol/L at 20° C, what is the turgor pressure within the cell of the developing fruit?

4.12. What is the pressure within the fruit cells of example 4.11 if a sudden rain storm raises the soil water potential to −0.3 atm?

4.13. The soil water potential is −3 atm at a depth of 0.3 m below the soil surface. If a plant with roots absorbing water at this location is in equilibrium with the soil water, what will be the water potential within the cell wall of the plant at a point 1 m vertically above the absorbing root surface? (Hint: Assume a datum for gravitational potential measurements through the point in the soil where absorption is occurring.)

4.14. Assume a redwood tree is absorbing water from soil which has a water potential of −5 atm. What is the water potential in a leaf cell wall located 90.9 m vertically above the point where absorption is occurring? For simplicity, assume there is no transpiration occurring so that no potential gradient due to flow exists.

4.15. Characterize the parameters in the perfect gas law, PV = nRT, as to whether they are intensive or extensive.

Chapter 5

Elementary Concepts of Green Plant Anatomy

Understanding the green plant is fundamental for agricultural, biosystems, and bioresource engineers. Plants are unique in that they possess the capacity to capture and convert radiant energy to the production of food, feed and fiber. Green plants are absolutely essential for life on this planet. The green plant is the foundation upon which all life is built since green plants serve as the primary source of food for most other living matter. The exceptions to this are microorganisms that can obtain energy from chemical reactions with sulphur and iron. These, however, are of insignificant importance.

Green plants maintain the level of atmospheric oxygen by using carbon dioxide and, in the process, produce gaseous oxygen. The production of oxygen occurs during photosynthesis, the process by which food is created using the energy in visible light.

Example 5.1. The role of photosynthesis
The overall chemical reaction of photosynthesis is:

$$6CO_2 + 6H_2O + 2.882 \times 10^6 \text{ kJ} \rightarrow C_6H_{12}O_6 + 6O_2 \qquad 5.1$$

The atomic weights of carbon, oxygen, and hydrogen are, respectively; C = 12, O = 16, and H = 1. How many grams of CO_2 and H_2O are used and how many grams of O_2 are given off when 1 mol of $C_6H_{12}O_6$ is produced?

Solution
The molecular weight of carbon dioxide is:

$$12 \frac{g}{mol_C} + \left(2 \times 16 \frac{g}{mol_O}\right) = 44 \frac{g}{mol_{CO_2}} \qquad 5.2$$

The molecular weight of water is:

$$\left(2 \times 1 \frac{g}{mol_H}\right) + 16 \frac{g}{mol_O} = 18 \frac{g}{mol_{H_2O}} \qquad 5.3$$

And the molecular weight of glucose is:

$$\left(6 \times 12 \frac{g}{mol_C}\right) + \left(12 \times 1 \frac{g}{mol_H}\right) + \left(6 \times 16 \frac{g}{mol_O}\right) = 180 \frac{g}{mol_{C_6H_{12}O_6}} \qquad 5.4$$

One mole of glucose is 180 g. This requires 6 mol of CO_2 and produces 6 mol of O_2, therefore, the number of grams of CO_2 used is:

$$g_{CO_2} = 44 \frac{g}{mol_{CO_2}} \times 6\ mol_{CO_2} = 264\ g_{CO_2} \tag{5.5}$$

The water required to support the reaction is:

$$6\ mol_{H_2O} \times 18 \frac{g}{mol_{H_2O}} = 108\ g_{H_2O} \tag{5.6}$$

and the amount of oxygen produced is:

$$6\ mol_{O_2} \times 32 \frac{g}{mol_{O_2}} = 192\ g_{O_2} \tag{5.7}$$

By consuming carbon dioxide during photosynthesis, the gradual increase of this gas in the earth's atmosphere is slowed. This occurs in spite of the daily contribution of this gas due to the consumption of fossil fuels by mankind.

Example 5.2. Production of CO_2 by transportation
Commercial fuels are a mixture of many hydrocarbons. An analysis of combustion gasses indicates one chemical reaction for an internal combustion engine is:

$$\begin{aligned} C_{12.8}\,H_{26.6} + 22.4\ O_2 + 84.1\ N_2 &\rightarrow 12.5\ CO_2 + 3.1\ O_2 \\ &+ 0.4\ CO + 84.1\ N_2 + 13.3\ H_2O \end{aligned} \tag{5.8}$$

If an automobile uses 15.6 L/(100 km) (averages 15 mi/gal) and travels an average of 19 300 km (12,000 mi) annually, how much CO_2 does the automobile produce in one year's time? Assume the density of the fuel is about 0.7 kg/L.

Solution
The molecular weight of the fuel is:

$$12.8 \times 12 \frac{g}{mol_C} + 26.6 \times 1 \frac{g}{mol_H} = 180 \frac{g}{mol_{fuel}} \tag{5.9}$$

The number of moles of fuel used in one year is:

$$\frac{19\ 300\ km}{yr} \times \frac{15.6\ L}{100\ km} \times \frac{0.7\ kg}{L} \times \frac{mol_{fuel}}{0.180\ kg} = 1.171 \times 10^4 \frac{mol_{fuel}}{yr} \tag{5.10}$$

and using, from example 5.1, the molecular weight of CO_2 as 44 g/mol yields:

$$\frac{12.5\ \text{mol}_{CO_2}}{\text{mol}_{fuel}} \times \frac{1.171 \times 10^4\ \text{mol}_{fuel}}{\text{yr}} \times \frac{0.044\ \text{kg}}{\text{mol}_{CO_2}} = 6440\ \frac{\text{kg}_{CO_2}}{\text{yr}} \qquad 5.11$$

The transportation output of CO_2 is about 6.5 metric tons of carbon dioxide per year. This is about 2 metric tons of carbon per year. These values agree with estimates by Clark (1989) who considered the production of carbon dioxide to exceed 3.75 t/person/year in the United States from all sources including heating and manufacturing.

Certain plants are particularly sensitive to the presence of specific air pollutants and have been successfully used as indicators for the presence of the pollutants in concentrations lower than those which would prove harmful to man, but could be harmful to other, more sensitive plants. Green plants are effective in the removal of pollutants from the air circulating through and over the plant canopy. Green plants cleanse the air and thus maintain the air quality at tolerable levels.

In addition to manufacturing food and oxygen while removing carbon dioxide from the air, green plants may serve mankind in other ways.

Plants of Importance to the Biosystems Engineer

The overall classification and study of plant anatomy and physiology is done by either biologists or botanists. For certain plants that are important to agriculture, the disciplines of agronomy and horticulture have specialized in the study of plant propagation and growth. In this chapter we will be concerned with certain aspects of the green plant that overlap all the above areas. We will direct our attention to those aspects of the green plants that can respond to modification by engineering means.

Many botanical classifications of plants have been developed. A classification that includes nearly all agriculturally significant plants is by life history. Here certain plants are classified as annuals, other as biennials, while still others are termed perennial.

Plants classified as annuals complete their entire vegetative cycle in a single growing season spanning the period from the germination of the seed to the completion of vegetative growth and production of seeds from which future propagation will occur. Many important agricultural crops fall into the category of annuals, such as corn, wheat, beans, etc.

Biennials require two growing seasons before a complete vegetative cycle is established. However, it is not uncommon to have the agriculturally significant portion of the biennial develop during the first growing season so it is harvested before the cycle is completed, i.e., before seeds have been formed. Carrots and beets are examples of biennial plants that are grown for harvest during a single growing season.

Perennial plants may require several years after planting and germination before they are mature and can reproduce. Common agriculturally significant perennial are orchard crops such as apples, etc. Perennials, in general, are characterized by woody stems that increase in diameter from year to year as new layers of cells develop and secondary growth occurs. Some herbaceous perennials, clover, alfalfa, etc., can mature and produce seed during the first year's growth although, typically, at least one year is required before seed production is significant.

In addition to classification by life cycle, plants are sometimes classified in relation to how they respond to their environment. Environmental factors include climatic factors such as light, precipitation, temperature, and air movement, and edaphic factors that act through the soil. Important soil factors that affect plant growth and development include both physical and chemical properties such as crust strength, soil water content and distribution, soil nutrient availability and movement, soil temperature, soil air composition and diffusion, and solute concentration of the soil-water solution. Biotic factors are also an important consideration in the classification. Biotic factors relate to competition between species and include developmental considerations of the plant itself.

Examples of biotic factors include the ease by which certain plants may be prevented from developing due to insufficient light caused by overcrowding or competition from weeds. One example is the sugar beet that is very sensitive to light deficiency early in its vegetative growing stage. A significant and ever present biotic factor are weeds that compete with the crop for available soil water and nutrients. Biotic factors also include those hereditary characteristics influencing the development of the plant's root system, the nature of the plant's mechanism for control of transpiration, its susceptibility to injury due to parasitic attack, or to detrimental physiological response resulting from the presence of atmospheric pollutants.

A classification by their environment is of particular interest to the engineer who is concerned with modifying the plant environment. The way climatic, edaphic, and biotic factors can be modified, and the effect modifications have on the physiological response of the plant or its agriculturally significant portion such as the seed form the crux of the need for the engineering approach.

A very important means of botanical classification of plants is by their structure. Certain types of seeds yield characteristically structured plants and this structure provides a clue to the environmental modifications that can be used to complement the physiological response.

All monocotyledon seeds have a single cotyledon that develop into a plant whose leaves are erect and possess parallel veins. This classification includes all grasses. Dicotyledon seeds have two cotyledons and develop into plants with flat leaves attached to a central stalk by a stem or petiole. The leaves possess a random network of veins running throughout the leaf structure.

In addition to differences in leaf structure important differences exist in regions where growth occurs as well as fundamental differences in the structure of the plant itself, particularly in the vascular tissue. The vascular system transports water and minerals from the soil to the leaves and transports food produced in the leaves throughout the plant to regions of growth and food storage.

Example 5.3. Effect of plant type on food production
According to the Statistical Abstract of the United States, in 1982 the following areas in hectares were devoted to annual grass-type crops:

Grass Type Crop	Area (ha)
Corn	33 140 000
Wheat	34 912 000
Oats	5 648 000
Grain Sorghum	6 489 000
Barley	3 866 000
Rice	1 334 000
Rye	1 026 000
Sugar Cane	300 000
TOTAL	86 715 000

The farmable area in the United States in 1982 was 189 879 000 ha with 155 061 000 ha in crop land. Another 34 818 000 ha were available for crop land but were not being used as such.

What percent of crop land was being farmed in annual grass-type crops? If the only grain type crops were biennials with the agriculturally significant yield occurring the second year of growth, what percent of crop land would be needed if the yield remained the same, to produce the same total amount of food?

Solution
The percent of farmland in monocotyledonous crops was:

$$\frac{86\,715\,000}{155\,061\,000} \times 100\% = 55.9\%$$

If these were biennial crops with the same yield we would need to double the 1982 hectares to produce the same yield, or 173 430 000 ha. Since the total cropable land is only 189 879 000 ha (155 061 000 ha plus 34 818 000 ha including the idle land and pasture land which could be converted to crop land) the percent of crop land needed would be:

$$\frac{173\ 430\ 000}{189\ 879\ 000} \times 100\% = 91.3\%$$

Note that most arable crop land would be needed for grain-type crops, with little area available for vegetable and fiber crops.

Space does not permit the study of plants from all aspects of their classification. This text will present only those that are common to the majority of agricultural crops. In particular, we are concerned with the agriculturally significant output of the green plant, the yield. In addition, we will attempt to stress aspects of plant anatomy and physiology that could significantly increase yield and may be possible by an engineering modification of the plant's environment.

This chapter is devoted to developing some examining concepts of green plant anatomy and physiology which are essential to the green plant in the soil-plant-atmosphere continuum. The terminology used is important and should be memorized because it will aid in communicating with plant scientists. This will enable the engineer to use the concepts discovered by botanists, ecologists, agronomists, and horticulturalists that can be modified to increase agriculturally significant output.

Example 5.4. Economic benefit of engineering modifications
Engineering modifications of the plant's environment can have impressive results on final yield. Dowling (1928) presented the results of sugar beet thinning in terms of the plant growth and the condition of the plant at the time of thinning.

Date of Operation	Operation Performed	Relative Yield
24 May	Only seed leaves present	1.0
31 May	Single pair rough leaves seen	1.0
6 June	Two pairs rough leaves seen	1.0
13 June	Two pairs rough leaves fully developed	0.91
20 June	Three pairs rough leaves fully developed	0.83
27 June	Four pairs rough leaves fully developed	0.67

When these experiments were conducted, hand-thinning was the only means of accomplishing the desired result. Today, mechanical thinners are being developed, although their action is imperfect. A comparison of total labor requirements for mechanically and hand-thinned beets in 1964 as presented by Reeve (1964) shows:

Hand-thinned 39.86 h/ha

Machine-thinned 31.45 h/ha (11.19 h by machine + 20.26 h by hand)

If labor costs are $4.25/h and a grower has 16.2 ha of beets, assume a 12-h work day and 3 workers, and calculate the time and cost required to hand-thin compared to machine-thin the 16.2 ha stand. What reduction in yield would he expected based on Dowling's data if the grower were unable to begin thinning until 6 June due to field conditions?

Solution
Hand-thinning only:

$$\frac{39.89 \text{ h}}{\text{ha}} \times \frac{1 \text{ man·d}}{12 \text{ h}} \times \frac{16.2 \text{ ha}}{3 \text{ man}} = 18 \text{ d} \quad 5.12$$

Cost for thinning = 39.86 × 16.12 × \$4.25 = \$2,744.36

If machine-thinning is used along with hand-thinning, the machine thinning requires:

$$\frac{11.9 \text{ h}}{\text{ha}} \times \frac{16.19 \text{ ha}}{\text{man}} \times \frac{\text{man·d}}{12 \text{ h}} = 16 \text{ d} \quad 5.13$$

If the machine costs are neglected,

Cost for machine-thinning = 11.19 × 16.12 × \$4.25 = \$770.43

and the hand thinning requires:

$$\frac{20.26 \text{ h}}{\text{ha}} \times \frac{16.19 \text{ ha}}{2 \text{ man}} \times \frac{\text{man·d}}{12 \text{ h}} = 14 \text{ d} \quad 5.14$$

Cost for hand labor = 20.26 × 16.21 × \$4.25 = \$1,394.90

Total cost for machine-thinning + hand-thinning = \$2,165.33

Since only 1 man is required to run the mechanical thinner the other 2 men can be hand-thinning behind the thinner, the time for machine-thinning + hand-thinning is reduced to 15 days. Therefore, the percent reduction in time is:

$$\frac{18 \text{ d} - 15 \text{ d}}{18 \text{ d}} \times 100\% = 17\% \quad 5.15$$

And the percent reduction in labor cost is:

$$\frac{\$2,744.36 - \$2,165.33}{\$2,744.36} \times 100\% = 21\% \quad 5.16$$

The yield reduction for hand-thinning beets at 18 days roughly can be calculated by noting that between 6 June and 20 June, the relative yield can be represented by the linear model y = 1.28x, where y = relative yield and x = days from 6 June. Therefore, the rate of reduction is 1.28%/day. Thus, the yield reduction due to hand-thinning is:

$$\frac{1.28\%}{\text{d}} \times 18 \text{ d} = 23\% \quad 5.17$$

The yield reduction due to mechanical (in addition to hand-thinning) is:

$$\frac{1.28\%}{d} \times 15\,d = 19.2\% \qquad 5.18$$

Therefore the reduction in yield due to hand labor only is:

$$23\% - 19\% = 4\% \qquad 5.19$$

Understanding the physiological needs of the plant and using appropriately designed equipment resulted in a 17% savings in time and a 4% less yield reduction while reducing labor costs 21%.

Structure of the Green Plant

An engineer who specifies parts for a machine must be concerned with the granular structure of metals and alloys because the strength, behavior, and durability of the machine part depends on the structure of the metal that comprises the part. The very fine grained structure of high carbon steel will exhibit significantly different qualities from the coarse, large grain structure of an iron casting.

In the green plant, an analogous situation exists. Cells are the fundamental building materials of plants, and the combination of cells with other materials that are produced within the cells. Since all physiological processes of plant development occur within cells, the engineer who modifies the environment must understand the basic anatomy of the plant cell. This is essential if he or she is to understand how the responses may be affected by the environmental modifications. In this section we examine some of the secretions produced by cells, the various cells themselves and their structural differences and functions—all of which are vital to the plant's existence.

The most important of the substances produced by the cells of the green plant include cellulose, hemicellulose, protein, starch, fat, pectin, cutin, lignin, and suberin. We will be concerned not so much with what the structure of the various components are as with what their role is within the plant.

Cellulose is the basic building element of the green plant, making up a substantial portion of the plant cell wall. It is the primary component that remains after the death of the plant and constitutes a major source of energy for ruminant animals. We will be concerned with the role of cellulose and hemicellulose within the wall of the green plant cell.

Proteins, starches, sugars, and fats play an important role in the diet of all animal life. We dwell briefly on these entities here, but we will delay discussing environmental concerns related to these substances until chapter 6 which deals with storage physiology of the green plant.

Pectin is the adhesive that holds cells together. It is a primary component of the material between individual cells. Pectin plays an important role in certain fruits, such as tomatoes, in preventing splitting of the product before harvest. There is some indication that the pectin component of the material between cells may be modified by appropriate environmental treatment. For example, spraying a calcium solution on tomatoes may help prevent cracking of the ripening fruit. This is mentioned to remind the student that all forms of environmental modification should be investigated as a potential remedy should the need arise.

Cutin, suberin, and lignin all play a significant role in the living green plant. Cutin and suberin are fatty waxes that form an outer layer over all green plants. Cutin, in particular, is found on the surface of all stems, leaves, and fruits of all plants. Cutin is responsible for the sheen that an apple displays after being polished. It is extremely valuable to the plant because cutin forms a waterproof covering on the plant and prevents the loss of liquid by evaporation, except through transpiration. Suberin is found mainly in cork cells that make up the outer layer of roots and some tubers such as potato. It also is waterproof and differs only slightly in structure from cutin since it is also composed of fatty waxes.

Lignin is found wherever secondary growth occurs in the green plant. It is the chief component of wood and is responsible for strengthening cell walls. Lignin itself is not cellulose, but cellulose is found imbedded within the lignin. Lignin is the chief component of sclerenchyma cells of peach, apricot, and cherry pits.

A wide variety of cells, both type and structure, are present in the green plant during growth and maturity. However, regardless of the final form of the cell, the early stages of its life are very similar. The very early stages of cellular development may be most affected by environmental modification. It is important to become acquainted with those developing cells that all plants have in common. We examine the general basic structure of the green plant cell, both as it is first formed, and as it matures into the various cells common to most green plants.

A Basic Cell

New cells are formed in the growing plant by the process of cell division that takes place in meristematic tissues. Meristematic tissues are special areas where new cells are formed by the division of existing cells. Figure 5.1 depicts the processes that occur during cell division. In figure 5.1, as cell division occurs, the nucleus of the cell splits and the development of a primary barrier composed mainly of pectic substances divides the original cell into two daughter cells, both of which contain a nucleus. As the daughter cells continue to mature, the barrier forms the middle lamella, or middle layer, between the adjacent cell walls. The middle lamella, which separates adjacent plant cells, is mainly composed of pectin, a jelly-like substance that is highly hydrated meaning that it possesses a strong affinity for water.

As the cell matures the cell wall is formed and composed of many individual strands of cellulose called microfibrils. The individual microfibrils are only about 0.01 μ and are made up of elementary fibrils about 0.0035 μ in diameter. Even through they are small, their failure stress is significantly higher than steel. During cell formation and early cell development, in response to turgor pressure, they allow the cell surface area to increase, probably by allowing individual microfibril to slip over each other. When the cell is fully mature and secondary cell wall development has taken place, the microfibril and other components making up the cell wall may resist stretching and extension while allowing deformation by bending to occur. This combination of properties makes the cell wall strong and very resistant to rupture.

Some means of fixing the individual microfibril in place is necessary since, during the cell wall formation, each microfibril is individually formed and deposited. This function is served in the cell wall matrix by hemicellulose, a cellulose substance that lacks the stranded orientation of the microfibril. The result is that the individual cell wall resembles the structure of fiberglass where individual glass fibers are bound together by resin. The result is much the same with the cell wall. Because of the more or less random orientation of the microfibril, it possesses great strength, and, because of turgor pressure, it has structure and gives form to the cell making up a tissue.

The hemicellulose with the pectic substance in the middle lamella has a strong affinity for water. The structure of a mature cell wall forms a vessel of constant surface area. The mature cell is then capable only of volumetric changes through deformation. In all cells, the walls are bathed with water that is supplied to the cell through the xylem vessels of the vascular system.

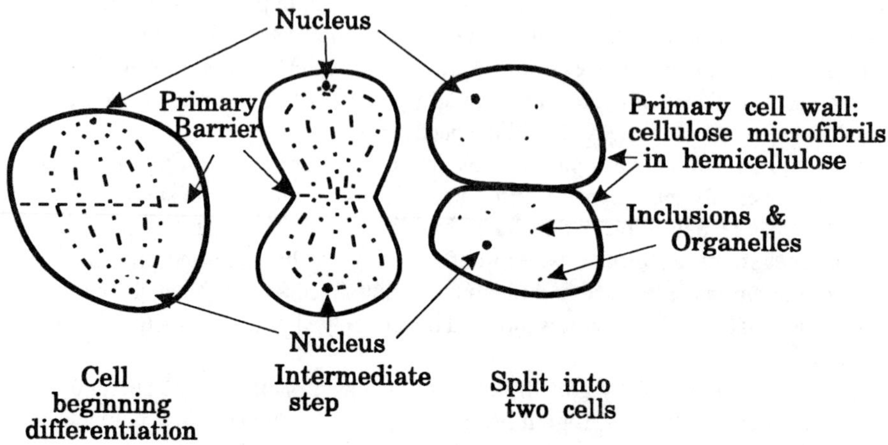

Figure 5.1. Stages in the duplication of a cell. The primary barrier coalesces into the middle lamella as the single cell splits into two cells.

Individual cells range in size from tenths of a micron in the meristematic area to as much as several millimeters in the conducting tissue of certain plants. There are enormous numbers of cells present in a single plant of a species.

Figure 5.2 shows the arrangement of cells in the petiole of a celery plant. Notice that the middle lamella is not continuous from one cell to another. As a result, there is a network of air spaces within the matrix of cells in the interior of a green plant. Each cell wall has a portion of the matrix of microfibril and hemicellulose exposed to the air in the intercellular spaces. The air spaces provide a convenient pathway for oxygen, carbon dioxide, water vapor, and other gases to diffuse throughout the collection of cells comprising the plant. While the network of intercellular air spaces is present to some degree in all plants, it is more well developed in some than in others. In particular, in plants that are grown in excessively wet soil, the intercellular spaces are often very pronounced.

Example 5.5. Cell numbers in a green plant leaf
Estimate the number of cells in a leaf 0.5 mm thick and 2500 mm² in area, if the cells are about 5 μ in diameter and 20 μ long with the long axis oriented in the direction of the thickness of the leaf.

Solution
To obtain the estimate, we first determine the number of cells lying in the leaf plane. The transverse cross-section of a single cell is:

$$\frac{(0.005\ \text{mm})^2 \times \pi}{4} = 1.96 \times 10^{-5}\ \text{mm}^2 \qquad 5.20$$

Therefore, the number of cells in one cell layer in the leaf is:

$$\frac{\dfrac{2500\ \text{mm}^2}{\text{leaf layer}}}{\dfrac{1.96 \times 10^{-5}\ \text{mm}^2}{\text{cell}}} = 1.27 \times 10^8 \ \frac{\text{cells}}{\text{leaf layer}} \qquad 5.21$$

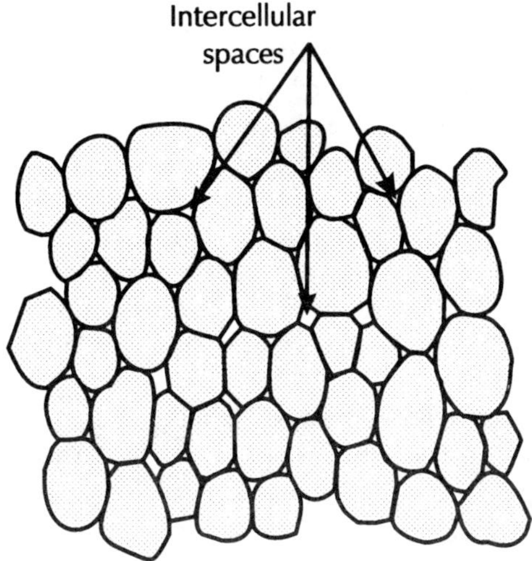

Figure 5.2. A schematic depiction of the intercellular spaces which are characteristic of the interior of most green plants.

Assuming each layer is composed of the same cells, the number of cell layers is:

$$\frac{0.5\ \text{mm}}{0.020\ \frac{\text{mm thickness}}{\text{layer}}} = 25\ \text{layers} \tag{5.22}$$

and:

$$25\ \text{layer} \times \frac{1.27 \times 10^{8}\ \text{cells}}{\text{layer}} = 3.17 \times 10^{6}\ \text{cells} \tag{5.23}$$

Although only an estimate, this example illustrates the great number of cells present in one organ of part of a plant.

Example 5.6. Intercellular surface area
If each cell in the above example has a surface area of 1000 μ^2 and 90% of this area is in contact with other cells, estimate the total surface area exposed within the leaf.

Solution
If 90% of the cell is in contact with other cells, 10% of the area is in contact with the air spaces. Therefore, since 1 μ^2 is 1×10^{-6} mm^2:

$$\frac{1.00 \times 10^{-3}\ \text{mm}^2}{\text{cell}} \times 0.1 \times 3.2 \times 10^{8}\ \text{cells} = 32\ 800\ \text{mm}^2$$

The area exposed in the intercellular spaces is about 1 order of magnitude larger than the external area of the leaf. The actual amount of air space will depend on the plant, and, in come cases, the life history of the plant since plants grown in waterlogged conditions have much larger air spaces than those grown under normal soil water situations.

Figure 5.3 presents the cross-section of an individual young developing cell. The cell wall encloses a thin, semipermeable membrane called the plasma membrane or plasmalemma containing the cytoplasm and the cell nucleus. The nucleus and cytoplasm together are called the protoplasm and comprise the living portion of the cell.

Contained within the cytoplasm are many cellular inclusions and organelles. One of the organelles is the nucleus. Other organelles include plastids, one type of which, the chloroplast, is where photosynthesis occurs. In addition to the chloroplasts there are mitochondria active in respiration, and ribosomes active in protein synthesis. Other microbodies include those which produce and contain various enzymes and organelles involved in the formation of the cell wall. In older cells, one may also find fats, oils, starch, and protein bodies. Within mature cells, vacuoles that contain cell sap and are bound by the vacuolar membrane or tonoplast will also be found. The nucleus contains the chromosomes by which hereditary traits are passed from generation to generation. The chromosomes contain individual genes. Manipulation of genes including selecting specific genes for insertion into a plant is referred to a genetic engineering. This is an area of biology in its infancy and largely the concern of the microbiologist.

Figure 5.3 also shows depressions in the cell walls, the pit pairs through which strands of cytoplasm termed plasmodesmata may be seen to communicate between adjacent cells. Observations of living cells reveal that the cytoplasm is often seen to be in rapid motion, a phenomenon called protoplasmic streaming that is thought to be essential to the physiological well-being of the cell and may be related to the existence of plasmodesmata.

As a cell matures, a second membrane may form. This membrane, the tonoplast or vacuolar membrane, encloses the vacuole. The tonoplast is similar to the plasma membrane or plasmalemma in that it is permeable to certain substances, chiefly water. The portion of the interior enclosed by the vacuolar membrane is comprised of

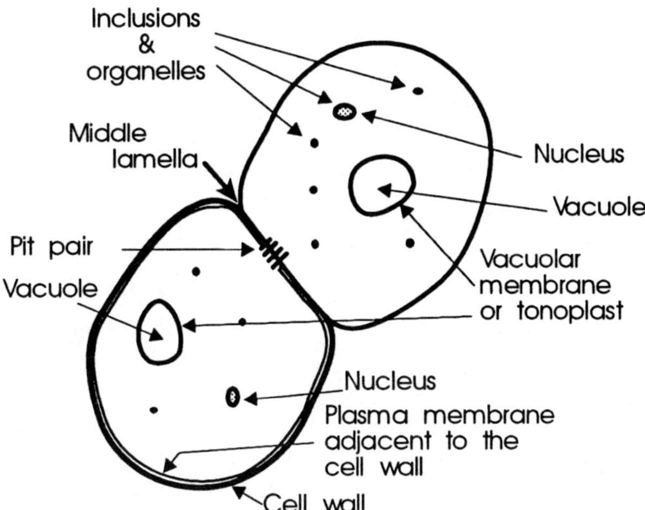

Figure 5.3. A maturing pair of cells showing the pit pair, which connects the four cells, along with the various inclusions common to the interior of a cell. Both the tonoplast and the plasma membrane are semipermeable membranes.

cell sap. The sap is assumed to be nonliving, consisting mainly of water in which salts, sugars, pigments, and other nonliving substances such as tannins and organic acids are dissolved. As the cell matures and becomes part of a plant tissue, the vacuole may enlarge and occupy most of the interior of the living cell or may disappear completely.

Cellular Maturity and Plant Tissue Formation

Obviously not all cells within a plant are identical. The process of maturation brings forth changes within the cells and cell walls that lead to the development of special cells that perform specific functions within the plant. A group of cells known as tissue, have undergone similar developmental changes. Since the characteristics of certain cells at maturity may serve to identify the function and purpose that the tissue serves within the plant, we must first examine the characteristics of the individual cells comprising the tissue.

Meristematic Cells

The meristematic cell is found in meristematic tissue wherever cell division occurs. Commonly, meristematic tissue is found at the apices of the stems and roots (apical meristem) and, in woody plants, immediately under the outer layer from which bark and vascular tissues are produced (lateral meristem). Meristematic cells are not fully mature cells. However, they may possess small vacuoles that are scattered throughout the cell. Since this is where cell development occurs, much of the food produced by the plant is utilized within the meristematic cells. In addition, the meristematic cells also produce various substances that are used in controlling growth or stored as output of the green plant system.

Parenchyma Cells

The most general cell to be found in a green plant is the parenchyma cell. The term parenchyma is applied to any cell that is not otherwise classified. This cell is thin-walled. It is always a living cell and may be capable of cell division. Therefore, meristematic cells are parenchyma cells. The main function of parenchyma cells is in the manufacture and storage of food. Parenchyma cells comprise most of the cells in the leaves and parts of the stem, as well as in the roots and in the pulp of the fruit. Parenchyma cells, when located in specific portions of the plant such as the leaf, also may be known by different names when specific plant tissues are discussed. The mature parenchyma is characterized by having a large vacuole and a very well developed microfibrilic structure for the cell wall. Often, secondary thickening of the cell wall may have occurred.

Collenchyma Cells

Collenchyma cells are living cells that differ from parenchyma in that they possess thickened walls and are usually elongated in the plane of the tissue in which they are found. The thickening is especially noticeable near the intercellular spaces. The chief function of collenchyma cells is to provide strength to the plant part. The value of collenchyma is most apparent in familiar plants like celery in which the thickened collenchyma and the turgidity imposed by the water in the parenchyma cells comprising the concave portion of the stalk provide strength and rigidity to the celery stalk. While collenchyma are strengthening cells, they are living and may contain chloroplasts and carry out photosynthesis. Some collenchyma cells are large and may attain lengths of 2 mm.

Sclerenchyma Cells

Sclerenchyma are mature cells with highly lignified walls whose chief function is to provide strength for the green plant. They differ from the collenchyma cells in that they are non-living, i.e., they have no protoplasm when mature. Of two types, sclereids and fibers, fibers are more elongated than the sclereids. In addition to forming the pit of stone fruits such as peaches etc., they may be found in many other locations throughout plants. For instance, the gritty texture of pear fruits is due to the presence of sclereids distributed throughout the flesh of the pear. Sclereids are found in the bark of certain plants where they provide strength. As mentioned previously, most sclereids are highly lignified.

Fibers

The group of highly lignified cells found in many plants includes fibers, elongated cells with thickened walls whose primary function is to provide strength. Fiber cells, like other highly lignified cells, are non-living. Fibers are commonly associated with the vascular system of green plants. In some plants they attain unusual lengths, 0.1 m or more in hemp and up to 0.55 m in Rami, a plant found in the Far East in Malaysia. Fibers are used extensively in the production of rope. Hemp, a plant which has long strong fibers, is used in the production of linen cloth. Other uses are found whenever the application requires an organic fiber of sufficient length and strength.

Tracheid Cells

Tracheids are elongated cells found in the vascular system of plants whose walls have undergone some secondary development with the secretion of lignin. These cells, when mature, are dead and possess no nucleus, cytoplasm, or vacuole. Their walls possess numerous pits, the remains of pit pairs, which serve to admit water into and out of the cell as the need arises. Tracheids are elements of the plant xylem, and provide strength and support by virtue of their thick walls.

Vessel Cells

The cell most common to the xylem tissue is the vessel cell. During maturation, vessel cells develop into an elongated tubelike form with prominent vacuoles that fill the interior of the cell. When completely developed the vacuoles of vessel cells disappear with the cell walls at the ends of the cells. Since vessel cells form a tube with the end of one cell in contact with its neighbor, the result is a hollow structure that may extend for meters within the xylem tissue of the vascular system. The lateral walls of vessel cells are highly thickened and lignified and have many pits. The pits extend through the cell wall and form passages through which water can move into and out of the vessel cell. Vessel cells are the primary water conducting tissue of the vascular xylem. Water also may move through the other cells such as tracheids that are part of the xylem tissue.

Sieve Cells

The vessel cell is fundamental to the function of the xylem. The sieve cell, with the companion cell lying immediately adjacent to it, is the fundamental cell of the phloem tissue. The sieve cell is a tubular *living* cell filled with a substance which resembles cytoplasm, but without the nucleus that is characteristic of other living cells. The ends of the sieve cells are perforated. The perforations enable the substance in each cell to communicate with its neighbor sieve cells. An important characteristic of sieve cells is the companion cell. The companion cell lies next to the wall of the sieve cell. Companion cells always have a nucleus and are thought to be vital to the function of the sieve cells in their ability to conduct food throughout the green plant.

Epidermal Cells

Epidermal cells cover the green plant from the leaves to the roots, forming the epidermis of the plant. Some modifications of form and structure may occur depending on where the epidermis is located. In general, epidermal cells are small in comparison to cells such as cortex or, vascular tissue such as vessels and sieve tubes. Epidermal cells occur in one or several closely packed layers. These cells have no intercellular spaces. The mature epidermal cell covering the leaf and stem is colorless. The outer cell wall is covered with a layer of cutin, the waterproof wax-like substance. When mature, the epidermis wall becomes covered with suberin, a substance that eventually transforms the cell into cork. Cork cells cover mature roots and also may cover the aboveground portions of plants. Cork is waterproof and also protects the underlying cells from injury.

Early in root development, the epidermal walls have no cutin or suberin and thus are permeable to water. In addition, the walls of individual epidermal cells may develop hair-like projections called root hairs. The diameter of an individual root hair is many times smaller than the diameter of the root. The root hair may be several millimeters long and can penetrate very small soil pores. Since the root hair is not covered with cutin, it will readily absorb water. Water absorption appears to be the primary function of root hairs since they form a radial extension of the root. This enables the root to absorb water that may be stored in soil pores several millimeters from the root itself. Because root hairs are so numerous, water does not need to move through the soil toward the center of the root. Instead, once absorbed by a root hair, the water molecule is in a medium through which movement is far easier than it would be in the soil. Root hairs do not exist for a very long period after their formation. In most plants, they will slough off and disappear after a period of from 12 h to several days although for trees, root hairs may last for weeks or even months.

Complex Plant Tissues of the Vascular System

The xylem and phloem are the tissues that form the vascular tissue of the green plant and extend throughout the plant from roots to leaves. Xylem tissue is composed of nonliving cells whose function is to form pathways through which water may be transported to all portions of the green plant. Xylem tissue is composed of vessel cells, fibers, and tracheids. The vessel cells and tracheids serve mainly as transport cells, while fibers furnish strength to the tissue structure.

Phloem tissue, on the other hand, is comprised of living cells, namely, sieve cells and companion cells. While transport of water through the xylem appears to be passive, i.e., movement occurs in response to a difference in water potential between two points, the movement of material through the phloem may require active transport. This is because plant food, namely sugars, moves *against* the energy gradient of the food medium. At this point we are mainly concerned with the structure of the tissues in the xylem and phloem, in particular, with the cells that comprise the tissue.

Xylem Tissue

The xylem tissue that is composed of vessel cells, tracheid cells, and some fibers may extend for several meters along a stem (Slayter, 1967). The cross-walls of these cells are perforated at intervals along their length by many pits through which water may move. The diameter of the vessel cells may range from 5 μ in grasses to more than 100 μ in some woody species. The pits, whose radius rarely exceed 1 μ, may be covered by perforated membranes whose perforations are in the order of 0.1 μ. Due to the pits in the crosswalls of the tracheids and fibers, if rupture of the xylem vessel cell occurs, water may drain out of the vessel, but it is stopped at the pit openings in the cell walls. The inability of water to drain completely from a cut stem may have implications in the processing of grasses and legumes for forage, related to the removal of excess water before storage.

Xylem tissue is located toward the interior of the stem aboveground, and toward the exterior of the stem below ground. This location is fortuitous since it places the xylem closer to the exterior of the root. Aboveground in the stem, however, the stem xylem tissue is placed as far away form the exterior of the plant as possible. This requires water leaving the xylem to be placed in contact with the plant cells before reaching the epidermis. Such an arrangement insures that all plant cells receive water.

Example 5.7. Anatomical features and agricultural operations
Would you expect either the leaves or the stems of forage to dry more slowly? Why? How might the drying process be accelerated?

Solution
The leaves dry more quickly, because they are flat and thus more surface area is exposed to air. The surface area of cells within the leaf is much greater than the leaf area, thus promoting moisture loss from the cells. In the stem, the path of water may be seriously impeded by the membranes that cover the pits and cross-walls of the tracheid vessels. Drying throughout plant is impeded by the cutin which forms on the epidermal layer. To speed up drying, crushing the stem and leaf is very effective. Crushing disrupts the cutin and epidermal layer, and breaks down the internal structure of the plant, releasing the water and providing more direct pathways to the atmosphere.

Phloem Tissue

As mentioned previously, phloem tissues are made up of sieve tubes and companion cells, although other cells such as phloem fibers and parenchyma cells may be found with the phloem cells. Sieve tubes may range up to 100 μ or more in length in potato and other species. The diameter of phloem cells will vary depending upon the location of the individual cell in the plant. A significant characteristic of phloem tissue is the presence of the companion cell. The sieve cell is *always* accompanied by a companion cell. It may be because only sieve cells, of all the types of cells within a green plant, do not possess a nucleus. Companion cells always have a nucleus and sieve cells are never found without companion cells.

Phloem cells, like xylem tissue, reverse their location between the above ground and below ground portions of the green plant. Aboveground, the phloem tissue appears toward the exterior of the stem. The lateral meristem is toward the exterior of the stem directly under the epidermis. This arrangement places the phloem near or in contact with the meristematic tissue. The plant food, produced by photosynthesis in the leaves and transported through the plant by the phloem, is located very near to the region where cell production takes place. Since the meristem is a region of high metabolic activity, i.e., a high requirement for food energy, meristematic cells are well supplied with their food.

In the plant root, however, the phloem tissue is located nearer the center of the stem. Phloem tissue, becomes functioning phloem quicker than xylem tissue. This is because xylem tissue must mature and then die before functioning as water conduction xylem tissue. This results in phloem tissue being mature most all of the way to the end of the root where the apical meristem is located. Because of this, the meristematic tissue is well supplied with the food it requires to function properly.

Developing Plant

A green plant may be cloned, i.e., it may develop from a cutting or a graft, or, more commonly, from a seed. Certain agriculturally significant crops are always cloned. An example is the yellow delicious apple, developed as a mutation on a red delicious apple tree, it is always sterile and cannot reproduce from seed. Therefore, the only propagation is by grafting or cloning. Actually, reproduction by cloning is practiced for most fruit crops. When propagated from seed, fruits often mutate and regress to some earlier undesirable trait. Cloning ensures the perpetuation of desirable traits. Furthermore, since the root system of some plant species plants with undesirable fruit are much more capable of resisting disease and stress, it is common practice to graft desirable fruits to hardy rootstocks to develop productive orchards.

Most agriculturally significant annuals and biennials are started from seed. The embryo within the seed from which the plant will develop is separated into two major divisions, the epicotyl from which the shoot, leaves and reproductive aboveground portions of the plant develops, and the hypocotyl, the extension of which is the radicle from which the root system arises.

As shown in figure 5.4, leaves are attached to a main stem by petioles, small stems that connect leaves with the nodes of the main stem. The length of main stem between nodes is the internode distance. In all plants the apical meristem is found at the tip of the stem and all its branches. The apical meristem consists of specialized cells in which the process of cell division and the formation of new cells is occurring. In addition to apical meristem in

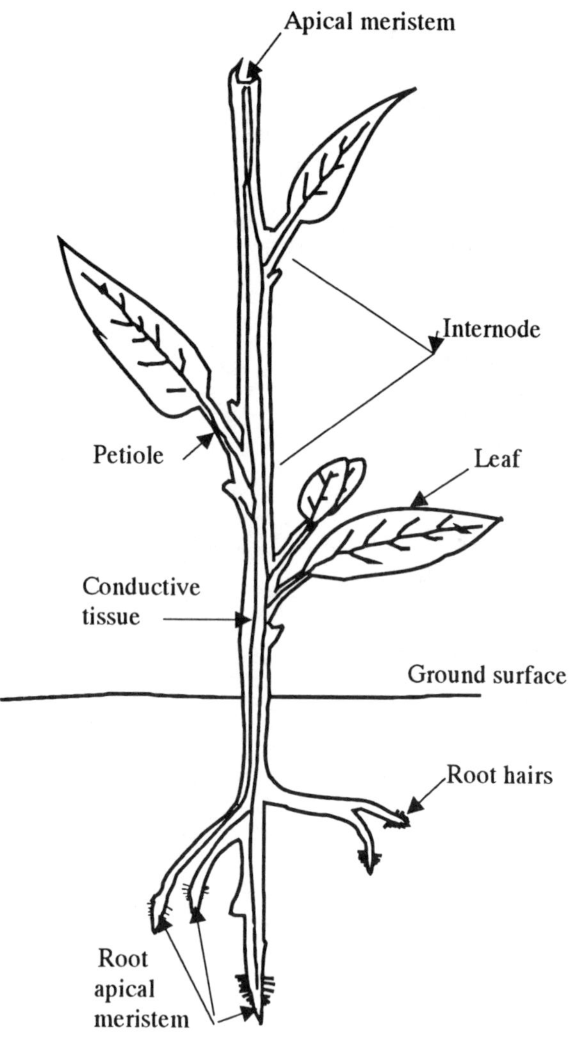

Figure 5.4. The young green plant. In the dicotyledonous plant, the meristematic tissue occurs primarily at the root and shoot tips. In monocotyledonous grasses, meristematic tissue occurs at the root tip, but at the base of the grass leaf rather than at the tip.

perennials such as trees, there is meristematic tissue at the circumference of the stems and roots termed lateral meristem. The lateral meristem enables perennials to expand in diameter as well as length. Because the meristematic tissue is located circumferentially just under the epidermis, it is susceptible to damage by mechanical abrasion, or by insects and disease, especially if the bark is disturbed.

The belowground portion of the plant comprises the root system with a few exceptions. While the carrot is a true food storage root, in some plants, such as the Irish potato, a tuber, a subterranean stem used chiefly for food storage develops. The bulb of an onion is a leaf, even, though it develops underground. As shown in these examples, variations in structure occur, however, the environmental requirements of these variations remain common to roots. Therefore, when a particular green plant has special requirement modifiable by engineering means, this will be noted. In general, however, the material that follows applies to most agricultural crops.

The nature of a plant root system varies from one species to another. The development of the radicle from the hypocotyl of the embryo is begin with the initial extension of the radicle downward into the soil. This is followed by the production of secondary branch roots.

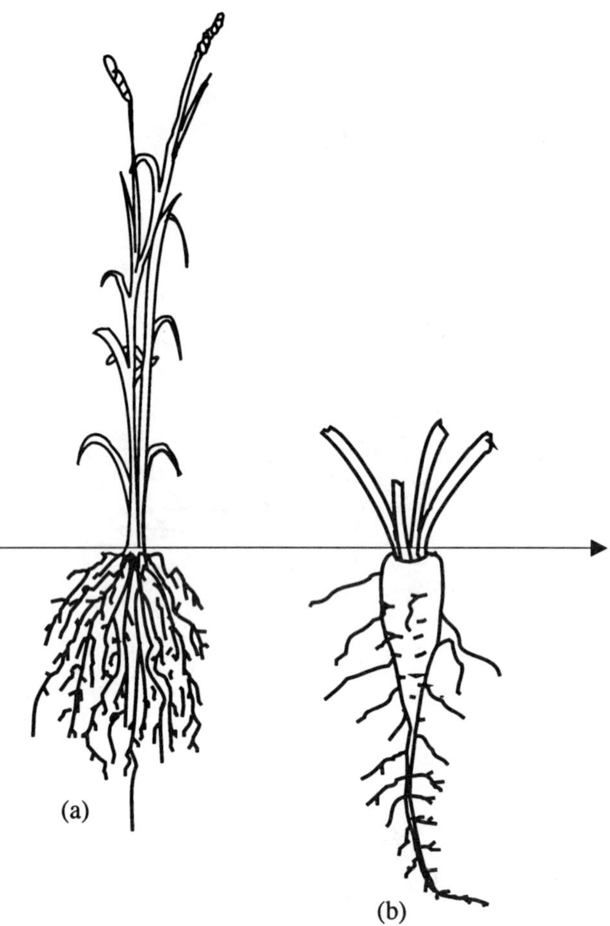

Figure 5.5. The fibrous root system of (a) is typical of many grasses as contrasted to the taproot system shown in (b).

Root systems, as illustrated in figure 5.5 may vary from the fibrous, characteristic of many grasses and shrubs, to a shape consisting of many roots extending outward from a central tap root. Tap root crops such as alfalfa may reach to great depths; in alfalfa, roots have been found growing at a depth of 3.6 m (12 ft) below the ground surface. One report stated that alfalfa roots were found extending through the side of a well 21 m (72 f) underground.

Regardless of the nature of the root system, careful examination of the roots of a growing plant will reveal many branches, the apex of which is covered by a protective tip called the root cap (see figure 5.6). The root cap protects the apical meristematic tissue, found immediately under the root cap. In the the shoot, the meristematic tissue is the site of cellular division resulting in the production of new cells that are essential to root growth.

Root growth occurs as a result of elongation of the cells in the portion of the root immediately behind the root cap in a region of cell development as shown in figure 5.6. This is the point in the root where the xylem tissue becomes fully developed and is, therefore able to supply water to the elongating cells. Here also, water absorbed by the root from the surrounding soil reservoir is taken into the root cells, probably by osmosis. The cells then expand and push the root forward into the soil matric. The root cap serves as a shield. However, it also prevents the root from penetrating into portions of the soil matric where the pores in the soil are smaller than the outer diameter of the root cap. The relationship of the root cap diameter to soil pore size has significant implications for the tillage process. Any action that compresses the soil making it more dense and decreasing the size of the soil pores obviously impedes the extension of the root system. Since the roots are the means by which water and minerals are taken in by the plant, anything that impedes root extension has a negative effect on plant growth. In the chapter on

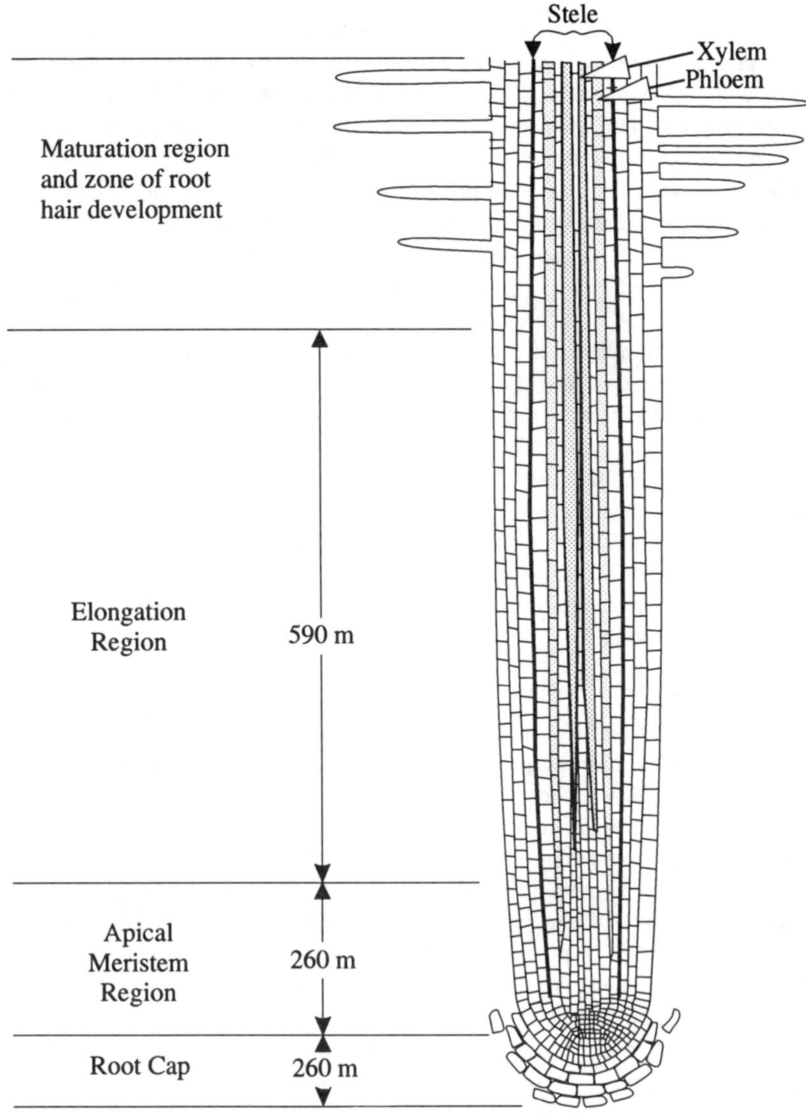

Figure 5.6. The developing root. The meristematic region is protected by the root-cap and is regenerated from the root apical meristem as the outer cells erode away due to abrasion by soil particles. Xylem tissue extends into the elongation region while phloem tissue is mature into the meristematic region where new cells are being produced.

water movement through the soil profile (Chapter 10), we will discuss the daily extension of the root into fresh soil during the growth of roots. This is the most significant means of providing the root with water for the following day's growth period.

Specific Functions of Parts of the Growing Plant

Stem

The shoot and root parts of the growing plant differ widely in form. Therefore, they also differ in gross function, although some functions so important to the physiological well-being of the plant are shared by both root and shoot. The common elements of function will be introduced during the discussion of individual structural relationships.

The stem provides support for the plant leaves in a manner that allows good exposure of the leaves to the incoming sunlight, the primary source of energy for food production by the plant. In addition, in most agricultural crop plants, the leaves, which constitute the primary food-producing regions of the plant, are maintained at varying distances above ground ensure better air movement through the plant leaf canopy. The inward movement of CO_2 into the plant leaf canopy and the outward movement of O_2, the by-product of photosynthesis, is thereby facilitated.

Significant amounts of water are lost from plant leaves during transpiration. The transpired water is carried out of the plant leaf canopy by air movement around the leaves so the structural qualities that enhance food production also favor water loss. Since water is an important factor in the food production, growth, and temperature regulation of the plant, excessive water loss is not desirable because the available supply in the soil reservoir may become depleted.

In addition to supporting the leaves, the stem with the petioles provide pathways through which the plant's vascular system extends. The vascular system consists of the xylem which serves as a conducting tissue carrying water and minerals from the soil to the leaves, and the phloem which is the primary pathway for the translocation of food away from the sites where it is produced to those where it is either stored or used. Both functions depend on the physiological needs of that part of the plant during the period under consideration.

In many plants the stem is green, so food production occurs in the stem as well as the leaves. The stem is also involved in the storage of food produced in the leaves. It also supports the reproductive organs and cell-producing regions of the plant. Since both reproduction in the form of seeds, and the production of new plant cells by the meristematic tissue are processes that require the utilization of food to obtain energy, the plant stem is an important temporary storage facility for the plant.

Roots

The roots of a growing plant perform several functions. One of the most important is to provide sites for the absorption of water and minerals from the soil. As will be shown later when we examine the anatomy of the root in more detail, the principal sites of water and mineral absorption are near the tips of the growing roots in a region where hair-like extensions of the individual epidermal cells, the root hairs, proliferate.

Extremely important to the well-being of the plant is the anchorage provided by the root system. Insufficient anchorage during periods of environmental stress, such as high winds, may cause the plant to become uprooted so lodging occurs. Lodging also results from insufficient strength in the plant stem, a condition that may be related to mineral imbalance and is independent of the anchorage provided by the root system. Severely lodged crops impede mechanical harvesting and are a challenge to the engineer's ingenuity.

Microscopic examination of the interior of the roots reveals the plant vascular system. The vascular system, consisting of xylem and phloem tissues, continues upward through the stem and petioles into the leaves. An important function of the plant root system therefore is to conduct water and minerals as well as plant food produced by the leaves and the stems throughout the plant. The roots also act as storage regions for plant food. In the sugar beet, for instance, plant food is stored as sugar, while in certain other crops such as potatoes, storage is primarily in the form of starches and other carbohydrates.

The location of meristematic tissue under the root caps at the apex of each rootlet is vital to the function of the roots. Cell division in the meristematic regions produces new cells that can then undergo growth by enlargement. This results in the continual extension of the root system through the pores of the soil. Because roots function in the unsaturated region of the soil where the water potential is less than zero, and because the resistance of the soil to water movement increases significantly as the water potential drops, the soil regions near the roots would soon be depleted of water without extension and the plant would become severely stressed. Only by continual extension of the roots can the plant root system penetrate regions of the soil environment where water is more readily available and where sufficient water to supply the physiological needs of the plant can be absorbed. Thus, any environmental condition that leads to cessation of growth and extension of the root system is almost certain to impair the aboveground development of the plant. If lack of readily available soil water continues long enough, the condition will inevitably result in irreversible damage and death to the entire plant.

Example 5.8. Roots and root surface area for soil interaction
A rye plant was found to have 6,400 roots (including main, secondary and all branches of the root system). The total length of the entire system was estimated at 64 m. If the weighted average root diameter was 250 μ, estimate the surface area of the root system in mm^2.

Solution
The system is equivalent to a cylinder 64 m long and 0.250 mm in diameter. Therefore, area (ignoring the end of the roots) is:

$$64 \text{ m} \times \frac{1000 \text{ mm}}{\text{m}} \times 0.25 \text{ mm} \times \pi = 5.0 \times 10^4 \text{ mm}^2 \qquad 5.24$$

Anatomy of Structural Parts of the Green Plant

The engineer must be familiar with the anatomical structure of the growing green plant tissues. In particular, the engineer should be cognizant of the relationship of the various tissues and parts of the plant to each other, to the metabolism and physiology of the plant as these depend on the plant tissues, and to the possible environmental influences that may affect the anatomical structure as well as the physiological response of the plant itself. We begin this part of our study with the root, particularly with that portion of the root where water absorption primarily occurs. We will then traverse the plant upward to the leaf. This approach should develop an appreciation of the continuity of the plant structure and of the need to consider the plant a complete entity in the soil-plant-atmosphere continuum.

Selected Anatomy of the Root

Figure 5.6 depicts the longitudinal cross-section of a young root. The apex of the root is covered with the root cap, a collection of cells that shields the cell producing tissue in the apical meristem immediately under the root cap as the root is forced through the soil. Cells in the root cap are continually being sloughed off as the root tip is pushed through the soil by the elongating region of the root immediately behind the apical meristem. The root cap protects the meristematic region and produces new cells to replace those lost by abrasion of the root cap.

The meristematic region is the region of cell development, it is not the region of root growth. While growth cannot occur without cell development, the presence of many tiny cells in the meristem is of no significance to agricultural production. Production is determined by volume or weight gain. For this reason, the region where growth occurs is in that portion of the plant where cell elongation takes place. Here is where true growth occurs since it is only by enlargement of the cell that an increase in the volume or weight of a plant or an individual plant part can occur. Growth is independent of the number of cells present and dependent more on the size of the cells.

Cell elongation forces the root tip through the void spaces in the soil. Cell elongation is therefore responsible for root extension. This creates a problem for roots growing in highly compacted soil like the dense layer of plow pan immediately below the plow layer of soil. Root extension is also a problem in soils with high clay content because voids in clay soils are normally very small. Fortunately, although clays possess a preponderance of voids smaller than 0.2 μ, there are usually some voids present whose diameter exceeds 60 μ. Most crop plants have root diameters ranging from 60 to 125 μ. Thus, even in clays there are voids present that are large enough to allow the root tip to penetrate. This is fortunate since it has been shown that roots could not penetrate a medium whose openings are smaller than the root tip. Even if the root penetrates a pore whose size is approximately that of the root tip, sufficient radial pressure cannot develop to enlarge the pore since most of the force of growing roots is axial. Figure 5.7 shows what happens when a root enters a pore that is smaller than the ultimate diameter of the root. While the tip of the root in figure 5.7 was able to enter the pore, radial pressures developed by the root were unable to enlarge the pore. The root cross-section deformed as the root diameter increased.

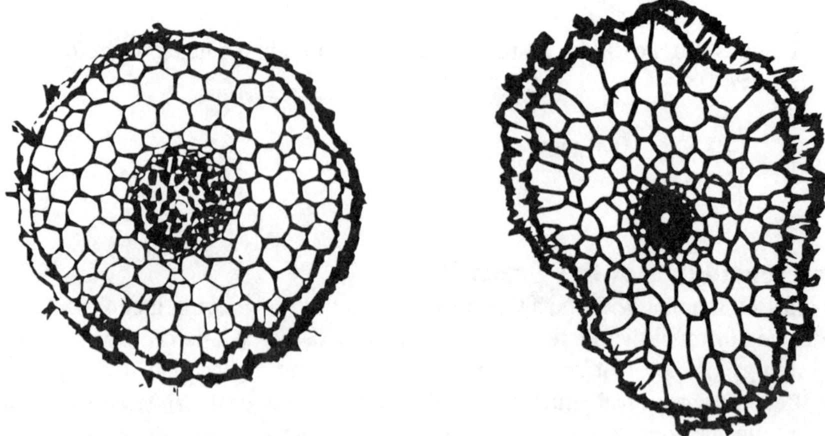

Figure 5.7. The root shown at the left grew in loose soil. The root on the right grew in compacted soil and was deformed because pressures in the root cells were not sufficient to deform the soil pore. (From *Soil Physical Conditions and Plant Growth*, Glinski and Lipiec, 1990, CRC Press)

Example 5.9. Stress in a cell wall
What is the stress within the wall of a spherical cell whose interior is at a pressure of 2500 kPa if the external cell diameter is 10 μ and the wall thickness is 1 μ?

Solution
A pressure of 2500 kPa must be maintained by a tension in the cell wall. Since the interior diameter is 10 μ and the thickness is 1 μ, the approximate cross-sectional area resisting the stress is:

$$\pi \times 10\ \mu \times 1\ \mu = 31.4\ \mu^2 \qquad 5.25$$

The force causing stress in the wall is:

$$F_R = 2500\ \text{kPa} \times \frac{(10-2)\ \mu^2 \times \pi}{4} = 1.26 \times 10^5\ \text{kPa}\cdot\mu^2 \qquad 5.26$$

The stress is the total force divided by the resisting area or:

$$\frac{1.26 \times 10^5\ \text{kPa}\cdot\mu^2}{31.4\ \mu^2} = 4 \times 10^3\ \text{kPa} = 40\ \text{atm} \qquad 5.27$$

Beyond the region of cell elongation is the region of cell maturation. Here the xylem and phloem conductive tissue is well developed. The xylem tissue, however, functions only into the region of cell elongation while phloem tissue is mature through the regions of cell elongation and into the meristematic zone of cell division. Therefore, the phloem is able to provide nourishment to the meristem, the region of high physiological activity, while xylem cells extend only into the region of cell elongation. This enables the xylem tissue to conduct water both downward to the elongating cells as well as upward from the root to the rest of the plant. Since water is not needed for cell division, the xylem does not penetrate the apical meristem.

In the region of cell maturation the root epidermal cell walls elongate to form root hairs. The zone of root-hair formation may begin as near as 0.5 to 2 mm from the meristematic region of the root and continue for 15 to 100 mm from the root tip. Beyond this point the epidermal cells begin to disintegrate as secondary growth and suberization of the external part of the root occurs. The root hairs increase the actual surface of the root making intimate contact with the soil. Root hairs are usually about 10 to 12 μ in diameter, considerably smaller than the root cap. Because of this, individual root hairs can penetrate soil voids much smaller than those into which the root can extend. The density of root hairs is enormous, exceeding 200 root hairs/mm^2 of root surface. Each root hair may extend up to 12.5 mm into the soil, depending upon the plant. Thus, the system of root hairs enormously expands the area of the root in contact with the soil particles and facilitates the absorption of water and minerals from the soil mass.

During the growth process as root elongation occurs, new root hairs continually form immediately behind the region of cell elongation. The older root hairs decompose and disappear as the root matures, sometimes within a day after forming. Root development maintains the water absorbing portion of the root in contact with that part of the

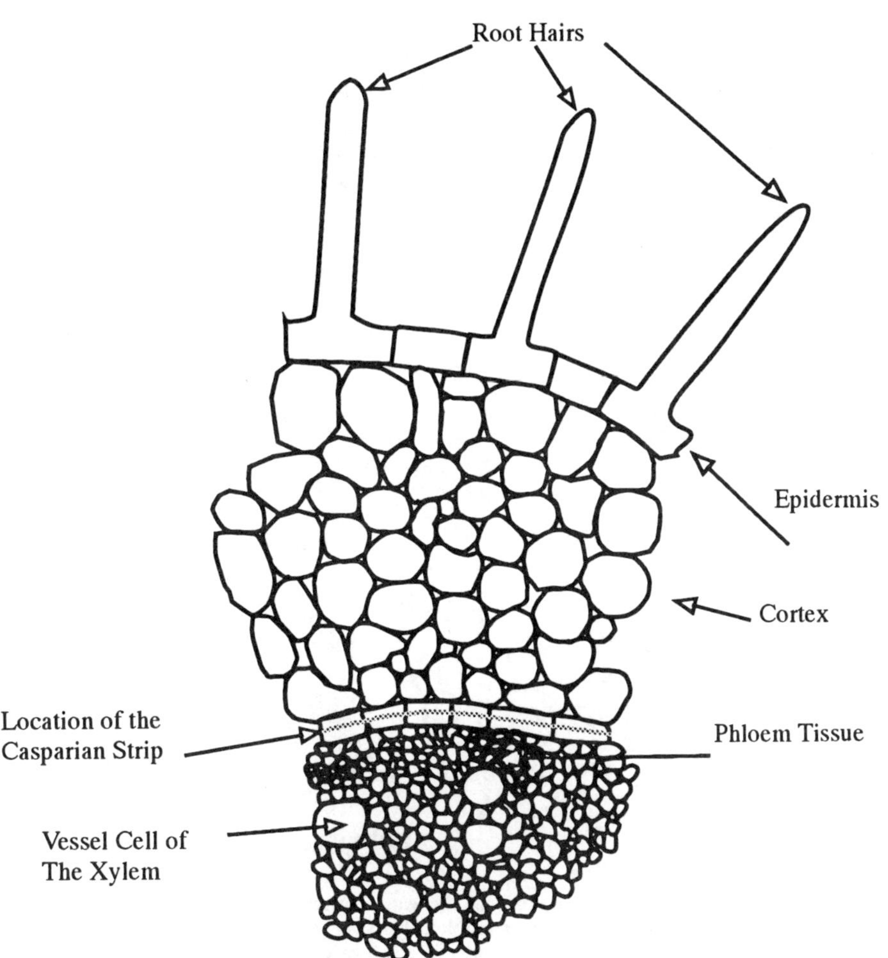

Figure 5.8. Cross-section of a developing root near the region of maximum water absorption.

soil mass in which the water content has not been depleted due to absorption by the root. In grasses and other monocotyledonous plants, and in some trees, the root hairs may persist for longer periods, often months at a time.

Figure 5.8 illustrates the cross-section of a growing root near the region of maturation and maximum water absorption. This is typical of the cross-sections of a growing root and will acquaint the student with the internal features of primary root growth. As is shown in figure 5.8, root hairs develop around the very young root in the epidermis, the closely packed layer of cells that is highly permeable to water. Although not all epidermal cells develop into root hairs, the number that do develop root hairs is large, possibly exceeding 200 root hairs/mm^2 of root surface area.

In figure 5.8, the stele, the center portion of the root, is enclosed by another layer of cells, the endodermis, within which lies the xylem and phloem tissue imbedded in miscellaneous parenchyma cells called pith. Within the endodermis is the casparian strip, a layer of a waxy waterproof substance imbedded in the walls of the endodermis. The casparian strip is unique in that water is forced to pass through the interior of the cells on which the casparian strip is deposited. Figure 5.9 depicts the pathways by which water may enter the xylem tissue within a root. There is uncertainty about the method by which the root accomplishes this transfer of water from the soil to the stele. One school of thought holds that the casparian strip disrupts the potential gradient for water transfer that exists throughout the plant, and that water through the casparian strip cells is due to *active transport* rather than passive transport that exists throughout the remainder of the plant. If this is the case, it helps to explain the intense metabolic activity that occurs in the developing root. Because of the metabolic activity, it is important for young roots to be well supplied with oxygen and that they be able to purge the carbon dioxide resulting from metabolism. The oxygen supply and the removal of carbon dioxide is accomplished by diffusion through the soil voids, thus it is very important that air-filled voids be present in the soil.

Exterior to the endodermis and immediately under the epidermis is the cortex tissue. It is composed mainly of parenchyma cells with large intercellular spaces. During root development, gaseous diffusion may occur through the cortex. In a mature root, however, the epidermis and cortex cells disappear while the walls of the endodermal cells become heavily suberized and waterproof and prevent water from entering or leaving the plant. Water absorption is therefore limited to the root hair region. Suberization prevents water from traveling up the xylem and exiting the root in a location of lower soil water potential. In this fashion, the transfer of water from wetter to drier regions of the soil through the root system is prevented.

Beneath the endodermis there is a layer of cells termed the pericycle. In some plants, the pericycle contain meristematic tissue resulting in secondary radial growth of the root as the plant matures. In the center of the young

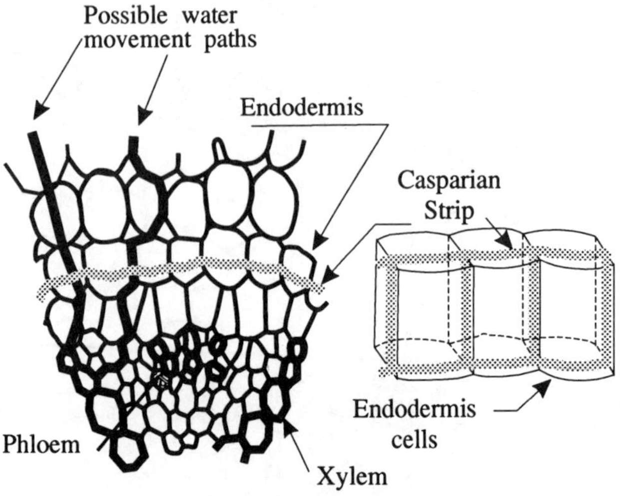

Figure 5.9. In the young root, the water movement is impeded by the casparian strip, a waxy condition that coats the lateral walls of the endodermis. For water to enter into the stele, the center of the root, water must somehow pass through the endodermal cells. Note that in the root, xylem tissue extends outward almost to the endodermis.

root, surrounded by the pericycle, the xylem tissue extends star-like up to the endodermis as depicted in figure 5.7 and 5.8. Within the rays of the starlike formation created by the xylem vessels the phloem cells are found. Both tissues these may be grouped together and surrounded by parenchymous pith cells but, more typically, the xylem and phloem occupy the greater portion of the center of the developing root.

Selected Anatomy of the Stem

The longitudinal anatomy of the plant is illustrated in figure 5.4. The tips of the stem are the apical meristematic regions that create new cells. The apical meristems are not protected by a cap as is the root apex. Cells formed in the apical meristems elongate, thus creating the internodal distances that separate the nodes, the points of attachment of the leaves.

The cross-section of a developing stem differs somewhat between monocotyledons and dicotyledons. In addition, considerable variation occurs between annual and perennial woody plants. We will restrict our examination to annuals of both monocot and dicots as these will serve to illustrate the basic functional relationships of interest to the engineer.

The cross-sectional structure of a herbaceous dicotyledonous stem is pictured in figure 5.10 and is characteristic of dicots in which the stem grows laterally. On the exterior, the stem is covered by the single layer of epidermal cells that are colorless. The epidermis, as it is everywhere on the aboveground portion of the plant, is covered by cutin which renders it waterproof.

The epidermal cells are characterized by the lack of intercellular spaces. A layer of collenchyma cells that gives strength to the structure of the stem is present under the epidermis and the exterior to the bundle fibers. The bundle fibers contain the phloem, meristematic cambium, and xylem and are imbedded in the cortex which is the center of the stem comprised of parenchymous pith cells.

In a typical monocot, such as corn pictured in figure 5.11, the meristematic cambium within the bundle fibers is absent. Any increase in stem diameter must take place in a monocot by cell expansion. The monocot stem, as the

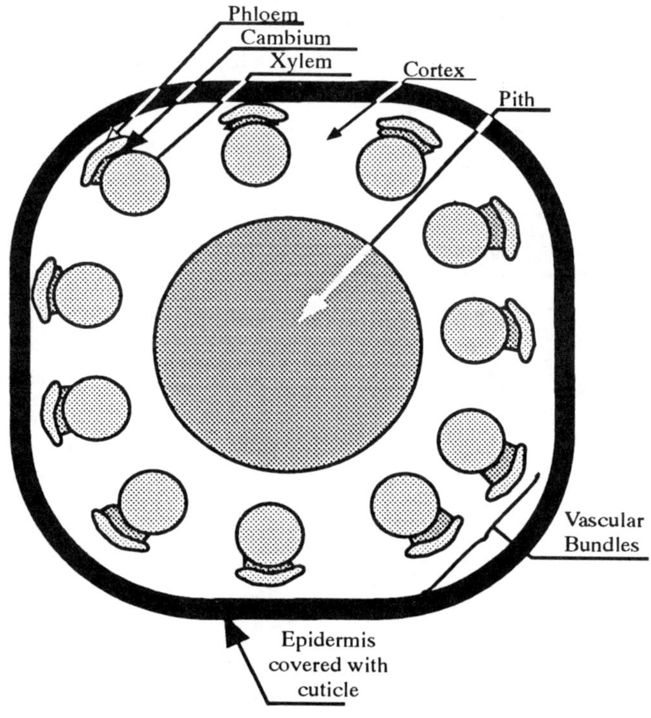

Figure 5.10. The cross-section of the stem of a dicotyledonous herbaceous plant showing the vascular bundles which contain the xylem and phloem. The phloem is toward the exterior and separated from the xylem by a layer of cambium cells.

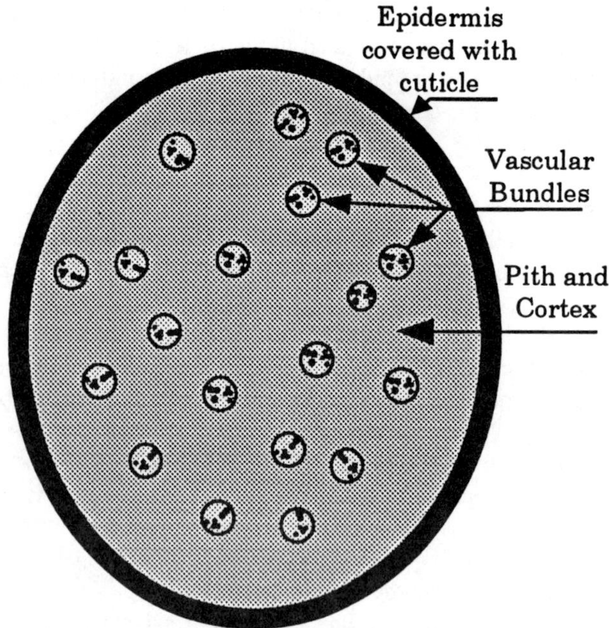

Figure 5.11. Cross-section of a corn stem: (a) epidermis, (b) sclerenchyma, (c) parenchyma, and (d) vascular bundle with (1) xylem, (2) phloem, and (3) bundle sheath.

dicot stem, is covered with epidermal cells that are heavily cutinized and colorless. The cortex tissue consists of parenchyma cells that store food. Scattered randomly throughout the cortex parenchyma are vascular bundles or bundle fibers that are comprised of the bundle sheath, strengthening cells that enclose both the phloem (toward the outside) and the xylem (toward the inside). In a monocot, no new formation of xylem and phloem can take place in the stem. All conductive tissue is differentiated and develops directly from the apical meristem.

Table 5.1. Distribution, number, and size of the stomata of selected plants

Plant	Stoma/upper mm^2 of the Epidermis	Stoma/lower mm^2 of the Epidermis	Stomatal size (μ) Length × width (lower epidermis)
Alfalfa	169	138	
Bean	40	281	17 × 13
Cabbage	0	226	
Cherry	141	249	
Corn	52	68	19 × 5
Oats	25	23	38 × 6
Pea	101	216	
Potato	51	161	
Tomato	12	130	13 × 6
Wheat	33	14	38 × 7

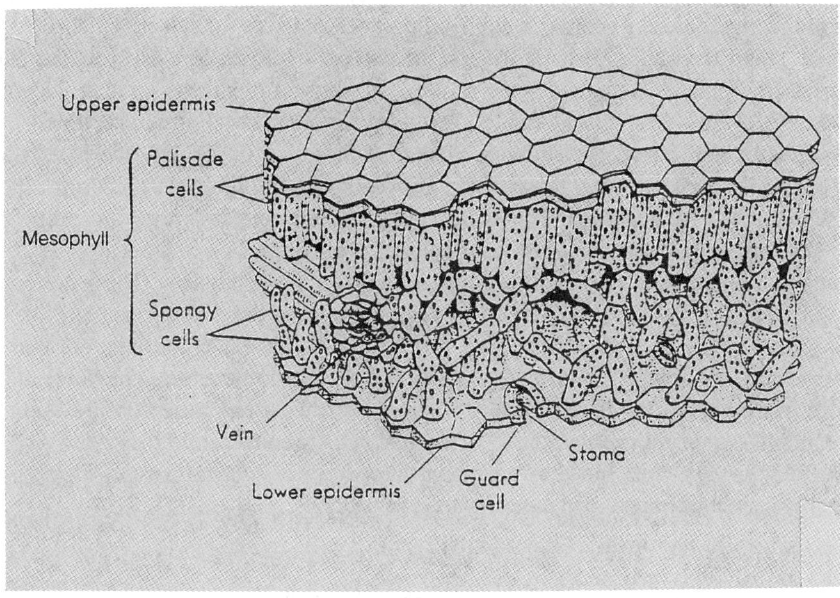

Figure 5.12. The cross-section of a leaf showing the arrangement of the cellular structure. The palisade layer containing chloroplasts lies immediately below the clear epidermal layer and above the spongy mesophyll which contains the veins that carry the xylem and phloem through the leaf.

Selected Anatomy of the Leaf

Many kinds of leaves exist depending on the species of plant being examined. Close examination of the leaves of several crops will show certain similarities and differences that are important since they enable one to characterize the anatomical structure and, to some extent, the function of leaves.

Gross differences exist between the external appearance of monocotyledonous and dicotyledonous leaves. These differences manifest themselves in the arrangement of the vascular bundles containing the xylem and phloem within the leaf. In monocots the leaves are parallel-veined, i.e., the vascular bundles run coaxially with the midrib as in a corn leaf. In a dicot, however, the veins or vascular bundles are highly branched and impart to the leaf the appearance of a net of supporting tissue filled with cellular tissue in between. Although the above differences enable one to characterize monocots and dicots easily, the internal structure of leaves from the two plants is similar, both in form and function. The form captures sunlight enabling the leaf to manufacture food for the plant. Figure 5.12 shows the main parts of the leaf structure.

The outer surface of all leaves is covered by the epidermis formed of the single layer of closely packed, transparent cells covered with a dense layer of cutin. The epidermis, except for the stomatal openings that will be discussed shortly, forms a waterproof layer over the entire leaf surface as it does over the remainder of the above ground portion of the green plant. The cutinized epidermal cells are the chief barrier to water loss from the leaf, and removal of the cutin with a solvent or by abrasion with the tip of a pencil eraser will cause rapid permanent desiccation of the leaf cells. Replacement of the removed cutin with an artificial agent, e.g., petroleum jelly, will prevent water loss although it will block the stomatal openings and prevent the interchange of air and water between the inside and outside of the leaf.

The stomata are openings existing within the epidermis. Their shape, position, and number vary with each plant. The stomatal openings are formed by a pair of guard cells and are usually accompanied by a second set of subsidiary cells that are a part of the epidermis. The stomata open as the guard cells absorb water and become turgid. When open, they provide a pathway from the outside air into the intercellular spaces within the leaf. Through the stomatal openings pass carbon dioxide, oxygen, water vapor, and, depending on the environment, gaseous pollutants present in the air. Table 5.1 compares stomatal number and size for selected plants. Although the size of an individual stoma is small, averaging perhaps 100 μ^2, the number of stoma on a leaf is enormous. From table 5.1, dicots have most of their stomata on the underside of the leaf while in monocots a rather even division is present between the upper and lower sides. This is related to the different internal structure which exists between the monocotyledonous grasses and the dicotyledonous plants. In dicots, a single palisade layer occurs immediately beneath the upper epidermis with a spongy mesophyll immediately under the lower epidermis. In grasses, however, the palisade layer may be less well defined. This is because grasses are erect so they are exposed to sunlight from either side and the chloroplasts are well distributed throughout the parenchyma of the leaf.

In some plants, the epidermal cells will develop leaf hairs or protuberances. These are common to potatoes and other plants and probably influence the action of air motion over the leaf surface and the removal of water vapor that has transpired. Under the epidermis of the dicots the palisade cells, consisting of one or more layers of parenchyma cells oriented at right angles to the epidermis, are arranged to capture a preponderance of the sunlight that passes through the colorless epidermis. Below the palisade layer is the spongy mesophyll, a layer of parenchyma with random organization so arranged that as the cells become turgid they exert a bridging action with many large intercellular spaces. This loose arrangement allows gases such as carbon dioxide, water vapor, and oxygen to circulate between the stomata and the palisade layer of cells.

Example 5.10. Force in a spongy mesophyll tissue due to turgidity
The spongy mesophyll of a leaf acts to support the leaf as its parenchyma cells gain turgidity due to the buildup of internal pressure within the cell. Estimate the force that a cell 10 μ in diameter and 25 μ long could exert if it carries an internal pressure of 25 atm and has a wall thickness of 1 μ. Assume the cell is oriented horizontally. If the cell is oriented vertically, how many cells of a similar size could the single cell support?

Solution
The microfibrilic structure of the cell wall without secondary thickening can carry no compressive load. Therefore, the only force that the cell can exert is that necessary to overcome the tensile stresses introduced within the walls due to the pressure within the cell. From example 5.9 we found that the stress in the cell walls was 6250 kPa in tension. Thus, assuming that the ends of the cell deformed until the force was distributed uniformly over the 10 μ diameter cross-section, the area over which the force acts is:

$$\text{Area} = \frac{\pi \times (10\mu)^2}{4} \times \left(\frac{\text{m}}{10^6\mu}\right)^2 = 7.8 \times 10^{-11}\text{m}^2 \quad 5.28$$

The force exerted is then:

$$7.8 \times 10^{-11}\text{m}^2 \times 6.25 \times 10^3\text{kPa} \times \frac{10^3 \frac{\text{N}}{\text{m}^2}}{\text{kPa}} = 4.9 \times 10^{-4}\text{N} \quad 5.29$$

In terms of mass being acted on by the acceleration of gravity, this is equivalent to:

$$\frac{4.9 \times 10^{-4}\ \text{kg}\cdot\text{m}}{\text{s}^2} \times \frac{\text{s}^2}{9.81\ \text{m}} \times = 5 \times 10^{-5}\ \text{kg}$$

The mass of a single cell of the size indicated is:

$$\left(\frac{4\,\pi\,(10\mu)^3}{3 \times 2} + \frac{\pi \times (10\mu)^2}{4} \times 15\mu\right) \times \frac{m^3}{(10^6\,\mu)^3} \times \frac{1000\ kg}{m^3} = \frac{3.27 \times 10^{-12}\ kg}{cell} \qquad 5.30$$

The number of cells that could be supported by a single cell of the dimension indicated is:

$$\frac{5 \times 10^{-5} kg}{3.27 \times 10^{-12} \frac{kg}{cell}} = 1.5 \times 10^7 cells \qquad 5.31$$

The cells within the palisade layer contain an abundance of chloroplast inclusions that contain chlorophyll and are the primary site of photosynthesis within the green plant. It is fortuitous that the palisade layer occurs immediately beneath the transparent epidermis so it is abundantly supplied with the light necessary for photosynthesis. The stomata and spongy mesophyll provide for ample diffusion of carbon dioxide, a raw material of photosynthesis, into the leaf, and allow for diffusion of oxygen, a product of photosynthesis, out of the leaf. Figure 5.12 shows that the vascular system of the plant containing the xylem and phloem occurs next to the palisade layer. This allows an ample supply of water, the primary substance in the xylem vessels, to be near food-producing cells. The phloem, on the other hand, are conveniently situated to conduct sugar, the food produced during photosynthesis, away from the sites of food production. There may be as many as 40 terminations of vascular bundles, sometimes called bundle ends, within a square millimeter of leaf area. Due to the abundance of vascular tissue and vascular terminations, the plant is adequately supplied with water the means to transport food necessary to life. However, little is known about the mechanism of food translocation from the cells to the phloem and through the phloem to the locations where food is stored.

Selected Anatomy of Fleshy Fruits

Botanically, any ripened or matured ovary containing a seed from which a new plant can develop is a fruit. This section, however, is restricted to fruits that are fleshy and which are the agriculturally significant parts of the plant. In particular, fruits that, when mature contain very little water and are stored in a quiescent state are properly called seeds while those that have a high water content when they are ready for consumption are known as fleshy fruits. The latter are not adapted to long term storage.

A fleshy fruit is comprised of the fruit along with a stalk or pedicel by which the fruit is attached to the plant. This fruit is composed of several layers. An outermost layer in most fruits is the pericarp which, as in the case of tomatoes as shown in figure 5.13(a), may comprise the primary agriculturally significant portion of the plant. In some fruits, the pericarp is separated into three distinct layers of tissues, the exocarp, which forms the outermost fruit wall, the mesocarp, a collection of cells that attains significant thickness, and an endocarp of stone-like sclereid cells that protects the actual seed. These are illustrated in figure 5.13(b).

A pome fruit is shown in figure 5.13(c). In pome fruits, the pericarp which envelopes the seeds is surrounded completely by remnants of the flower that preceded the fruit and from which the fruit developed. The flower remnant is comprised mainly of calyx tissue and constitutes the edible portion of the fruit. Apples and pears fall into the category of pome fruits, while peaches and cherries are primarily drupes whose edible portion consists of the actual ovary enclosing the seed.

The anatomy of fruits as presented here is simplified. The student interested in a specific fruit is directed to a text on anatomy such as that of Esau (1965).

Selected Anatomy of the Seed

Almost all agriculturally significant green plants start their development as seeds. Some exceptions exist, e.g., the potato that is started by sprouting the tuber. We should become familiar with the differences between types of seeds that the plants produce, and from which they initiate their life cycle. This information will become especially

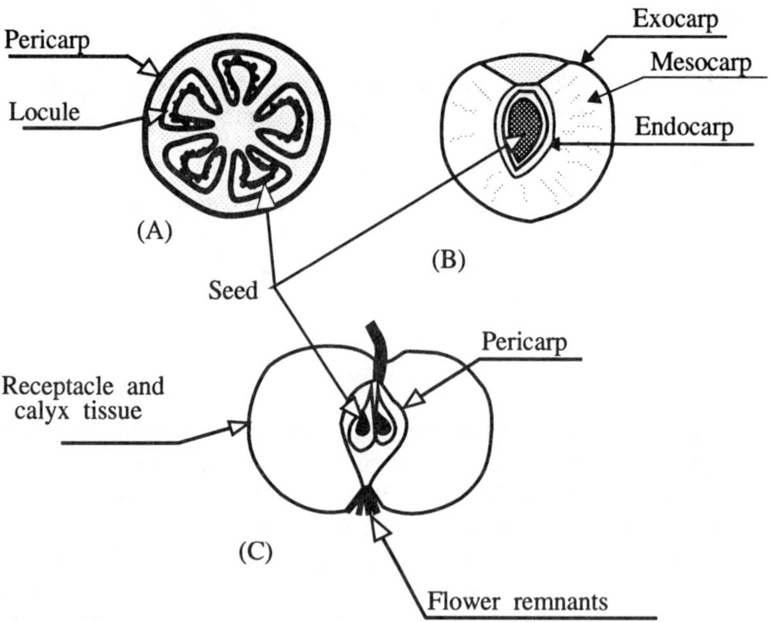

Figure 5.13. Various kinds of fruit (a) tomato, (b) peach, and (c) apple. In each, a different portion of the fruit can be eaten.

valuable when we consider the physiological processes of germination, as well as those processes that occur during periods of storage, which, for most seeds, entails periods of quiescence, relative physiological inactivity.

Because of differences in structural composition and germinating behavior between plants developed from the two major types of seeds, monocotyledons and dicotyledons, we will examine these in some detail. The similarities in structural parts will first be described, and then differences that are important in the plant's development will be discussed.

The outer layer of the seed, the seed coat or testa, may vary from plant to plant as regards thickness and structural strength, but it performs an important role in the life of the seed. It is composed of several layers depending upon the seed and characteristically is very tough and waterproof. In most seeds the seed coat performs several important functions. It serves to reduce the rate of water loss from the seed, it forms a natural barrier to insect penetration and therefore plays an important role in preventing insect damage to seeds. The seed coat's mechanical strength resists abrasion and cracking during seed harvesting, handling, processing, and storage. The latter property is especially important for agricultural crops such as corn, wheat, etc., for which the seed is the agriculturally significant component, since the seeds are subjected to considerable mechanical handling during harvesting and processing.

In certain cases these properties of the seed coat are so pronounced that special treatment is necessary to enhance seed germination. This is particularly true of clover for which the seed coat should be scarified or mechanically injured to ensure that water can penetrate the seed coat and promote germination. Cracking of the seed coat during harvesting and storage, however, may injure the embryo and prevent germination. This is especially true where the engineering practices related to the handling of seeds are conducive to cracking or checking of the seed coat such as may occur during the shelling and storage of corn and handling and shipment of dried beans and similar seeds. Therefore, it is important for the design engineer to be acquainted with the anatomical and physical properties of the seed to ensure that the product is in good condition for its intended use. The central portion of any seed is the embryo which lies within the seed. Careful examination of the embryo shows it to be a miniature, inactive, undeveloped plant that remains in an inactive state until germination. The embryo consists of the cotyledon(s) or seed leaves, the epicotyl (from the Greek prefix "epi", meaning on or above) from which the plumule and shoot or aboveground portion of the develops, and the hypocotyl (from the prefix "hypo", meaning below) from which the radicle and ultimately the entire belowground root system develops. In addition, especially

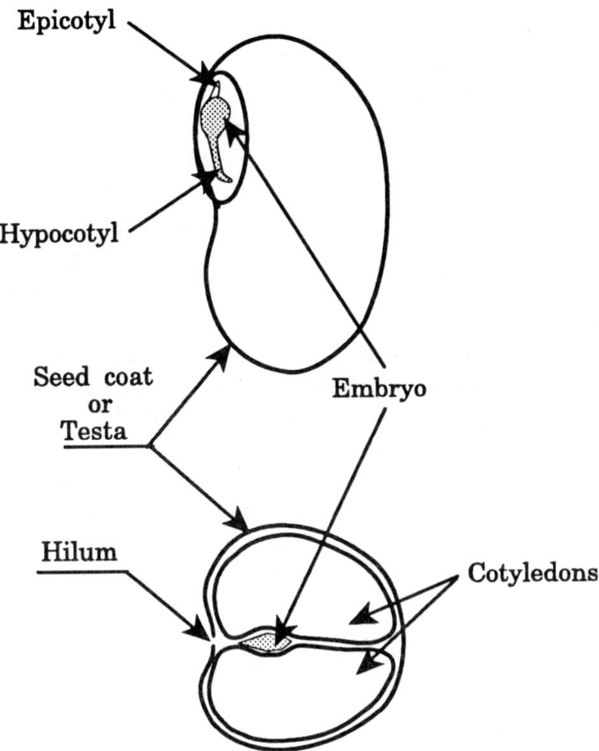

Figure 5.14. A schematic cross-section of a bean. The testa or seed coat is waterproof so that moisture can only enter the seed through the hilum. The cotyledons nourish the embryo as soon as water reaches the interior of the seed. The emergence of a dicotyledon may occur within one to four days of planting.

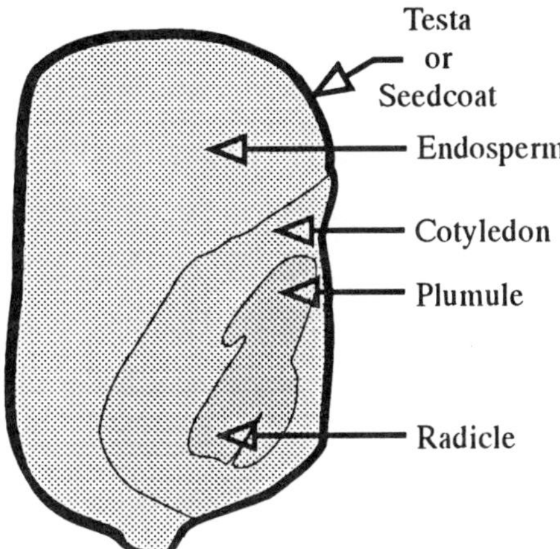

Figure 5.15. Cross-section of a corn kernel. The testa is waterproof and water enters or leaves the seed only through the hilum at the base of the seed. The cotyledon is small and most of the nourishment for the developing embryo comes from the endosperm, the fatty tissue occupying most of the seed. Because the endosperm must be digested before it can be converted into usable nourishment for the embryo, monocotyledons require a much longer time before the seed emerges from the ground.

in monocotyledons, there may be a quantity of tissue called endosperm which is a food source for the developing embryo. The endosperm is composed of vegetable fats that must be digested before being absorbed by the embryo. Endosperm tissue is an important energy source because it is composed primarily of fats and stores approximately twice as much the energy per unit of mass as the cotyledon, a carbohydrate. In some plants, particularly dicotyledons, the endosperm may be absorbed by the embryo during seed maturation while in others, mainly monocotyledons, absorption does not occur until the initiation of germination.

Figure 5.14 shows the structure of a dicotyledonous seed while figure 5.15 depicts a monocotyledonous seed. Dicotyledonous garden crops such as beans, peas, peanuts and pumpkins (the list is not exhaustive), produce seeds in which the endosperm tissue is absorbed by the embryo while the seed is maturing and is absent in the harvested seed. In crops where the endosperm is absent, such as beans, seed germination and emergence of the developing plant occurs more quickly after planting than for monocotyledonous seeds such as corn, oats, and wheat that contain significant volumes of unabsorbed endosperm. In monocotyledons, the embryo must undergo further development during germination. Typically, monocotyledons in which endosperm occupies a significant volume of the seed germinate slower after planting then do dicotyledons because the embryo is immature and must absorb the digested endosperm and complete development before emergence can take place.

The cotyledon(s) are structurally leaves and function in the processes of digestion, absorption and storage of food derived from the endosperm. In certain dicotyledons such as beans, the cotyledons will be pulled out of the soil by the arched, developing epicotyl and will become green when exposed to sunlight. When this occurs, the cotyledons themselves function as leaves, becoming food-producing organs of the plant, thus aiding in the early growth of the developing shoot.

Example 5.11. Seed properties and mechanical processing
Navy beans with a cracked testa often possess cracked cotyledons and exhibit poor germination. Narayan and Stout (1972) reported that a mathematical model relating kinetic energy at impact with seed damage is $E = [7.7 \times 10^{-7} J \cdot s^2/(mg \cdot m^2)] m v^2$, where m is the mass of the seed in milligrams, v is the relative velocity of the seed and the impacting surface, m/s, and E is the critical kinetic energy in Joules. The mass of a bean is closely related to its moisture content, and studies indicated the following values at various moisture content:

Moisture Content (%)	Seed Mass (mg)	Critical Energy (J)
17.4	269	4.35×10^{-2}
15.6	261	5.69×10^{-2}
13.4	253	5.65×10^{-2}

What is the maximum permissible peripheral velocity to minimize damage if beans are being harvested at 14% moisture?

Solution
By interpolation from curves drawn through the graph of mass and KE as a function of moisture content at 14% moisture the average seed mass is 255 mg and the critical KE is 5.59×10^{-2} J. Using Narayan and Stout's model:

$$v = \sqrt{\frac{5.66 \times 10^{-2} J}{\frac{7.7 \times 10^{-7} J \cdot s^2}{mg \cdot m^2} \times 255\ mg}} = 16.87 \frac{m}{s} \qquad 5.32$$

Particularly important are dormancy and quiescence. During dormancy, the embryo which may be incompletely developed at the time of harvest has a chance to complete development. Dormancy may be essential to produce a viable seed capable of germinating. The dormant period occurs when a very low water content, approximately 4 to 5% by mass, is present in the seed. Low water content is advantageous because it reduces the seed's susceptibility to temperature and mechanical injury.

Quiescence is an induced rest period due to environmental conditions that are unfavorable to seed germination but not unfavorable to the embryo, which is at a very low level of physiological activity. Unfavorable conditions during quiescence, such as sudden extremes of temperature, injury to the seed coat, or excessive humidity, may initiate seed germination that rapidly ceases causing the death of the embryo. The engineer should be familiar with the needs of the seed during periods of dormancy and quiescence so appropriate storage facilities can be provided. A well-planned storage facility will be one in which quiescence is assured and will fulfill the physiological needs of the seed being stored.

Example 5.12. Moisture, seed storage, and seed germination
A seed stored at 5% moisture (dry weight basis) weighs 88 mg. How much water is present? During germination the moisture content may rise to 150%. How much water is present at germination?

Solution
The dry weight of the seed is:

$$\frac{W_{wet} - W_{dry}}{W_{dry}} = 0.05$$

and

$$W_{dry} = \frac{88 \text{ mg}}{1.05} = 84 \text{ mg}$$

Therefore, during storage the amount of water present is:

$$84 \text{ mg} \times 0.05 - 4.2 \text{ mg} \qquad 5.33$$

At germination, after the seed has imbibed all the water needed for germination, the water within the seed increases to:

$$84 \text{ mg} \times 1.5 = 126 \text{ mg} \qquad 5.34$$

Note that a 30-fold increase in moisture was necessary to sustain germination.

Example 5.13. Storage temperature and product condition
If the rate of respiration doubles for every 10° C rise in temperature according the Q10 law of chemical reactions, what would you conclude about the desirable storage temperature? What atmosphere should be maintained during storage?

Solution
Since respiration is a biochemical process that uses stored food, a higher storage temperature would promote more rapid respiration. Also, if the storage were such that energy could not rapidly be dissipated, increased respiration would result in still higher temperatures, using stored food at a greater rate. Thus, low storage temperatures are desirable. Also, respiration is a process requiring O_2, so that availability of O_2 to the stored product should be reduced. This concept is utilized in controlled-atmosphere storage of apples to reduce weight loss due to utilization of stored sugars during the respiration process.

Exercises

5.1. How many kilograms of CO_2 are used by the green plant in producing 1 metric ton (1000 kg) of $C_6H_{12}O_6$ glucose? [1.47×10^3]

5.2. How many kilograms of H_2O are needed to produce 1 metric ton of $C_6H_{12}O_6$ (glucose)? [6.0×10^2]

5.3. If the yield of sugar beets is about 37 t/ha of which 20% is crystallizable sugar, how much water would be needed per acre to produce the sugar if it were sucrose which has the formula $C_{12}H_{22}O_{11}$? What depth of water does this represent on a hectare (assume the density of water is 1000 kg/m^3)? Is this a realistic water requirement for a growing season? [4.3×10^3, 0.1, No]

5.4. In example 5.2 how much water (kg) is produced per kilometer of travel? What volume of liquid water (give your answer in mm^3) does this represent? [0.149, 149×10^3]

5.5. What does example 5.3 indicate regarding the delicate balance between nature and man?

5.6. If navy beans are to be harvested at 12.5% moisture, what cylinder speed would you recommend to the operator for minimum damage? (Hint: Plot the mass as a function of percent moisture to determine the bean mass, then plot the critical energy as a function of percent moisture and fit a line through the points by eye to determine the critical energy.) [≈15.7 m/s]

5.7. What is the total surface area of the root system in example 5.8 if the system contained 12.5×10^6 root hairs with an average length of 1.320 mm each and an average diameter of 14 μ? Assume that the end of each root hair tip is a hemisphere. [7.8×10^5 mm]

5.8. A 12.7 mm (1/2 in.) rope made of manila hemp fibers is capable of carrying 8340 N = 1900 lb. If a particular fiber cell is 100 μ in diameter and consists almost entirely of thickened lignified walls, what tensile force in Newtons would such a fiber support? What is the ultimate tensile stress in a manila hemp fiber kPa? Assume that the fibers are densely packed, i.e., that lines joining the centers of adjacent fibers form a triangle, and further assume there is the same percent of void space present in the rope as there would be in a circle encompassing seven smaller circles. [87 000 kPa]

5.9. In grasses, the meristematic tissue in a leaf is at the base of the leaf near the point of attachment to the stem. In a corn leaf, which part of the leaf would you expect to be composed of older tissue?

5.10. In dicots, the leaves possess little or no meristematic tissue. What does this imply concerning growth of a bean leaf?

5.11. Assume a bean leaf has an area of 0.20 m^2. Estimate the surface that stomatal openings occupy. You may assume the stomatal opening is a rectangle. [1.4×10^{-2} m^2]

5.12. The testa of legumes often needs to be scarified or broken to facilitate water absorption. Sketch a machine to accomplish this.

5.13. Assume a cylindrical cell of length L has radius r, wall thickness t, and hemispherical ends. If the pressure within the cell is P, what is the longitudinal stress in the cell wall? What is the transverse stress in the cell wall? Which is larger?

5.14. A balloon 76 mm in diameter and 300 mm long with hemispherical ends is resting on two flat supports, each 25 mm in diameter. The supports are 100 mm apart. The balloon has an internal pressure of 30 kPa. What mass of diameter 25 mm placed midway between the supports will the balloon "beam" support?

Chapter 6

Elements of the Green Plant Physiology

This chapter will examine some important basic concepts of the physiology of the green plant. The physiology of the green plant is important to the engineer for many reasons. The complex physiological processes that occur within a plant are not completely understood, even by those who have dedicated their efforts to the study of these processes. However, certain aspects are well enough understood to allow engineering modifications that affect the physiological responses can be performed. These include the mechanism that triggers the action of stomata, the nature of the processes involved in moving food from the site of photosynthesis to parts of the plant where it is stored or utilized, the influence of the utilization of plant food on certain allied processes such as water absorption and mineral uptake, the approximate requirements of plants for minerals and the action of these minerals in the physiological processes.

Some well-defined physiological processes may be of little interest from an engineering viewpoint because they cannot be influenced by engineering modification except in an indirect fashion. Photosynthesis, one of the processes we shall discuss in its most elementary form, is actually a complex series of biochemical reactions involving energy absorption, electron transfers, and the production and use of complex chemical products before food is produced. If the accepted scheme of photosynthesis were completely understood, there would still be little or nothing the engineer could do to increase its efficiency except by modification of environmental parameters that enter into the process. Thus, it is more important to understand photosynthesis from the viewpoint of its role in the whole plant as influenced by the parameters of the soil-plant- atmosphere continuum than to delve into the complex chain of events that occur on a single chloroplast within a plant cell. We will concentrate on what effect the environment has on the photosynthetic output rather than on the individual steps detailing the production of a molecule of sugar from the interception of a photon of radiation.

The student will become acquainted with those physiological phenomena that affect the plant in the soil-plant-atmosphere continuum. It is important to develop an appreciation for how the parameters of the continuum relate to the physiological processes. In this chapter we will examine the influence of temperature on processes involved in seed germination, respiration, photosynthesis, and water balance, as well as related topics such as the importance of light, oxygen, and carbon dioxide to the well-being of the entire plant. Considerable emphasis will be place on water and its role, since water is a primary parameter of the soil-plant-atmosphere continuum, and one for which engineering modification is possible.

The intent of this chapter is to make the engineer aware of basic physiological processes so that the engineer can turn to other specialists, i.e., plant physiologists, geneticists, soil scientists, and crop scientists, to obtain the best possible engineering approach to modify the plant environment and complement the agriculturally significant plant response.

Elementary Physiological Concepts

Photosynthesis

Photosynthesis is the process by which the green plant, using the raw materials of water, carbon dioxide and energy from the sun, can manufacture sugars that are used by the plant in the processes of growth and production of significant agricultural output. It is the link between solar energy and the energy required by man. In the photosynthetic process an almost minuscule amount of solar energy is used within the chloroplast of the green plant. The amount is less than 1.5% of the total energy arriving from the sun. The paucity of the amount is such that the energy used in photosynthesis can be ignored in the computation of an energy balance for a plant canopy. This fact will be demonstrated in chapter 8 which treats the energy balance of the plant environment.

When the energy balance of a single leaf is examined, however, a considerable portion of the incoming energy is actually used by the photosynthetic process. Estimates of this amount range to 30% with the remainder of the energy either reflected, transmitted, or converted into heat.

The chemical equation that can be used to represent the photosynthetic process is:

$$6CO_2 + 6H_2O + 2.82 \times 10^6 J \rightarrow C_6H_{12}O_6 + 6O_2 \qquad 6.1$$

Equation 6.1 states that 6 moles of CO_2, combined with 6 moles of H_2O and solar energy in the visible range, will yield 1 mole of $C_6H_{12}O_6$, a simple sugar (glucose) along with 6 moles of O_2.

Example 6.1. Energy requirements of photosynthesis
Assume that the immediate products of photosynthesis are glucose ($C_6H_{12}O_6$) and oxygen (O_2). In equation 6.1, 2.82×10^6 J are required per mole of glucose. How much energy must be used to produce one metric ton of glucose. How much incoming solar energy is needed?

Solution
One metric ton is 1000 kg of sugar. The gram molecular weight of sugar is 180 g/mol. From equation 6.1, the number of moles of glucose required to comprise a metric ton is:

$$\frac{2.82 \times 10^6 J}{mol_{C_6H_{12}O_6}} \times \frac{5.56 \times 10^3 \; mol_{C_6H_{12}O_6}}{t_{C_6H_{12}O_6}} = \frac{1.57 \times 10^{10} J}{t_{C_6H_{12}O_6}} \qquad 6.2$$

The energy required per ton is then:

$$\frac{1.55 \times 10^7 kJ}{0.015} = 1.04 \times 10^9 kJ \qquad 6.3$$

If the photosynthetic reaction uses only 1.5% of the total incoming solar radiation, the required incoming radiation is:

$$\frac{1.55 \times 10^7 kJ}{0.015} = 1.04 \times 10^9 kJ \qquad 6.4$$

Equation 6.1 represents the simplest form of the photosynthetic reaction, however, the process is considerably more complex. The first chemically identifiable substance in the photosynthetic process is phosphoglyceric acid which contains phosphorus. Through a complex series of electron transfers and intermediate chemical reactions, a simple sugar such as glucose appears. Depending upon the plant and the environmental conditions, glucose may be transformed into other substances such as fructose (fruit sugar), sucrose (cane sugar), starch and other carbohydrates, fats, and proteins. Of these substances, only the sugars are readily soluble in water. Food translocation within the phloem vascular structure of the plant occurs as solution of sugars. Starches, carbohydrates, and fats are important because they furnish the modes of energy storage within the plant. Sugars are transformed into starches for temporary food storage until the insoluble starch is converted back into sugar. Other sugars are used for the production of cellulose, a polysaccharide which may be empirically represented chemically as $(C_6H_{10}O_5)_x$, where the x subscript denotes that the cellulose molecule is a long chain of glucose molecules.

Cellulose fibers or microfibril made up of glucose chains comprise the wall of the individual mature plant cell. The proportions of hydrogen and oxygen in cellulose and in the sugars are the same as those in water. The difference is that cellulose and sugars have different numbers of carbon atoms in each molecule.

Fats, another by product of photosynthesis, have varying chemical formulas but contain less oxygen than sugars or carbohydrates. Fats store approximately twice as much energy per unit weight as sugars. In plants, fats are always vegetable oils that can be converted back to sugars for translocation and utilization. Fatty substances are present in cytoplasm in varying amounts, probably in temporary storage. Maturing seeds are often high in fats that are excellent concentrated sources of energy for the maturing embryo. The endosperm in a corn seed is an excellent example. Seeds with large cotyledons and little or no endosperm do not contain a high proportion of fat. In seeds such as beans, very little stored fat can be found.

Example 6.2. Energy in foods
Proteins are a result of the photosynthetic process. Protein molecules may be very large. For example, two proteins isolated from corn and wheat, respectively, are zein ($C_{736}H_{1161}N_{184}O_{208}S_3$) and gliadin ($C_{736}H_{1068}N_{196}S_5$). Proteins are not produced directly in the photosynthetic process, but occur as a result of the production and recombination of amino acids. Amino acids, the precursors of proteins, are thought to be formed in either the leaves or in the younger roots of plants, depending upon the particular plant. Amino acids are soluble and therefore translocatable. Protein synthesis takes place where the protein is needed by the plant, depending on whether amino acids are present at the site. In most plants, the actual site of protein synthesis is the meristematic or cell-producing region of the plant where *protoplasmic proteins* are constructed from amino acids. In plants where seed maturation is occurs, protein synthesis occurs mainly in the seeds, thus explaining the high protein content of certain seeds, such as soybean, which may reach 40% or more. The process of protein formulation requires certain other processes, which include respiration, digestion, and assimilation, all of which contribute to the metabolism of the plant. We shall discuss these processes briefly.

Example 6.3. Gram molecular weight of large molecules
It has been shown empirically that the oxidation of protein yields approximately $1.8 \times 10^7 J/kg$. Calculate the molecular weight of a molecule of zein. How much energy would be available in 30 kg of corn containing 8% protein if all of the protein was zein?

Solution
The molecular weight is:

$$
\begin{array}{lcrcrcr}
C_{736} & \rightarrow & 736 & \times & 12 & = & 8832\ g \\
H_{1161} & \rightarrow & 1161 & \times & 1 & = & 1161\ g \\
H_{184} & \rightarrow & 184 & \times & 12 & = & 2576\ g \\
O_{208} & \rightarrow & 208 & \times & 16 & = & 3328\ g \\
S_3 & \rightarrow & S & \times & 32 & = & 96\ g
\end{array}
$$

To find the energy stored in protein:

$$\frac{1.8 \times 10^7 J}{kg} \times \frac{8\%}{100\%} \times 30kg = 4.3 \times 10^4 J \qquad 6.5$$

Respiration

Respiration, the process by which sugars react with oxygen to yield water and carbon dioxide plus energy, can be thought of as the reverse of photosynthesis. The simplified chemical equation for this phenomenon is:

$$C_6H_{12}O_6 + 6O_2 \rightarrow 6H_2O + 6CO_2 + Energy \qquad 6.6$$

Actually, respiration is a very complex reaction. Some 50 major chemical reactions have been identified as involved in respiration. Respiration is composed of two sets of reactions. In the first set, glucose undergoes a series of reactions that do not require additional oxygen, and is converted to pyruvic acid. This part of the process is glucolysis.

In the second set of reactions, the end products may differ depending on whether or not atmospheric O_2 is available to affect the breakdown of pyruvic acid. With free O_2 available, aerobic respiration occurs with enzymes that catalyze the reaction. This process involves the formation and breakdown of several organic acids until the result, water carbon dioxide and energy is attained.

Without free O_2, a second set of reactions occurs. This is anaerobic respiration in which O_2 from the organic acids is used. The result of this set of reactions yields a different group of compounds. Energy again is liberated, but instead of CO_2 and H_2O, alcohols are produced. In most green plants, the appearance of alcohols results in severe injury and death to the plant. This is especially true of respiration occurring in roots that are submerged in water the roots have difficulty obtaining O_2 since the rate of diffusion of free O_2 through water is about four orders of magnitude less than the rate of diffusion through air. Submerged roots will undergo anaerobic respiration unless some other mechanism has developed in the plant to supply O_2. The alcohols and resulting toxic products of anaerobic respiration cause the death of that portion of the root through which water is absorbed. The result, which may occur within 48 hours or less, is death due to severe water deficiency brought on by the plant's inability to obtain water.

Respiration, chiefly aerobic respiration of sugars throughout the plant, is the green plant's means of obtaining energy for complex growth processes that include protein synthesis. According to equation 6.2, about 2.8×10^6J of energy is liberated for every 180 g of glucose oxidized. This much energy can manifest itself as heat energy that can often be detected. In sprouting seeds, temperatures of 45° C above the ambient temperature have been found. Heat energy of this magnitude may inhibit and destroy the vitality of poorly stored seeds whose environment is periodically allowed to reach conditions conducive to the respiration occurring when the seed has an excess of moisture and available free oxygen.

A similar and often disastrous situation occurs when high-moisture hay is confined. Bacterial respiration can produce temperatures high enough to convert free sugars into caramel causing a brown color in the hay. If the respiratory activity is high enough, sufficient heat will be generated to ignite the material. Spontaneous combustion is said to occur under these circumstances.

Respiration is not limited to roots. Respiration continually occurs throughout the green plant, especially in the meristematic regions and within living cells. During daylight hours under aerobic conditions, the CO_2 produced by respiration in the leaves is used in the photosynthetic process, while in the roots CO_2 is liberated and diffuses through the soil. In the dark, respiration continues to occur throughout the plant and any available O_2 within the intercellular spaces is rapidly consumed. At night CO_2 is produced by the leaves of the plant as well as by the roots. In most green plants, CO_2 escapes from the leaf through imperfectly closed stomata. That which has not escaped during respiration is rapidly used again in photosynthesis when light strikes the leaves.

There are plants that have evolved a unique method of recapturing the CO_2 produced by respiration. These are the plants that possess a C4 pathway. In C4 plants the stomata will close when the climate becomes hot and dry, thus reducing excessive water loss through transpiration, and preventing CO_2 from reaching the chloroplast from the atmosphere. In C4 plants, however, CO_2 goes through intermediate steps in the mesophyll cells ending up as malic acid which is transferred to the bundle sheath cells that surround the vascular system. Here the malic acid breaks down into CO_2 that can then be reused by the plant. In this fashion, C4 plants are able to reuse CO_2 produced during respiration. Thus, the C4 plants that include corn and sugarcane, have become adapted to conditions that would impede food production in plants which undergo the more common photosynthetic reaction.

There is one more collection of plants which exist under arid conditions. These are the succulents such as cacti. In the succulents, stomata are normally closed during the daylight hours. This prevents the loss of water from the plant, and also cuts off the source of CO_2. However, the stomata open at night and allow CO_2 to enter the plant. Here it is converted into malic acid and stored. When daylight comes and the stomata again close, the malic acid produces CO_2 that is then fixed in the photosynthetic cycle.

Digestion

Digestion is the process in which foods that are insoluble in water are broken down into simpler water-soluble substances and can be translocated throughout the plant. Digestion is necessary to ensure that sites of high respiratory activity within the green plant (roots, seed-forming regions, area of meristematic activity, etc.) can obtain the raw materials for respiration regardless of where the raw materials were produced.

Digestion is a biochemical process and requires complex substances called enzymes for its initiation. Enzymes are catalytic in their action. Their presence is necessary for the reaction to occur, but they themselves are not used up during the reaction. There are many enzymes and each enzyme is specific in its action for a particular food substance. Enzyme production occurs within plant cells and is controlled by genes located within the living cell and important in the transmission of hereditary characteristics, specific properties that are peculiar to a given species.

A physiologically significant property of enzymes is that, in a liquid medium, they lose their ability to function as temperature increases. At temperatures above 50° C almost complete inactivation and destruction of most enzymes occurs. At lower temperatures there is less inactivation. In dry tissues, however, such as seeds, periods of temperatures above 100° C can be tolerated for some time if the moisture content remains low.

Example 6.4. Water's role as a solvent
Why is a high moisture content necessary for respiration to occur when water does not appear on the left side of the basic equation describing respiration, but as a product of the respiratory process?

Solution
Water, the universal solvent, forms the medium in which all chemical processes within the plant occur. Digestion, activated by enzymes in solution, is a prerequisite in the conversion of insoluble products to sugars. Therefore, sufficient water is necessary to support the digestive processes that produce the substance that is respired. Thus, although respiration is chemical in nature in which the rate is approximately doubled for every 10° C rise in temperature, it is only a step in the overall metabolic process, and depends on many other factors, such as moisture, oxygen, nature of the material being digested and reduced to sugars, etc.

Since enzyme production is an intracellular process, its action could be expected to be largely intracellular. In general this is true; digestion of fats and carbohydrates occurs largely within the plant cell, breaking down complex foods into simpler, translocatable foods. Digestion also occurs within the seed as endosperm is used to develop the embryo. As seedling development takes place in the early stages of germination, digestion must occur in the cotyledon or cotyledons of the seed also. As anticipated, digestive activity is directly related to enzyme activity and thus digestive rates will be dependent upon temperature

Assimilation

Probably the least understood of the physiological-chemical processes within the green plant is the conversion of nonliving substances into the living cytoplasm in the plant cell. This process is assimilation and is continually at work wherever cell division, enlargement, and general cell development are occurring. In the synthesis of protoplasm, the foods assimilated are largely proteinaceous while cell-wall formation requires the assimilation of carbohydrates.

Obviously, since assimilation must be preceded by digestion and respiration must supply energy needed for these processes, the same factors that inhibit respiration and digestion will also inhibit assimilation. Assimilation is essential for plant growth so there is an ever-increasing link between environmental parameters and plant development.

Metabolism

Up to this point we have considered individually some processes involved within the growing, developing plant. In living entities, these processes must occur continuously and simultaneously. The totality of such processes is grouped under the term *metabolism*, and any specific activity is termed metabolic activity or, for short, metabolism. Thus, the metabolic activity of a tissue or group of tissues includes photosynthetic activity (if light and chloroplast are present), respiratory activity, as well as the processes of digestion and assimilation.

Certain areas of the plant, such as apical meristems in which cellular division and cell elongation are occurring, are areas of very high metabolic activity. Water- and mineral-absorbing regions of root systems exhibit considerable metabolic activity as do the leaves, the sites of food production for the plant. Metabolic activity is necessary for food transfer throughout the phloem tissue even though the mechanism of such transfer is as yet poorly understood. Likewise, the opening and closing of stoma may require metabolic activity on the part of the guard cells and companion cells.

Among processes that involve little or no metabolic activity is the movement of water up the stem of the plant through the xylem vessels and tracheids. The water movement process is not completely independent of metabolic activity, however, since the overall water balance of the plant is closely tied to the metabolism of individual tissues. The demands for water result from physiological processes and the overall water demands imposed by the energy balance of the plant itself. Some concepts related to plant water use are discussed in the following sections.

Water Within the Cell

Since the process of photosynthesis occurs in chloroplast within the cytoplasm of living cells, first we first investigate the nature of cytoplasm and examine its role in the physiology of the whole plant.

Cytoplasm, the protoplasm exclusive of the cell nucleus, is a mixture of liquids and solids dispersed in water. The size of individual particles of this mixture (in which are suspended larger particles such as the nucleus, plastids, chloroplast, or other cellular entities) is in the range of from 0.1 μ to as small as 0.001 μ. While the mixture is liquid, as a fluid, cytoplasm is more properly a member of that class of colloidal substances termed a sol. Under certain conditions, the mixture may become gelatinous and is more properly termed a gel.

The chemical analysis of living protoplasm is impossible since the very act of analysis will destroy it. From a careful chemical examination of dead protoplasm, however, the following can be inferred. The chief elements in protoplasm are oxygen, carbon, hydrogen and nitrogen, which make up 95 to 98% of the protoplasm. Much of the oxygen and hydrogen are in the form of water which alone has been estimated to comprise 70 to 97% of the liquid within a cell. The role of water is immediately apparent since it is the chief constituent of protoplasm. Nitrogen occurs largely in the form of proteinaceous substances whose gram molecular weights may range from 20 000 up to 1 000 000 g/mol and probably average 22 000 to 25 000 g/mol. In part, proteins may be responsible for the sudden transformation of the "sol" form of cytoplasm to the "gel" form as coagulation of the protein occurs. In some cases, such as under the influence of high temperatures, the coagulation is permanent, and deprived of its liquid form the protoplasm dies. This phenomenon may be related to plant injury resulting from excessive environmental temperatures.

Example 6.5. Effect of high temperatures on enzymes
On one occasion the heating control in an office failed and the temperature within the office rose to over 43° C (110° F) overnight. The following morning a Philodendron plant appeared flaccid. Within two days the plant had lost all its leaves and died. Can you suggest several possible causes for this?

Solution
With increasing ambient temperature, increasing respiration rates occurred. Based on the 10° C rate-doubling concept, the respiration at 43° C was four times greater than it was at 21° C, the normal room temperature at the time. At 43° C some of the enzymes necessary for the digestive process lose their ability to catalyze, and thus available raw materials for respiration were soon exhausted. In addition, coagulation of cytoplasm may have taken place, thus stopping all metabolic activity essential to the well-being of the plant and inducing permanent plant injury.

Additional chemical elements are detected in cytoplasm in much lesser amounts. These include sulfur (S), phosphorus (P), calcium (Ca), magnesium (Mg), potassium (K), iron (Fe), and, in addition, probably yet smaller amounts of chlorine (Cl), zinc (Zn), aluminum (Al), copper (Cu), boron (B), molybdenum (Mo), and manganese (Mn). While the latter group of elements are present in lesser amounts than the elements C, N, O, and H, they appear to be physiologically significant in processes essential to cell life though not in the photosynthetic process. In particular, these elements may help form specific entities within the plant cell such as the chlorophyll molecule, or they may be essential in the production or action of enzymes.

Example 6.6. Role of trace elements in the cytoplasm
List several of the substances in which S, P, Ca, Mg, K, and Fe are to be found in the plant.

Solution
Some types of chlorophyll are known. The two most important in plants are:

$$\text{Chlorophyll a} = C_{55}H_{72}O_5N_4Mg$$
$$\text{Chlorophyll b} = C_{55}H_{70}O_6N_4Mg$$

It is seen that both species contain 4 atoms of N and 1 atom of Mg. Additionally, phosphorus, P, is important in the energy-transfer process of photosynthesis by which chlorophyll produces sugar using sunlight as its energy source. Calcium is an ingredient of calcium pectate, of which the middle lamella of cellular formation is composed. Ferrodoxin, a water-soluble protein, contains Fe, as does cytochrome, which is insoluble but may accept and donate electrons, primarily by changing the atomic structure of its Fe atom. S and N are both constituents of protein molecules in general. Finally, K is found as an ion in the guard cells and surrounding companion cells and acts in enzyme activation.

Example 6.6 demonstrates that while carbon, oxygen, and hydrogen form the primary building blocks of the green plant. Other elements are essential, though secondary, components of the green plant.

Since only CO_2 is obtained from the atmospheric environment, the remaining elements must be obtained by the plant from other sources. These elements are usually called trace elements since they occur in such small quantities (less than 1%). The soil mass via the root system of the plant is the chief source for many of trace elements. The processes by which mineral and even water absorption by the roots occurs are not completely defined, these elements must be present in proper amounts for satisfactory physiological response to the soil component of the soil-plant-atmosphere continuum.

Not all the particles in cytoplasm are in the form of a solar or a gel. Some elements may be present as true solutions with the water as the solvent and molecules or ions of compounds or elements as solutes. Variability of the components of the solution and sol constituents of cytoplasm, with the semipermeable properties of the plasma and vacuolar membranes, results in the transfer of liquid water through these membranes via osmosis. Since osmosis plays an important role in the amount of water present in a cell (and therefore in the cell's volume and turgidity, the pressure buildup within the cell as water enters a confined volume) its study in the water balance of the soil-plant-atmosphere continuum can lead to a better appreciation of the effects of the osmotic process on the physiology of the green plant.

Depending on the way the chemical elements are combined, in addition to proteins, components such as sugars, starches, fats, organic acids, salts, and pigments will be found within the cell cytoplasm; all these elements are formed from the products of photosynthesis by physiological processes. Usually they will be used by the plant during the growth process as metabolism takes place. Any solids within the cytoplasm must undergo change and be dissolved in water before they are transported from the site of production to the point of utilization.

Thus the role of water in the physiological activity occurring within the cytoplasm cannot be ignored. Indeed, water begins to appear as one of the single most important components of the plant. The cytoplasm usually surrounds the vacuole completely. The vacuole, comprised mainly of a solution of salts and minerals, may act with the semipermeable vacuolar membrane and the osmotic process, as an important reservoir of water dependent upon the needs of the cytoplasm. Little is actually understood about the processes of water transfer into and out of the vacuole. Although during periods of extended water deficit, the water demand from the vacuolar sap may be so great that a significant portion of the water is withdrawn leaving the cell limp and flaccid. This can lead to the death of the cell if extended over a prolonged period.

Water in the Cell Wall

Since all water entering the cytoplasm or vacuole of the cell must pass through the cell wall, the role of water in the cell wall must be considered. Water in the cell wall will occur in one of two forms. Bound water is chemically fixed in the cell wall matrix. Free water is water that is able to move either along the cell-wall surfaces or through the walls into the interior of the cell. Movement will occur in response to differences in the energy level of the water at various points.

As was shown in chapter 5, the structure of cell walls is that of a matrix of cellulose microfibril imbedded in hemicellulose. The hemicellulose is highly hydrated and the cell wall may be comprised of up to 50% water. It is estimated that some percentage of water in the cell wall is absorbed on the solid surfaces of the hemicellulose matrix. This water is very tightly bound. It represents less than 10% of the total water in normal, turgid cells. The remainder of the water in the cell wall is probably retained by the surface tension acting across the liquid-air menisci of the interfibrillar spaces and can be free to move in response a gradient of water potential.

Example 6.7. Bound and free water in a leaf
Assume that a cell has a surface area of 1000 μ^2 and a wall thickness of 1.5 μ. If a leaf contains 4×10^9 cells, how much water should one expect to find in the walls of the cells comprising the leaf? How much of the cell wall water is bound water?

Solution
The volume of the above cell wall is about:

$$1.5\mu \times 1000\mu^2 \times \frac{\text{mm}^3}{1 \times 10^9 \mu^3} = 1.5 \times 10^{-6}\text{mm}^3 \qquad 6.7$$

Therefore in all the cells there is:

$$4 \times 10^9\text{cells} \times 1.5 \times \frac{10^{-6}\text{mm}^3}{\text{cell}} = 6 \times 10^3\text{mm}^3 \qquad 6.8$$

If 50% of the cell wall volume is water:

$$\frac{50_{H_2O}\%}{100\%} \times 6 \times 10^3\text{mm}^3 = 3 \times 10^3\text{mm}^3_{H_2O} \qquad 6.9$$

Then the bound water would be:

$$3\text{mm}^3_{H_2O} \times \frac{1\text{g}_{H_2O}}{\text{mm}^3} \times \frac{1\%}{100\%} = 0.3\text{g}_{H_2O} \qquad 6.10$$

Bound water also may occur on inclusions within the cytoplasm or the vacuole. The amount of water held depends on the nature of the inclusions and their number. From an engineering standpoint bound water is of little value to the plant. Before it is removed from the binding surface, water deficits within the plant are so severe that death has occurred.

Free water, on the other hand, is found not only in the interfibrillar spaces but throughout the plant. Obviously, water is found within the xylem vessels where it is continuous from the interfibrillar spaces of individual mesophyll and palisade layer cells, throughout the vascular system to the terminations of the xylem vessels in the roots.

Free water constitutes an important element in the physiological and metabolic processes of the plant. In the cell wall and within the cell, it passes freely through semipermeable membranes and plays an active role in the cell metabolism. However, a small portion of the total water use by the plant occurs in this fashion. The majority of the water moves freely through the xylem to the surfaces of the intercellular spaces within the plant leaves and stem. There it absorbs heat energy and evaporates into the intercellular spaces. Providing there is sufficient water in the guard and companion cells of the stomata to cause the stomata to open, the evaporated water passes through the stomatal openings and out into the atmosphere. Thus the heat removed from the plant cools the plant ensure favorable temperatures for the complex chemical reactions of growth to take place. The spongy mesophyll and palisade layers of the leaf cells are conveniently arranged for this purpose. In these areas it has been estimate that the surface available for evaporation is at least one order of magnitude greater than the exterior exposed surfaces of the leaf.

Example 6.8. Water use comparisons, photosynthesis and transpiration.
In example 6.1 it was shown that 1.57×10^{10} J of energy were used to produce 1 metric ton of glucose by photosynthesis. If this represents 30% of the total energy absorbed during the photosynthetic process and the remainder is used to evaporate water from the cell walls. How much water is evaporated and transpired from the leaves during the production of 1 ton of glucose? Assume that 2.47 kJ is needed to evaporate 1 g of water. How does this compare with the total water required to produce 1 ton of glucose?

Solution
The amount available for evaporation is:

$$\frac{1.57 \times 10^7 \text{ kJ}}{\text{t}} \times \frac{100\%}{30\%} = 5.22 \times 10^7 \text{kJ}_{\text{absorbed}} \qquad 6.11$$

$$5.2 \times 10^7 \text{kJ} - 1.57 \times 10^7 \text{kJ} = 3.6 \times 10^7 \text{kJ} \qquad 6.12$$

The water evaporated per metric ton of glucose is:

$$\frac{3.7 \times 10^7 \text{kJ}}{2.47 \times 10^3 \frac{\text{kJ}}{\text{kg}_{\text{H}_2\text{O}}}} = 1.5 \times 10^4 \text{kg}_{\text{H}_2\text{O}} \qquad 6.13$$

To produce 1 metric ton of glucose 3.3×10^3 moles of water are used:

$$3.33 \times 10^4 \text{mol}_{\text{H}_2\text{O}} \times 0.018 \frac{\text{kg}_{\text{H}_2\text{O}}}{\text{mol}_{\text{H}_2\text{O}}} = 600 \text{kg}_{\text{H}_2\text{O}} \qquad 6.14$$

The total water used is transpired and that fixed in photosynthesis or:

$$1.5 \times 10^4 \text{kg} + 600 \text{kg} = 1.56 \times 10^4 \text{kg} \qquad 6.15$$

The percent of water transpired in the production of 1 metric ton of glucose is:

$$\frac{1.5 \times 10^4 \text{kg}}{1.56 \times 10^4 \text{kg}} \times 100 = 96\% \qquad 6.16$$

Partially closed stomata are ineffective in reducing transpiration since an individual water molecule is approximately 4.54×10^{-4} μ in diameter. A typical stomatal opening far exceeds 1 μ, even when partially closed. A partially closed stoma is still many times larger than the diameter of the water molecule and considerable movement of the water vapor out of the leaf can occur.

Example 6.9. Comparison of stoma opening and water molecule diameter
If the stoma of a bean plant when fully open is 7 μ wide, at 90% closure how many times larger than a water molecule is the stomatal opening?

Solution
At 90% closure, the width of the stomatal opening is:

$$7\mu - \left(7\mu \times \frac{90\%}{100\%}\right) = 0.7\mu \qquad 6.17$$

The water molecule is 4.54×10^{-4} μ in diameter, so:

$$\frac{0.7\mu}{4.54 \times 10^{-4}\mu} = 1542 \qquad 6.18$$

The stomatal opening, even at 90% closure, is still 3 orders of magnitude larger than a water molecule.

The rate of transpiration depends on the rate at which escaping water molecules are carried away from the leaf. If the leaf is in still air, the difference in the concentration of water molecules between the inside and outside the leaf is less than it is for leaves exposed to moving air carrying away the diffusing water molecules. Therefore, one could expect a leaf in a rapidly moving airstream to experience a large difference in the concentration of water molecules from the interior to the exterior of the leaf. Figure 6.1 illustrates the influence of air movement and stomatal aperture on transpiration.

Water in the Xylem

Water in the xylem is almost all free water except for a small amount that may be bound to the cell-wall surfaces. Water within the xylem transports minerals to all parts of the plant where they will be utilized. Movement of water through the xylem is passive, that is, it occurs mainly as a result of the "pull" exerted on the water this results from a lowering of the water potential in the xylem as water is evaporated from the cell walls in the leaves.

Some aspects of xylem water are not clearly understood. Specifically, in some plants during periods of darkness when transpiration demand is least, liquid is observed on leaves at the points where xylem vessels terminate. This phenomenon, called *guttation*, occurs in a wide variety of plants. An accepted explanation for guttation, perhaps in response to an osmotic process, involves active absorption of water through the root system. However, since the pressure that causes guttation to occur is not sufficient to force water from the roots into the leaves of trees more than 10 m tall, this pressure cannot be considered as an active means of distributing water and mineral salts through the green plant.

While water enters the xylem vessels in the roots in response to the transpirational pull of evaporating water in the leaves, many researchers consider that the presence of minerals in xylem water is evidence of *active* transport occurring at the point of water entry into the root. Some feel that mineral entry itself is an active transport phenomenon. From an engineering viewpoint, the important concept is that the xylem vessels and tracheids are

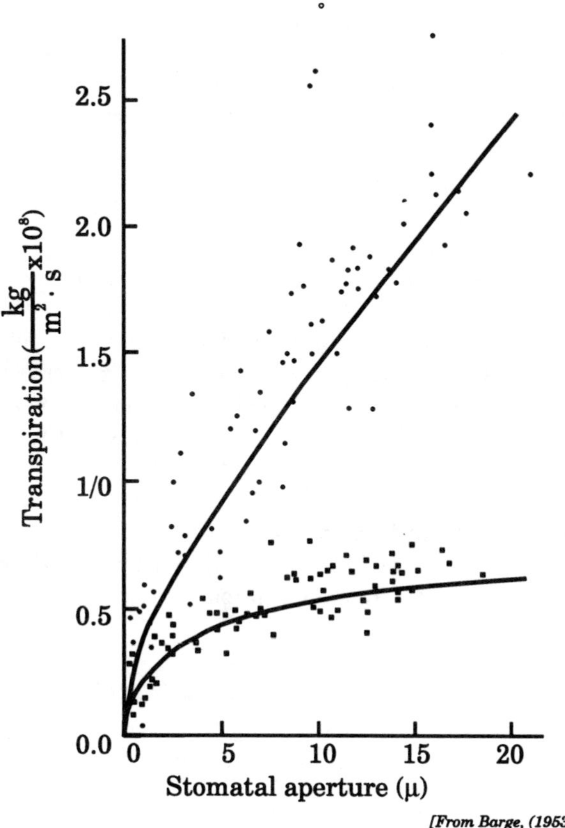

Figure 6.1. The rate of transpiration is influenced significantly by the velocity of air passing over the stoma. The circles indicate transpiration measurements made in rapidly moving air while the squares indicate measurements made in still air. (From Barge, 1953)

filled with water that acts as a solvent for a low concentration of mineral solutes. The solution of minerals is distributed via the xylem vessels throughout the plant. Note that not all minerals that enter the xylem end up in leaf cells. Mineral transfer directly from the xylem into the phloem has been observed. The destination of most minerals within the xylem stream, however, is believed to be the cells wherein metabolic activity takes place.

Water in the Phloem

Physiologically, the primary function of the phloem tissue is the transfer of soluble food products from the site of their production or storage to the location where they are being utilized. These are either the location of metabolism associated with cell division and growth or with food storage, such as the seed. Water in the phloem is different from water in the xylem. A major component of phloem solution is sucrose. This is particularly true in orange trees and probably in most other fruits as well. The phloem solution in apples and plums, sorbose, another fruit sugar, is the primary solute.

In addition to containing dissolved sugars, an analysis of solution from phloem shows that minerals that have been organically combined into soluble substances are also present. Thus, while water in the xylem is relatively pure, water in the phloem is in the form of a solution. The chief function of phloem water is a solvent for the transport of food throughout the plant. Indeed, the concentration of solutes in phloem is so great that the solution has the appearance of slime and is commonly identified as such in many texts.

Physiologically, metabolic activity associated with living cells is necessary for transport of food in the phloem tissues. Thus, any conditions that would be detrimental to processes associated with metabolism (respiration, digestion, etc.) will disrupt the flow of food throughout the plant. However, actual processes involved in moving plant food from one phloem cell to another through the phloem tissue is unknown.

Water in the Absorbing Zone of the Root

Two theories exist concerning the movement of water into the plant root. We shall adopt current usage and identify these as active intake and passive intake.

Active intake of water is water movement against the gradient of concentration, i.e., from a region of low concentration to a region of higher concentration. Active transport is thought to occur as a result of physiological and metabolic processes that take place in the root at the zone of absorption. As evidence for active intake of water, many authors cite guttation which shows pressure build-up within the xylem vessels. No doubt some active intake of water does occur, just as there exists evidence that ion and mineral absorption by the root is an active process requiring a high rate of metabolic activity in the absorbing region. However, it is not possible to assign the entire process of water absorption by roots of growing plants to active transport.

Example 6.10. Active intake—theory and observation
Redwood trees are known to reach heights of 122 m (400 ft). What would the pressure within the xylem vessels in the roots have to be if active absorption was responsible for water intake?

Solution
One atmosphere of pressure is equivalent to 101.325 kPa or 10.33 m of water, therefore:

$$\frac{122\text{mm}}{10.33\frac{\text{m}}{\text{atm}}} = 11.8\text{ m}$$

$$11.8\text{m} \times 101.325\frac{\text{kPa}}{\text{atm}} = 1190\text{kPa} \qquad 6.19$$

Several investigators have measured the guttation pressure of roots by measuring the pressure buildup in a cut-off stem. This pressure has seldom been found to exceed 2 atm. Thus, though liquid within individual parenchyma cells has been observed to attain pressures of 20 to 25 atm, such pressures may not exist within the xylem. Researchers have found that stem diameters decrease during transpiration. If the movement of water were due to pressure, increases in diameter should have been found. This is further evidence that water movement through the xylem is passive rather that active.

The passive transport theory of water absorption suggests that water enters the root in response to the pull of water exerted in the xylem vessels by the transpiration in the leaves. In passive transport, evaporation of water in the leaves creates a water deficit which is corrected by water withdrawn from the xylem. The removal of water from the xylem, however, in turn creates a deficit which is transmitted throughout the plant vascular system to the root zone where water is available from the soil outside the plant root. The deficits in the root xylem vessels is then corrected by the intake of water in the absorbing zone of the root.

Example 6.11. Passive movement of water
A corn plant is 182 cm tall to the topmost leaf from the ground surface. If the topmost leaf is losing water very slowly by evaporation and there is no frictional loss as water travels along the xylem vessel supplying water to the topmost leaf, what is the tension at the bottom of the xylem vessel 0.12 m below ground?

Assume the friction loss in the xylem vessel is 1 atm/m and is independent of the flow velocity. What is the tension in the xylem vessel at the ground surface? Assume that ample water is available to the root but no active absorption of water by the root is occurring.

Solution

Disregarding frictional losses, the total length of the water column is 1.82 m + 0.12 m = 1.94 m. This is the maximum tension. It will occur in the topmost leaf and will progressively decrease down the water column in the xylem vessel. At the lower end of the vessel, if there is no frictional loss the tension will be zero.

However, if frictional losses are to be considered, at the ground surface and with a frictional loss due to the movement of water in the xylem vessel, the tension is:

$$0.12\,\text{m} \times \frac{1\,\text{atm}}{\text{m}} = 0.12\,\text{atm} \tag{6.20}$$

This tension must be added to the tension resulting from the weight of the water column below the ground surface:

$$0.12\,\text{m} \times \frac{1\,\text{atm}}{10.33\,\text{m}} = 0.012\,\text{atm} \tag{6.21}$$

The tension inside the xylem vessel at the ground surface is:

$$0.012 + 0.12 = 0.13\,\text{atm} \tag{6.22}$$

Very little is known about the frictional effects due to water movement within thc xylem vessels. It is probable that much higher tensions develop within the xylem vessels during periods of high evaporative water loss by the leaves.

The most likely pathway of water into the xylem vessels is via the root hairs and along the walls of the cortex to the endodermis. The casparian strip, the zone of cutinized tissue surrounding the radial and transverse walls of the endodermal cells, probably impedes the flow of water along the endodermal cell wall. Therefore, the water must pass through the outer endodermal cell wall, through the cytoplasm of the endodermal cell, and through the inner endodermal wall since these walls are not cutinized. When the water overcomes the endodermal wall barrier, it can quickly make its way into the xylem vessels which, in the absorbing zone of the root, extend to the endodermal region thanks to the star-like arrangement of the phloem and xylem vessels in the root (see figs. 5.8 and 5.9).

The soil itself may play an important role in controlling the absorption of water by the roots since one would rarely expect to find the entire plant root system growing in a soil that has a uniform soil-water content and soil-water availability. A root entering a region of higher soil-water availability would absorb water most readily in that region. However, since water must move through the cortex of the root before entering the endodermal cells containing the casparian, it is possible for water to move longitudinally along the root cortex to a zone where the soil-water deficit is at a lower water potential. In response to purely passive movement, water should than exit the root and return to the soil to satisfy the lower potential.

Undoubtedly some longitudinal movement of water does occur along the external cell layers of the root, soil water being transported from regions of high soil-water content, i.e., high soil-water availability, to regions of lower soil-water content, i.e., low soil-water availability. The roots could then act as a “short circuit” moving water through the soil to equilibrate the soil water potential, but this does not usually occur. Roots may grow through regions of very low water potential into regions where water is available without recharging the low water potential regions. This is most likely because roots become covered with suberized tissue immediately beyond the region of water absorption. Since the suberized tissue is relatively waterproof, a barrier to the entry or loss of water by the root cortex is present.

While we will defer the treatment of water movement through the soil and treat the subject in some detail in chapter 10, we note here that continued root growth is essential so that the absorbing regions of plant roots can continually enter regions of high water availability. It is very likely that suberization of the root begins very close behind the water absorption region preventing roots from transferring water through the soil profile into regions from which the water has been exhausted.

As in the stem, the growing region of the plant root is a region where metabolic activity is very high, with digestion, respiration, and assimilation occurring to support the growth processes. Thus, the root zone of a plant system is a region of high metabolic activity, requiring a source of free O_2 for respiration and a sink or place where the CO_2 (formed as a by-product of respiration) can be given off.

The zone of water absorption in the roots, as already noted, requires that O_2 be available with the soil water, and that CO_2 be transported away from the metabolizing region. These processes occur by diffusion in the soil mass or, for some crops such as rice, by diffusion up and down the stem of the plant via large intercellular spaces. It is also possible that metabolic mechanisms adapt to existing conditions dependent upon the plant. It is obvious that physiological activity and water movement into the root are closely related. It is not clear what influence active absorption plays in the water-absorption process, or it if acts only at some particular stage in the overall process such as in the control of flow through the endodermis. It is an accepted fact, however, that metabolism is necessary for water absorption by the plant root system to occur.

Water and Plant Development

Water movement into, through, and from a plant obviously has an influence on the metabolic activity of plants and thus on their growth and development. At this point it is only possible to speculate on the full significance of the relationship between water balance and plant growth. A schematic representation has been developed by Weatherly (1963) to depict the effect of water stress on plant development. An adaptation of this representation is given in

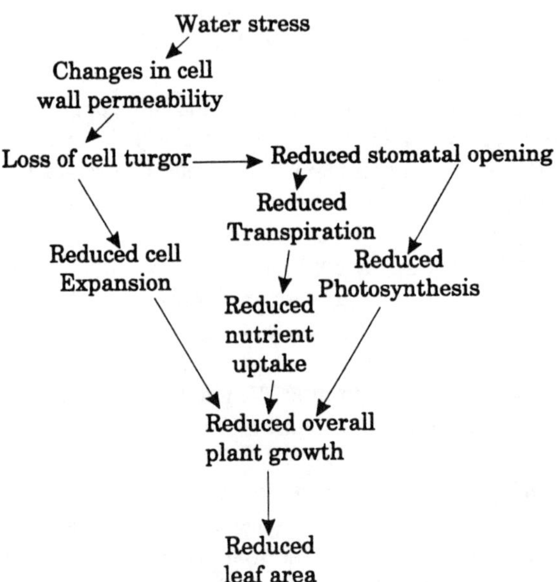

Figure 6.2. The effects of water deficit are illustrated in the flow diagram. Deficits influence both growth as a result of lower turgor and photosynthesis through reduced stomatal opening.

figure 6.2, where water stress is shown as causing an internal strain that influences various physiological activities of the plant.

Little is known about the effect of water deficiencies on cell permeability because of the difficulty of studying the membrane and its structure under stress conditions. It is obvious that lack of water will reduce the pressure within cells, thereby causing all expansion to cease and overall lack of growth to occur. A considerable water deficit must occur, however, before there will be appreciable reduction in the movement of water molecules through the plant leaves and out the semi-closed stomata. Once water loss by transpiration is reduced, it is certain that a corresponding reduction in CO_2 available for the photosynthetic process will occur as a result of the stomatal closure, causing less production of plant food and reduced overall metabolic activity, followed by reduced overall growth.

Since most water flow from the soil into the plant is caused by transpiration, one should anticipate a reduction in ion uptake from the soil. This must take place for two reasons: (1) less overall water flow occurs, and perhaps more important; (2) root growth is also reduced so less root extension takes place, resulting in reduced contact between root surface and soil particles. Since some ions are absorbed directly from soil particle surfaces, a reduction in ion availability can be expected.

Environmental Factors Affecting Seed Germination

It was pointed out earlier that within a viable seed there exists a miniature but inactive or quiescent plant. In this section we discuss in general terms the germination of the seed and the role played by water in the young, developing seedling.

The germination of seeds requires ample oxygen, water, and the proper temperature. The temperatures required are those which might be expected 10 to 70 mm below the ground surface. However, the precise temperature will vary with the plant. Wheat will germinate at temperatures as low as near 0° C and as high as 35° C. Maize, a corn plant of subtropical origin, prefers a narrower range, from 5 to 10° C, and has an upper limit of 45° C although the optimum temperature, 17 to 20° C lies between these limits. It is not possible to designate any exact temperature as the optimum for germination because this varies with the other prevailing environmental conditions and with the criterion selected as the index of germination. For instance, the most suitable temperature for elongation of the hypocotyl and primary root does not correspond to the most suitable temperature for the development of the plumule or epicotyl. Finally, in some species, the permeability of the testa or seed coat to oxygen and water is increased by exposure of the seed to wide fluctuations in temperature. Example are found within several millimeters of the soil surface as the diurnal variation of temperature due to solar radiation occurs.

Water plays a central role in the germination of all seeds since water acts as the solvent for the many enzymes that trigger digestion of the stored food reserves and initiate respiration, providing energy for the growth processes of cell division and cell enlargement.

Most seeds possess a very strong affinity for water and absorb soil water by imbibition, a term implying the movement of water to and through the seed coat. Water movement will occur as liquid soil water at a higher soil-water contents such as those favorable to plant growth. Water movement also may occur under conditions of very low soil-water content as the seed imbibes water in the vapor phase. Imbibition of water and germination may occur in many seeds even though the seed is surrounded by soil so dry that the plant, once it emerges from the seed, cannot absorb sufficient water to continue to grow.

The initial stage of seed germination involves a rapid increase in both seed volume and mass. It is not unusual for a germinating seed to double its weight after only two days in the soil. The initial weight gain is due wholly to water absorption, since no food is manufactured by the seed until the seed plumule emerges.

Example 6.12. Moisture imbibition by germinating seeds

A bean seed containing 5% moisture and weighing 0.2 g is found to weigh 0.4 g after planting has occurred and the seed has remained in the soil two days. What is the percent moisture content if the base is the dry mass of the seed? With seeds, the moisture content is usually based on the wet mass of the seed. What is the percent moisture content if all values are based on the wet mass?

Solution
The moisture content on a dry mass basis is:

$$\frac{\text{mass}_{\text{wet}} - \text{mass}_{\text{dry}}}{\text{mass}_{\text{dry}}} \times 100\% = 5\% \quad 6.23$$

Therefore, at 5% moisture, a seed weighing 0.2 g has a dry mass of:

$$\frac{0.2\,\text{g}}{1.05} = 0.19\,\text{g} \quad 6.24$$

The moisture content after two days in the soil is:

$$\frac{0.4\,\text{g} - 0.19\,\text{g}}{0.19\,\text{g}} \times 100\% = 110\% \quad 6.25$$

Therefore, the dry mass is:

$$\frac{\text{mass}_{\text{wet}} - \text{mass}_{\text{dry}}}{\text{mass}_{\text{wet}}} \times 100\% = 5\% \quad 6.26$$

The amount of water in the seed is:

$$\text{mass}_{H_2O} = 0.2\,\text{g} - 0.01\,\text{g} = 0.19\,\text{g} \quad 6.27$$

At a wet weight of 0.4 g, the moisture content on a wet weight basis is:

$$\frac{0.4\,\text{g} - 0.19\,\text{g}}{0.4\,\text{g}} \times 100\% = 52.5\% = 53\% \quad 6.28$$

Example 6.13. Availability of soil water for seed germination
Initially the seed in example 6.12 was planted in a dry soil with water content on a dry mass basis of 11%. If all the water imbibed by the the seed came from a spherical mass of soil 20 mm in diameter with a mass of 3.14 g, what was the soil moisture content in the spherical mass of soil after two days? (Note: While seed moisture content is usually computed on a wet basis, soil is always computed on a dry basis.)

Solution
The dry mass of the soil was:

$$\begin{aligned}\text{mass}_{\text{soil}_{\text{dry}}} &= \frac{3.14\,\text{g}}{1.11} \\ &= 2.83\,\text{g}\end{aligned} \quad 6.29$$

The mass of water initially in the soil was:

$$\text{mass}_{\text{water}} = 3.14\,\text{g} - 2.83\,\text{g}$$
$$\text{mass}_{\text{water}} = 0.31\,\text{g} \qquad 6.30$$

If 0.2 g was imbibed by the soil, the water remaining in the soil is:

$$m_{H_2O} = 0.31\,\text{g} - 0.2\,\text{g}$$
$$= 0.11\,\text{g} \qquad 6.31$$

The water content in the soil on a dry basis is:

$$\frac{0.11\,\text{g}}{2.83\,\text{g}} \times 100\% = 3.9\% \qquad 6.32$$

The initial stage of water absorption by imbibition is important because an increase in water content by the seed markedly increases the permeability of the seed coat to O_2, a raw material of respiration, and CO_2, a by-product of the metabolic processes. The initial stage also renders the seed coat permeable to water. A higher water content of the seed enhances enzymatic activity and digestion, enabling food stored as a solid to become soluble and moved to the rapidly developing parts of the embryo where it can undergo assimilation into the growing embryo.

As germination develops, two types of seed development can be observed: (1) seeds with emerging cotyledons; and (2) seeds in which the cotyledons remain within the soil mass. For seeds with emerging cotyledons, the following sequence of events occurs if adequate soil water is available.

From the hypocotyl of the embryo, a primary root develops and pushes through the seed coat downward into the soil mass. From this primary root secondary roots develop and push into the soil mass laterally and absorb water. The hypocotyl continues to enlarge upward from the primary root and forms an arch that penetrates the soil surface. Once the soil surface is penetrated, the hypocotyl continues to enlarge and the cotyledons are literally pulled upward through the surface of the soil. At this point the plant arch straightens The cotyledons may then turn green and participate in food manufacture, supporting the food-manufacturing capabilities of the plumule, or epicotyl, which rapidly develops leaves and forms the shoot or food-manufacturing portion of the plant. Figure 6.3 illustrates a typical germination sequence for a seed with emerging cotyledons.

For seeds in which the cotyledons remain within the soil mass, the hypocotyl of the embryo emerges from the seed and a primary root rapidly elongates, followed by the development of secondary roots. Thus, embryo is thus provided with a root system through which water absorption can take place. The embryo can obtain moisture from regions distant from the germinating seed. Once the hypocotyl and primary and secondary roots have developed, the epicotyl develops. The epicotyl lengthens as the cells within the epicotyl divide, elongate and mature, using food stored in the cotyledons that remain in the soil mass. The epicotyl then pushes the plumule vertically out of the soil mass, leaves appear, and food production by the leaves augments the food being provided by the cotyledons. Figure 6.4 illustrates steps in the germination of a red oak following the sequence of events described above.

In monocotyledons the epicotyl develops a sheath called the coleoptile which pushes its way through the soil. As the coleoptile emerges, the plumule or shoot, enclosed by the coleoptile while the coleoptile was submerged within the soil mass, breaks out of the coleoptile and the leaves unfold, enabling food production to take place to supplement the food stored in the cotyledon. Figure 6.5 depicts the behavior of a germinating corn that is typical of the behavior just described.

Germination is affected by anything that alters the environmental parameters of temperature, water, and O_2. Particularly important is the plant's need for O_2. Seeds germinating in a soil whose pores are nearly filled with water will suffer from lack of O_2, since the primary source of O_2 is diffusion from the soil surface through the soil mass. Anything that impedes the diffusion of O_2 through the soil mass may negatively affect the growth process. For instance, corn growth is limited at soil O_2 contents of 5.2% which is considerably below the 20% commonly

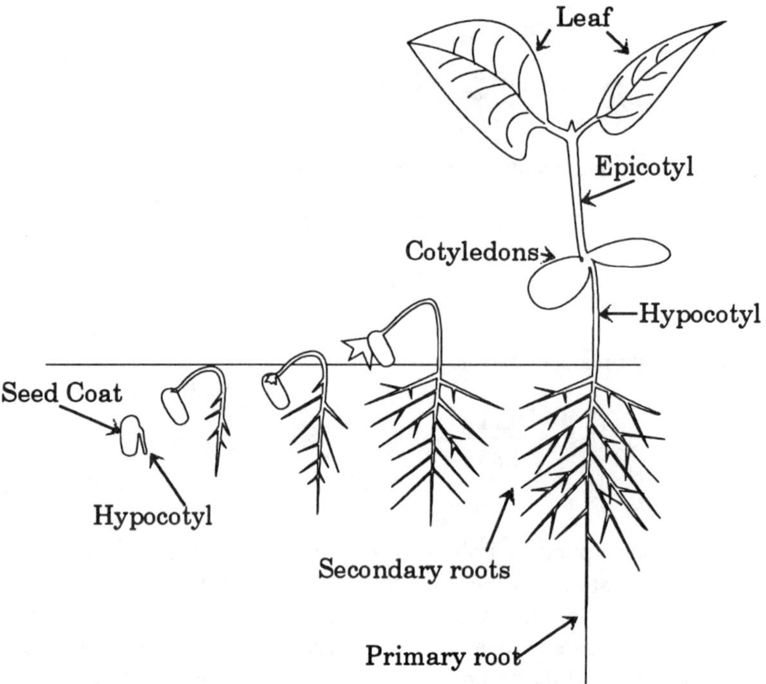

Figure 6.3. Stages in an emerging dicotyledonous plant. The cotyledons or seed leaves become food-producing structures when they emerge from the soil.

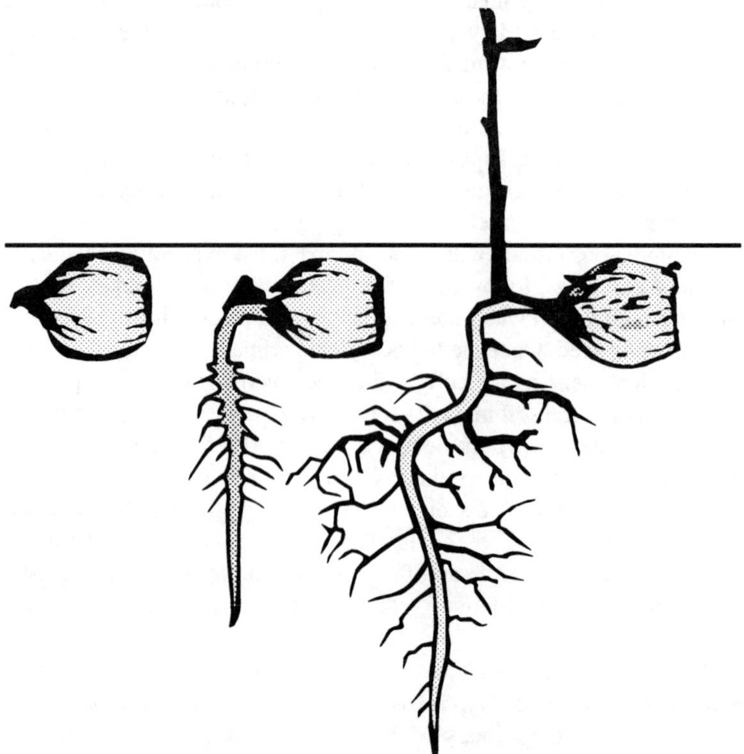

Figure 6.4. A developing and emerging oak sprout. The acorn remains belowground and furnishes nourishment for the young sprout.

found in atmospheric air. Inadequate pore space will inhibit O_2 diffusion as will a highly compacted crust on the soil. In addition to inhibiting diffusion of O_2, soil crusting may provide mechanical obstruction such that the enlarging hypocotyl or epicotyl cannot penetrate the crust. Obviously, such conditions are not advantageous to seed germination.

Example 6.14. Oxygen flux and soil environment interaction
The growth of wheat seedlings was found to be inhibited when the O_2 flux to the germinating seed was less 6×10^{-5} g/(cm^2·s). Assuming a soil crust thickness of 70 mm, and further assuming that wheat seedlings will suffer if the O_2 content of the soil falls below 5.2%, compute the diffusion coefficient of the soil crust for O_2 diffusion. The expression:

$$F_{O_2} = \frac{-D\left(C_{soil} - C_{air}\right)}{\Delta x} \qquad 6.33$$

applies where F is the O_2 flux, C is the concentration of O_2 in g/cm^2, Δx is the thickness through which diffusion occurs, and D is the diffusion coefficient. T = 300 K.

Solution
Atmospheric concentrations are given as volume of gas per volume of air, both taken at atmospheric pressure. Therefore, to obtain the concentration as mass per unit volume, it is only necessary to determine the mass of oxygen per volume of air. By the perfect gas law, PV = nRT, so:

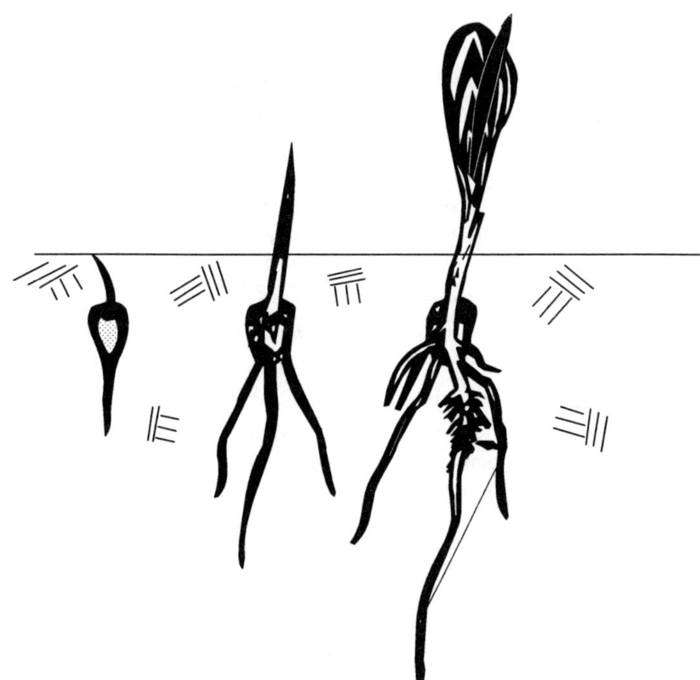

Figure 6.5. An emerging corn plant. The seed contains considerable endosperm which must be digested before it can be used as food for the developing plant. This is the reason a corn plant takes longer to germinate and emerge from the soil than does a bare cotyledon.

$$\rho_{O_2} = \left(\frac{m_{O2}}{V_{O_2}}\right)_{20\%} = C_{\%} \times \frac{P_{atm} \cdot M_{O_2}}{R \cdot T} \qquad 6.34$$

The concentration of O_2 at 20% is:

$$\rho_{20\%} = \frac{20\%}{100\%} \times \frac{32 \frac{g_{O_2}}{mol} \times 100 \text{ kPa}}{8.314 \times 10^{-3} \text{ kPa}\cdot\text{m}^3 \times 300\text{K}} = 257 \frac{g_{O_2}}{m^3} \qquad 6.35$$

Similarly:

$$\rho_{5.2\%} = 67 \frac{g_{O_2}}{m^3} \qquad 6.36$$

Therefore, the O_2 diffusion coefficient is:

$$6 \times 10^{-1} \frac{g}{m^3 \cdot s} = -D \times \frac{67 \frac{g}{m^3} - 257 \frac{g}{m^3}}{70\text{mm} \times \frac{m}{1000\text{mm}}}$$

$$D = 2.2 \times 10^{-4} \frac{m^2}{s} \qquad 6.37$$

The above problem illustrates the mathematical reasoning required to obtain unknown values required for engineering calculation.

Seeds will imbibe water from very dry soil but the roots that develop from the hypocotyl cannot. Thus, soil near developing seeds should be sufficiently porous to allow root penetration, and should contain adequate moisture to supply the young plant's needs. Often, supplemental irrigation during the germination process is necessary to ensure adequate moisture.

For seed germination minerals from the soil are unnecessary as the endosperm or cotyledon contains sufficient nutrients and minerals for seedling development. However, once germination has occurred, the young roots must supply the plant's needs for minerals by absorption through the root surfaces and transmission through the vascular system. Thus, supplying supplementary minerals in the form of fertilizers may be essential. The needs should be dictated by soil testing and the advice of crop scientists. This subject is discussed in more detail in chapter 7. At this point it is sufficient to be aware that the needs of the whole plant must be considered in seed-bed preparation and management to obtain optimum plant responses.

Fruit Development

Fruit development in of this text will be limited to fleshy fruits in which the high water content pericarp forms the edible portion.

The physiology of a developing fruit can be separated into three segments: (1) flowering and fertilization; (2) cell division; enlargement and maturation; and (3) fruit ripening. It is especially important to the engineer to appreciate the physiological processes involved in fruit development, especially from the aspect of the role of water in the fruiting process.

The period of highest metabolic activity (respiration, translocation of foods, digestion and assimilation) occurs during the second developmental segment for many fleshy fruits such as apples, cherries, and citrus. During this developmental segment, the cells of the pericarp are rapidly changing in tissue structure and, more important, in volume. The volumetric changes occur mainly as a result of the storage of solutes within the cells of the developing fruit. Extension of the cell walls is due to the development of turgor pressure.

During the period of fruit development, the overall water balance of the plant is critical. Water is lost from fruit during periods of water stress. In response to high water demand by the transpiration of water from plant leaves, daily changes in fruit diameter can easily be observed even during periods of adequate soil moisture. Figure 6.6 illustrates the changes in the diameter of orange fruit during a period of adequate soil water availability. As transpiration peaked during the daylight period, the plant root system was unable to supply adequate soil water to satisfy the deficits created by evaporation from the leaf cell walls and transpiration. Because the xylem vessels and tracheids throughout the plant were continuous, water was withdrawn from the fruit, thus causing an abrupt reduction in diameter.

During the drought period when the soil water was inadequate to supply the transpiration demands, a decrease in fruit diameter occurred, and the diameter continued to decrease so that the fruit never regained its original size during the period of record, as shown in figure 6.6. Splitting of fruits following irrigation or rainfall after a prolonged period of drought, probably due to rapid cell enlargement associated with increased availability of soil water, is a serious problem. Significant losses occur in susceptible crops such as cherries.

Physiologically, fleshy fruits have a high demand for water, especially during maturation while growth and expansion are occurring. Ample water supplied to even a small portion of the root zone during the enlargement phase can increase the overall diameter of fruits such as tomatoes. One must take care to ensure that the entire root zone is not saturated for extended periods as physiological damage to the plant may occur.

The final stage of fruit development is concerned with ripening. The nature of the ripening process varies with the type of fruit. Usually, it is a period of high enzymatic activity during which water plays an important role. In

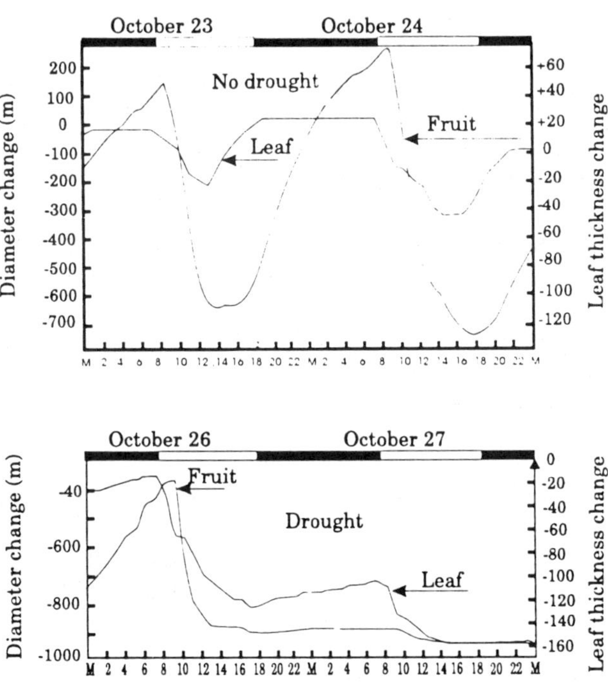

Figure 6.6. Evapotranspiration and drought on orange fruit.

many fruits a partial breakdown of cell-wall structure occurs with the accumulation of fruit sugars and increases in pectinaceous substances. Here the relationship between absorption of water by the fruit in response to irrigation, or conversely, withdrawal of water from the fruit due to high transpiration levels and low availability of soil water, is controlled by abscission layer development that disrupts the continuity of the vascular tissue throughout the plant.

As an example of the possible influence of the abscission layer, sweet cherries grown and ripened under low soil water availability may split badly after a rain that occurs immediately before harvest time. Sour cherries are not as susceptible to splitting as a result of increased soil-availability, probably because an abscission zone forms at the point of attachment of sour cherries to the tree, this zone is absent in sweet cherries.

The subject of physiological relationships of fleshy fruits is one requiring considerably more information before sound engineering practices related to the management and control of water and its effect on fruit quality and structure can be stated. Until more information is available, engineering practices related to fruit development and quality will have to be inferred from a knowledge of the anatomical and physiological behavior of plants.

Exercises

6.1. In the production of 1 metric ton of glucose, how many kg of O_2 are produced? [1.07 × 102]

6.2. The combustion reaction for propane in an internal combustion engine is approximated by:

$$C_3H_8 + 5O_2 \Rightarrow 3CO_2 + 4H_4O + 2.283 \times 10^7 \text{ J}$$

How many kg of O_2 are used to produce 1.614×10^7 J, the amount of energy required to photosynthesize 1 metric ton of glucose? [0.1]

6.3. How many kg of CO_2 are produced during the liberation of 1.6×10^7J by the combustion of propane? How does this compare to the CO_2 used by photosynthesis during the storage of 1.6×10^7 J. [16.2 more used in photosynthesis]

6.4. An individual was shown to require 1.24×10^2 J/kg-day to sustain itself. If a 21-day-old bean plant has a mass of 35 g, a leaf area of 800 cm^2, and uses CO_2 at an average rate of –85 g/(cm^2·s) over a 12-h day, what energy input is necessary to sustain the plant.[1.0×10^5] (Reduce your answer to the same terms as those given for the individual cited.)

6.5. Can humans use green plants as their sole source of food? If the answer is yes, would there be any fear of a food crisis in the densely populated sectors of the world?

6.6. A 40 kg sheep ingests 8.79×10^3 J in a day. Corbett (1969) showed that the sustenance relationship for sheep was related to the body mass to the 0.75 power. What is the daily energy requirement for sheep in J/(body mass)$^{0.75}$-day? [5.52×10^3]

6.7. Based on example 6.8 and the estimate of percent of water in the cytoplasm of a cell, if the cells in example 6.8 are roughly cubical, what mass of water would you expect the leaf to have? [5 to 6.1]

6.8. Assume that a hectare of green crop will produce 10 metric tons of glucose. If 30% of the energy entering a leaf is used in photosynthesis, and 6 kg H_2O are used to produce 1 metric ton of glucose, compute the total water requirement of the crop on a per hectare basis? [211]

6.9. If 30% of the energy striking the bean plant in exercise 4 is used in photosynthesis, calculate the rate of water use per day of the bean. Express your answer in cm^2 water/cm^2 leaf. [4.9×10^5]

6.10. It has been estimated that the rate of water loss through a leaf when all stomata are closed may approach 5% of the daily rate. What rate of water loss could be expected from the plant in exercise 9, if this estimate is valid? This is known as cuticular transpiration. [0.157 cm^3]

6.11. Calculate the stress (tension) that must exist in the xylem vessels at the top of a tree 100 m tall if the tension in the soil at the point of absorption by the root is –500 kPa. [–1470 kPa]

6.12. If the seed in example 6.14 was 0.5 cm in diameter and spherical, assume that the water absorbed by the seed was done uniformly over a 48 h period and calculate the flux through the testa in $gH_2O/(cm^2 \cdot s)$. [3.68×10^{-7}]

6.13. Carnes (1934) expressed soil crust strength in terms of its modulus of rupture R, where:

$$R = \frac{3\,PL}{2\,bd^2}$$

and P is the force required to fracture a crust with dimensions of length L, width b, and thickness d. Assume a clay with a rupture modulus of $8 \times 10^4\ N/m_2$. An arched bean epicotyl is found to have emerged and fractured a soil crust whose dimensions were 8 × 2 × 0.9 cm. What force was the bean epicotyl exerting on the crust? If the contact area of the arch was about 1 × 0.2 cm, what pressure in kPa was exerted by the seedling? [1.08, 54]

6.14. Assume a cell in a sweet cherry suffering from water stress is roughly cubical and has dimensions of 0.2 μ. If, due to rainfall, the cell expands to a spherical shape without increasing in surface area, what is the new diameter of the cell? What percent increase does this represent over the cubical edge dimension? [0.276, 38]

Chapter 7

Soil Environment of the Green Plant

Soil is that part of the earth's crust consisting of rock that has been weathered and altered by climate, plant and animal life, and time. Table 7.1 gives the general composition of the lithosphere or solid part of the earth's crust from which all soils were formed.

Time plays an important role in the formation and presence of soils. The processes of weathering by which soil is formed take place slowly but continuously, forming soil by bringing about changes in the upper layers of the earth's crust. Weathering embraces the following factors, all of which involve chemical reactions: (1) oxidation (the combining of oxygen with other elements); (2) reduction, the reverse of oxidation, resulting in the loss of oxygen from an elemental combination; (3) hydration, the union of water with minerals; (4) hydrolysis, in which a positively charged hydrogen ion changes places with a cation, forming a new ion; and (5) carbonation, which occurs when CO_2 combines with other bases.

The result of weathering is a mixture of substances that is similar to other soils formed from substantially different materials. The similarity prevails over wide soil groups and will become more evident as we examine the mineral constituents of soils. The minerals in soil are similar for all soils. The only differences are excesses and deficiencies of specific minerals. The identification of excesses or deficiencies a job for competent soil scientists. The correction or mineral prevention of the creation of an excess or deficiency is often the responsibility of the agricultural or biological systems engineer.

The soil composition that is ideal for plant growth is different from the composition of the lithosphere given in table 7.1. An ideal soil consists of approximately 45% mineral matter, 5% organic matter, 25% water and soil minerals solution, and 25% soil air. A productive soil contains many voids. The voids comprise approximately 50% of the total composition of the soil.

The role of soil in the soil-plant-atmosphere continuum is unique because soil is not essential for plant growth. Plants are often grown without soil or hydroponically, in a liquid culture adjusted to provide the nutrients required by the plant. Problems are associated with hydroponic culture, however. These include providing support for the plant and maintaining adequate supplies of minerals in the proportions necessary for growth.

Table 7.1. Composition of the lithosphere

Element	% Present	Element	% Present
Oxygen	48.30	Hydrogen	0.20
Silicon	27.70	Carbon	0.20
Aluminum	7.80	Phosphorus	0.10
Iron	4.50	Sulfur	0.10
Calcium	3.50	Magnesium	2.20
Sodium	2.50	Manganese	0.08
Potassium	2.50	Other	0.82
Titanium	0.50		

Source: Kossovich (1916) and Miller et al. (1965)

In the biosphere, one of the roles the soil plays is to provide support and anchorage for the plant through the root system. In addition, the soil also provides for storage and release from the soil mass of water and minerals essential to plant growth. Furthermore, by storing energy and transmitting heat through the soil mass, the soil provides a thermal environment suitable for the physiological processes of plant growth.

For a large class of plants of agricultural significance, the soil is the primary source and sink for gases that support the physiological processes of the plant. The soil voids diffuse gases and maintain a balance of the components of soil air that are suitable for the plant. Exceptions to the role of soil as a supplier or diffusing medium for gases exist for plants such as rice that have become adapted to an aquatic existence. In rice, for instance, gaseous diffusion occurs through the stem in the voids created by the extensive and well-developed system of intercellular spaces. In this chapter several important physical and chemical soil concepts will be discussed. This chapter is not intended replace a study of soil physics and soil chemistry. It is intended to acquaint the student with some properties of the soil biosystem which are important to the physiological well-being of the plant. The material is chosen to familiarize the engineer with the terminology of soil scientists. This chapter will introduce those aspects of the soil that are amenable to engineering modifications of soils. Complementing the physiological responses of the plant to enhance the desirable agriculturally significant output is a primary role of the agricultural or biological systems engineer.

Important soil properties may be physical, chemical, biological, or a combination of these. We focus attention on the physical and chemical, but remain aware that the biological aspects are not independent and, therefore, must be considered with both physical and chemical properties.

Physical Aspects of the Soil

Particle size is an important parameter of the composition of a soil. While soil comprised of large particles will provide anchorage for the plant roots, soil water held in the spaces between soil particles may soon be exhausted by the plant. This is because a soil comprised of large particles has fewer void spaces than that containing many small particles. In addition, the volume of water held by capillary forces between large particles is small. Minerals so necessary to plant growth are held as anions or cations, on the surfaces of soil particles, or are joined by a weak chemical bond to the particle surfaces. Large particles present considerably less surface area for the retention of minerals than the same volume of soil comprised of small particles. Small soil particles retain a significant amount of water hygroscopically as a film of water completely surrounding the particles. Finally, the small particles retain water in the pore spaces between the particles and a significant amount of this water can be used by the growing plant. Some aspects of the importance of particle size are easily illustrated by the following example.

Example 7.1. Particle number as a function of particle size
Assume that a cube of silica 1 m on a side is decomposed into cubes measuring 2 μ on a side. How many particles will be obtained from the parent material?

Solution
The number of particles in one edge of a cube of silica 1 m/side is the same as the number of times 2 μ can be divided into 1 m or:

$$\frac{\text{No. particles}}{\text{edge}} = \frac{1 \times 10^6 \mu}{\text{edge}} \times \frac{\text{particle}}{2\mu} = 5 \times 10^5 \text{particles} \qquad 7.1$$

Therefore, the number of particles is:

$$\text{Total No.} = \left(5 \times 10^5\right)^3 = 1.25 \times 10^{17} \text{particles} \qquad 7.2$$

An examination of sand and silt particles that can be seen under an ordinary microscope will show that the particles are not perfect cubes. They are somewhat irregular in shape. The irregularity is significant in creating the void spaces essential to the transport of water and gaseous products through the soil needed by or expelled by the plant.

Example 7.2. Particle number as a function of particle shape
How many spherical silica particles 2 μ in diameter can be obtained from 1 m^3 of silica. If the particles were stacked so each sphere touched 6 other spheres (i.e., the maximum possible void space is created), estimate the volume of soil that would result from the silica spheres obtained from 1 m^3 of silica?

Solution
Volume of a 2 μ diameter sphere is:

$$V_{2\mu} = \frac{\frac{4}{3} \times \pi \times (1\mu)^3}{\text{particle}} = \frac{4.189\mu^3}{\text{particle}} \qquad 7.3$$

The total number of spheres required to produce 1 m^3 of silica is:

$$\text{No. particles} = \frac{\left(1\text{m} \times \frac{1 \times 10^6\mu}{\text{m}}\right)^3}{\frac{4.189\mu^3}{\text{particle}}} = 2.387 \times 10^{17}\text{particles} \qquad 7.4$$

The number of particles on each side of a cube formed of the spheres in the above equation is:

$$\frac{\text{No. particles}}{\text{side}} = (2.387 \times 10^{17})^{\frac{1}{3}}\,\text{particles} = \frac{6.20 \times 10^5\text{particles}}{\text{side}} \qquad 7.5$$

Therefore, the length of each side is:

$$\frac{6.20 \times 10^5\text{particles}}{\text{side}} \times \frac{2\mu}{\text{particle}} = \frac{1.24 \times 10^6\,\mu}{\text{side}} = \frac{1.24\text{m}}{\text{side}} \qquad 7.6$$

This yields a volume of soil of:

$$(1.24\text{ m})^3 = 1.9\text{m}^3 \qquad 7.7$$

It is apparent from example 7.2 that there is almost as much pore space created by the reduction of 1 m^3 of solid into 2 μ diameter loosely packed spheres as the solid had originally. Particle size, shape, and arrangement all contribute to the physical properties of soil. Particle size forms the basis of the system of soil textures by which soils are identified and classified. Soil is used for many purposes other than plant growth, such as road building, construction material, etc. Therefore, not surprisingly, there is more than one method of classifying soil according to the size of its particles. Table 7.2 gives the two most important textural classifications in use today with the terminology used to identify the particle-size designation.

Table 7.2. Classification of soil texture by particle size

Particle Designation	USDA Classification System Diameter* (mm)	International Soil Science Society System Diameter (mm)
Very coarse sand	2.00 – 1.00	– –
Coarse sand	1.000 – 0.50	2.00 – 0.20
Medium sand	0.50 – 0.25	– –
Fine sand	0.25 – 0.01	0.20 – 0.02
Very fine sand	0.01 – 0.05	– –
Silt	0.05 – 0.002	0.02 – 0.002
Clay	0.002 –	0.002 –

* Particle diameter is defined in terms of the diameter of the sphere which would pass through a screen with square openings the size of the diameter.

Determining the percentage of larger particles present in a sample of soil is not difficult. For sand particles (or sand separates) sieves are used to screen out particles larger than those passing through the sieve. For silt and clay particles, however, a different method must be used because of the small particle size. A common technique is to use the speed at which particles of a given diameter will settle out of a suspension of particles. The time required to remove all particles of the given diameter or larger is calculated. The sample is allowed to settle and the suspension remaining is heated to remove the water. The residue is then weighed to determine the mass of the particles remaining.

The mathematical model upon which this technique is based was developed by G. G. Stokes, and is based on Stokes' observation that the frictional drag exerted on a spherical particle in a fluid is directly proportional to the particle diameter, the rate of fall, and the fluid viscosity. Experimentally, Stokes verified that the force opposing the motion of a small, spherical particle in a fluid could be expressed as:

$$F_f = 3\pi\eta dv \qquad 7.8$$

where η is the viscosity of the liquid, d is the diameter, and v is the velocity of the particle.

Example 7.3. Viscosity
If the dimension of F_f is Newtons and the diameter of the particle is given in m with velocity expressed in m/s, what are the dimensions of η, the fluid viscosity?

Solution
By rearranging equation 7.8:

$$\eta = \frac{F_f}{3\pi dv} \qquad 7.9$$

After substituting the appropriate dimensions:

$$\eta(\ ) = \frac{F_f\left(\frac{kg \cdot m}{s^2}\right)}{3\pi d(m)v\left(\frac{m}{s}\right)}$$

$$\eta(\text{Pa}\cdot\text{s}) = \frac{F_f\left(\frac{\text{kg}\cdot\text{m}}{\text{s}^2}\right)}{3\pi dv\left(\frac{\text{m}^2}{\text{s}}\right)} \qquad 7.10$$

The collection of units for η, the viscosity of the liquid in the SI system has the dimension given above. The magnitude of η is a function of the liquid and its temperature (for Newtonian fluids).

Example 7.4. Development of Stokes' Law
By equilibration of forces, what is the gravitational force causing particles of density δ to settle out of a liquid density ρ?

Solution
The gravitational force on a particle of density δ and diameter d in air is:

$$F_g = \delta g \frac{\pi d^3}{6} \qquad 7.11$$

The particle is buoyed upward by the fluid buoyancy force:

$$F_\beta = \rho g \frac{\pi d^3}{6} \qquad 7.12$$

The resultant force opposing motion under the effect of gravity is:

$$F = F_g - F_\beta = (\delta - \rho)\frac{\pi d^3 g}{6} \qquad 7.13$$

By equating the frictional drag expressed in equation 7.8 with the gravitational force causing particle fall in a liquid, equation 7.3, one can easily solve for the velocity of fall of a suspended particle. The result is:

$$3\pi\eta dv = (\delta - \rho)\frac{\pi d^3 g}{6} \qquad 7.14$$

So that the velocity can be obtained from:

$$v = (\delta - \rho)\frac{d^2 g}{18\eta} \qquad 7.15$$

The velocity of a falling particle is proportional to the square of the particle diameter, d, the acceleration of gravity, g, the difference in densities between the particle and the fluid ($\delta - \rho$), and inversely proportional to the viscosity of the liquid η. Table 7.3 gives a brief description of the viscosity of water as a function of temperature.

Table 7.3. Dynamic viscosity of water as a function of temperature

Temperature (°C)	Dynamic Viscosity (Pa·s)
0	0.001 792 1
10	0.001 307 7
20	0.001 005 0
50	0.000 549 4
100	0.000 283 8

Example 7.5. Stokes' Law
Compare the rate of fall in water at 20° C in mm/s of a silt particle at the upper range of particle diameter, and a silt particle at the lower range of particle diameter. What is the rate of fall of a clay particle 1 μ in diameter? Assume the density of the particles is approximately that of silica (2650 kg/m³) while the density of water is 1000 kg/m³.

Solution
According to the international system of classification, the velocities corresponding with the particle diameters are:

$$v = (2650 - 1000)\frac{kg}{m^3} \times \frac{9.81\frac{m}{s^2}}{18 \times 1.0 \times 10^{-3} Pa\cdot s} \times d^2$$

$$= 8.99 \times 10^5 \frac{kg}{m^2 \times \frac{kg \times m}{m^2 \times s^2} \times s^3} d^2 \quad 7.16$$

$$= \frac{8.99 \times 10^5}{m \times s} \times d^2$$

For the upper silt size:

$$v = \frac{8.99 \times 10^5}{m \times s} \times \left(\frac{0.02\ mm \times 10^{-3}\ m}{mm}\right)^2 = 3.6 \times 10^{-4}\frac{m}{s} \quad 7.17$$

For the lower silt size:

$$v = 8.99 \times 10^5 \times \left(0.002 \times 10^{-3}\frac{m}{s}\right)^2 = 3.6 \times 10^{-6}\frac{m}{s} \quad 7.18$$

The clay particle 1 μ in diameter settles at:

$$v = 8.99 \times 10^5 \times \left(1\ \mu \times 10^{-6}\frac{m}{\mu}\right)^2 = 8.99^{-7}\frac{m}{s} \quad 7.19$$

Example 7.5 illustrates that coarse silt settles 100 times faster than silt at the lower range of particle diameter, and 400 times faster than clay. The model developed in equation 7.3 can be applied to the separation of various particle sizes in a solution of soil and water after the sands have been separated and determined by sieving.

Soil Texture

Soil scientists make use of soil textural classifications that are dependent on the relative amounts of sand, silt, and clay in the soil. Three broad groups of soils, sands, loams, and clays, are recognized, with additional variations within the groups. Sandy soils are comprised of soils containing 45 to 50% sand. Within the sandy soils, the textural classifications include sand, loamy sand, sandy clay loam, and sandy clay.

Soils with less than 50% sand and 20% clay, or 45% sand and 28% clay, comprise the group of loam soils and include silt, silt loam, and loam.

Any soils containing over 28% clay have the term clay in their name. These include silty clay loam, clay loam, silty clay, clay, sandy clay loam, and sandy clay. The physical properties of the soil that influence plant growth, especially those associated with water, are closely related to the textural classification grouping, especially to the proportion of silts and clay present.

After a mechanical analysis has been performed on a soil to determine the mass percentages of sand, silt, and clay present, it is easy to determine the proper textural group to which a soil belongs by making use of a textural diagram (or textural triangle) illustrated in figure 7.1. In the textural triangle, the percent of sand silt and clay at a point within the triangle must total 100%. Thus, point A represents a soil with 15% clay, 65% sand, and 20% silt and would be classified as a sandy loam soil.

Example 7.6. Soil texture and textural grouping
The mechanical analysis of a particular soil indicated 33% clay, 33% silt, and 34% sand were present. To what textural group should this soil be assigned?

Solution
From figure 7.1, point B designates a soil with the indicated proportions. It is a clay loam.

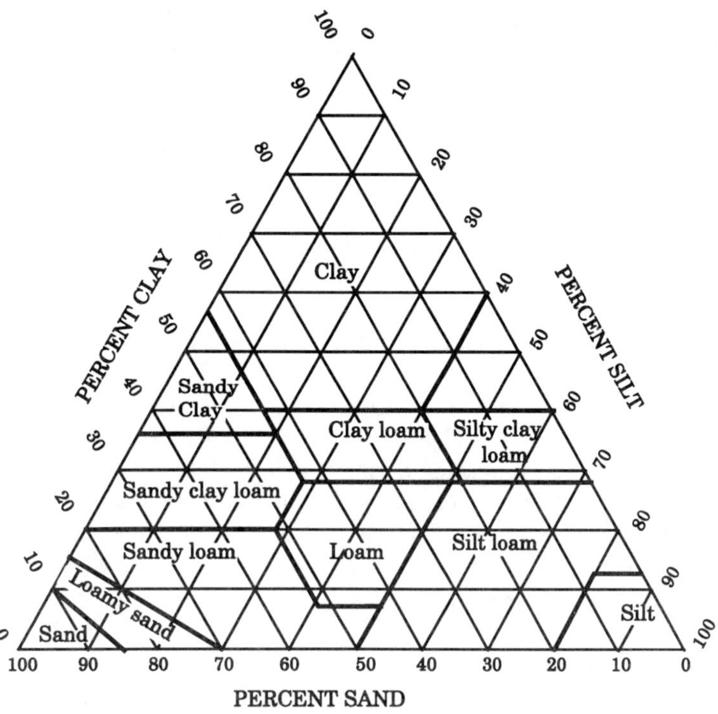

Figure 7.1. The triangle shows the textural designations accompanying the percents of sand, silt, and clay in a soil sample.

Physical Nature of Soil Particles

Soil is seldom composed of spheres. Only the sand fraction of a soil is spherical and only for particles that have undergone weathering by mechanical abrasion. This might occur by prolonged contact with running water.

Silt and clay particles, despite their small size, are not spherical. Silt particles are fragmental in shape and might be thought of as micro-sand-particles. Silt may be sticky when wet, but this is due to the presence of coatings of clay that often cover the silt, rather than to the silt itself. On the other hand, clay is plate-like. Due to its size which places it in the range of colloidal materials, as well as to its chemical composition, clay is extremely important to the soil. The percent and type of clay present in a soil will affect its water-holding capacity, its ability to hold and release minerals necessary for plant growth, and its structure. Clay promotes the coherence of individual particles into larger structural units that, in turn, promote diffusion of soil air and the movement of soil water.

In the field, determinations of sand, silt or clay soils are often made by touch alone. Soils that are predominantly sandy feel gritty when rubbed between thumb and forefinger. Silty soils, when dry, feel like flour or talcum powder. A better test occurs when the soil is slightly wet. Wet soil containing clay, when rolled between the palms into a rope about 3 mm in diameter will hold together. Silt soils, on the other hand, will separate. With increasing clay, the rope can be bent into a 25 mm diameter curve without breaking.

The reason that chemical, structural, and water-holding capacities or soils are dependent on the amount of clay present is due in large part to the chemical structure of clays that are complex crystals composed of silicon, aluminum, potassium, oxygen, and, depending on the degree of weathering, water. Adsorption of cations is possible on the outer faces of the clay particle. This is due to negative charges that attract the positively charged cations. The negative particles exist due to the presence of electrons on the edges of the particles or to negative charges that result from the exposure of incompletely bonded oxygen atoms on the edges of the particles. The ability of a particle to absorb or release cations is characteristic of colloidal particles and is identified as the property of cation exchange.

Several types of clays exist. The most common and important clays are kaolinite, montmorillonite, and illite. Kaolinite is composed of alternating layers of alumina and silica held together by oxygen atoms shared in common bonds by both the silica and alumina. The alumina and silica are in a 1:1 ratio and form a nonexpanding crystal lattice.

Illite is a 2:1 type lattice clay, except that about 15% of the silica atoms have been replaced by aluminum. Potassium ions are present between the layers. This has the effect of stabilizing the 2:1 lattice structure so little water can find its way between the lattice layers. Thus, illite is stable, and is a non-expanding clay. The cation exchange capacity of illite clays is two to three times greater than that of kaolinite.

In montmorillonite clays, two sheets of silica atoms with a sheet of aluminum atoms in between are bonded by mutually shared oxygen atoms. The oxygen-to-oxygen bond and is weak. The result is that water molecules may enter between the O–O bonds. Therefore, the lattice structure of montmorillonite is therefore bellows-like. Montmorillonite clays with their 2:1 type lattice will absorb considerable water and expand when wetted. Also, montmorillonite clays are noted for their swelling and shrinking properties. Finally, due to the nature of the lattice structure, montmorillonite possesses 10 to 15 times the cation-exchange capacity of kalonite clays.

Although other types of clay exist, kaolinite, montmorillonite, and illite, are the main clays found in temperate humid regions.

A constituent of soils unmentioned up to now is the organic matter (O.M.). Organic matter is important in a soil because usually it is present as a colloid and possesses properties of cation exchange. Organic matter adds to a soil's water-holding capacity, as well as furnishing a cementing action so soil structure is improved. Mineral soils rarely contain more than 5% O.M. However, there is a group of soils formed primarily from decomposing organic material. These soils are termed peat or muck soils, and contain 60 to 90% organic matter. Such soils are usually localized in nature and have properties that differ significantly from mineral soils. Organic soils will be discussed briefly later in this chapter along with nutrient availability.

Soil Structure and Soil Characterization

In addition to the physical properties of pore space and particle size, another important property, namely, soil structure, deserves attention. Due to the presence of colloidal particles and natural cementing agents exuded by plant root hairs, soil bacteria, earthworms, etc., individual soil particles may adhere to each other and form larger structural units. Under the influence of freezing and thawing, as well as the penetration and expansion caused by growing plant roots, the structural units are broken into smaller units called aggregates. A soil made up of the smaller units is a well-aggregated soil and possesses good aeration properties, has excellent permeability to water, and is favorable to plant growth.

As a result of the processes of weathering, the upper several meters of the earth's surface form distinct layers or horizons. Each of the horizons comprising the profile has its own characteristics by which it is identified. Soil scientists refer to these horizons in classifying a soil and characterizing it for plant growth.

The uppermost layer, commonly called the plow layer, is the A horizon and extends through the plow depth about 0.2 m (8 in.). Occasionally, deeper A horizons are found. The A horizon is characteristically high in organic matter, and, especially under sod vegetation, is usually well aggregated due to the presence of colloidal organic matter and cementing agents resulting from bacterial activity and root action. The well-aggregated structure in the A horizon promotes water movement. The water acts as a solvent carrying minerals and colloidal material to locations deeper in the soil profile. As a direct result of this "leaching" action, colloidal clay is often removed from the plow layer and carried downward into the second distinct layer, the B horizon.

Example 7.7. Bulk density
A typical loam soil may have a bulk density (in situ mass of soil per unit volume) of approximately 1300 kg/m³. Compute the total mass of soil in 1 ha of the plow layer (0.2 m deep)? What is the weight of the mass in metric tons?

Solution

$$\text{mass} = 1300\,\frac{\text{kg}}{\text{m}^3} \times 10\,000\,\frac{\text{m}^2}{\text{ha}} \times 0.2\text{ m} = 2.6 \times 10^6\,\frac{\text{kg}}{\text{ha}} = 2.6 \times 10^3\,\frac{\text{t}}{\text{ha}} \qquad 7.20$$

Example 7.8. Soil voids under field conditions
If the bulk density of the soil in example 7.7 is 1300 kg/m³ and the particle density is 2650 kg/m³, what is the volume of voids in the plow layer? What percent of the plow layer is void space?

Solution
The total volume of the plow layer in 1 ha is:

$$10\,000\,\frac{\text{m}^2}{\text{ha}} \times 0.2\text{ m} = 2000\,\frac{\text{m}^3}{\text{ha}} \qquad 7.21$$

The volume of solids present is:

$$\frac{2.6 \times 10^6\,\dfrac{\text{kg}}{\text{ha}}}{\left(2650\,\dfrac{\text{kg}}{\text{m}^3}\right)_{\text{solids}}} = 980\,\frac{\text{m}^3_{\text{solids}}}{\text{ha}} \qquad 7.22$$

Therefore, the percent of voids is:

$$\frac{2000\,\dfrac{\text{m}^3}{\text{ha}} - 980\,\dfrac{\text{m}^3_{\text{solids}}}{\text{ha}}}{2000\,\dfrac{\text{m}^3}{\text{ha}}} \times 100\% = 51\% \qquad 7.23$$

Example 7.9. Surface area of soil particles
Assume the plow layer in example 7.7 is 30% clay with a surface area of 5×10^5 m²/kg. What is the total area in hectares available for surface reactions such as water adsorption, cation exchange, etc.?

Solution
The surface area of clay per hectare of plow layer is:

$$980 \frac{m^3_{solids}}{ha} \times \frac{30\%\ clay}{100\%} \times 2650 \frac{kg}{m^3} = 7.79 \times 10^5 \frac{kg_{clay}}{ha}$$

and

$$7.79 \times 10^5 \frac{kg_{clay}}{ha} \times 5\times 10^5 \frac{m^2}{kg} = 3.9 \times 10^{11} \frac{m^2_{clay}}{ha} \qquad 7.24$$

or

$$\frac{3.9 \times 10^{11} \frac{m^2_{clay}}{ha}}{10\,000 \frac{m^2}{ha}} = 3.9 \times 10^7\ ha_{clay}$$

Examples 7.7 through 7.9 illustrate the role of the clay colloidal fraction in an ideal soil (50% voids). The amount of surface area available due to the presence of 30% clay presents a startling picture of the effect of colloidal properties in a soil, since each clay particle may have many negative charges that must be filled by positively charged cations.

The second layer, the B horizon in a typical soil profile, is a zone of accumulation. Clays leached out of the A horizon accumulate in the B horizon. Minerals, particularly calcium, also accumulate here. Root activity although present in the B horizon is developed to a lesser degree in this horizon than in the plow layer. Inhibition of root development may be further influenced by soil compaction resulting from tillage operations. In particular, the formation of a soil layer of high bulk density known as the "plow pan" often inhibits the penetration of roots into lower horizons. Thus, even though minerals pertinent to plant growth may be present in the lower horizons, plants may not penetrate to a sufficient depth to obtain the necessary nutrients available at the lower depths. The soybean plants in figure 7.2 are the same age. The plant on the left has a root system which developed over a highly compacted soil layer. Contrast this development with that of the plant on the right in figure 7.2 in which the soil structure did not impede the root development.

Example 7.10. Soil voids and bulk density
In the B horizon of certain soils, accumulations of clay with chemical cementing agents as iron oxides create soil bulk densities often attaining values of 1700 to 1800 kg/m³. What is the volume of voids in a soil with a bulk density of 1800 kg/m³? What is the percent pore space? Assume the individual soil particles have a bulk density of 2650 kg/m³.

Solution
In 1 m³ of soil the volume actually occupied by particles is:

$$V = \frac{1800\ kg}{2650 \frac{kg}{m^3}} = 0.68\ m^3 \qquad 7.25$$

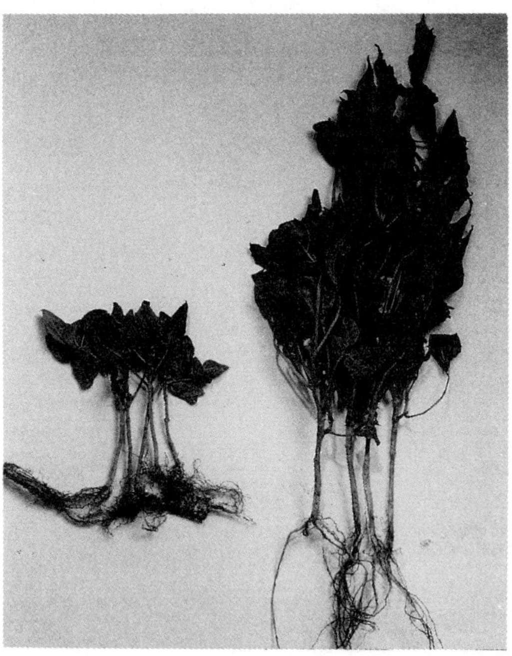

Figure 7.2. Two soybean plants of the same age have developed in soils containing minerals and water adequate for plant growth. However, the plant on the left developed over a highly compacted layer that the roots could not penetrate. The roots of the plant on the right easily penetrated the soil voids to obtain water and minerals essential to growth and development. (Figure courtesy of H. W. Foth, Michigan State University)

Therefore, the volume of voids is:

$$1m^3 - 0.68m^3 = 0.32m^3 \qquad 7.26$$

Therefore, the percent void space is:

$$\%\ \text{void} = \frac{0.32m^3}{1m^3} \times 100\% = 32\% \qquad 7.27$$

In a soil with a bulk density of 1350 kg/m^3 in which one half of the voids are filled with water, about 0.25 m^3 of voids space would remain for oxygen diffusion. However, if the soil had a density of 1800 kg/m^3 as shown in example 7.10, the same amount of water would reduce the void space for gaseous diffusion to 0.32 m^3 - 0.25 m^3 or 0.07 m^3 of void space.

Example 7.11. Soil porosity
If the voids in the B horizon with a bulk density of 1800 kg/m^3 were spherical, what would the average volume of a void be if the number of voids equaled the number of soil particles and the soil consisted of 60% clay, 30% silt, and 10% sand? Assume the average particle density is 2650 kg/km^3.

Solution
In 1 m³ of soil there would be:

$$0.6 \times 1800\,\text{kg} = 1080\,\text{kg clay}$$

$$0.3 \times 1800\,\text{kg} = 540\,\text{kg silt}$$

$$0.1 \times 1800\,\text{kg} = 180\,\text{kg sand}$$

If it is assumed that the average particle diameters for sand, silt, and clay are 0.1 mm, 0.01 mm, and 0.001 mm, respectively, then each particle will have a volume of:

$$7.78 \times 10^{17} + 3.89 \times 10^{14} + 1.30 \times 10^{11} = 7.78 \times 10^{17}\text{particles}$$

The number of particles of each texture will be:

$$\text{sand}: \frac{\left(0.1\ \text{mm} \times 10^{-3}\,\frac{\text{mm}}{\text{m}}\right)^3 \times \pi}{6} = 5.24 \times 10^{-13}\text{m}^3_{\text{sand}}$$

$$\text{silt}: \frac{(0.01 \times 10^{-3})^3 \times \pi}{6} = 5.24 \times 10^{-16}\text{m}^3_{\text{silt}} \qquad 7.28$$

$$\text{clay}: \frac{(0.001 \times 10^{-3}) \times \pi}{6} = 5.24 \times 10^{-19}\text{m}^3_{\text{clay}}$$

The total number of particles is then:

$$\frac{180\text{kg}_{\text{sand}}}{\dfrac{5.24 \times 10^{-13}\text{m}^3_{\text{sand}}}{\text{particle}} \times 2650\,\dfrac{\text{kg}}{\text{m}^3}} = 1.3 \times 10^{11}\text{particles}_{\text{sand}}$$

$$\frac{540\text{kg}_{\text{silt}}}{5.24 \times 10^{-16} \times 2650} = 3.89 \times 10^{14}\text{particles}_{\text{silt}} \qquad 7.29$$

$$\frac{1089\text{kg}_{\text{clay}}}{5.24 \times 10^{-19} \times 2650} = 7.78 \times 10^{17}\text{particle}_{\text{clay}}$$

The number of particles is the same as the number of voids, so:

$$\frac{0.32\text{m}^3}{7.78 \times 10^{17}\,\text{particles}} = 4.11 \times 10^{-19}\,\frac{\text{m}^3}{\text{particle}}$$

So the diameter of an equivalent spherical void is:

$$\left(\frac{4.11 \times 10^{-19}\text{m}^3 \times 6}{\pi}\right)^{\frac{1}{3}} = 9.23 \times 10^{-7}\text{ m} = 0.92\mu$$

Example 7.11 illustrates why root penetration into soils of high bulk density is difficult. Pore size in a soil containing a high percentage of silt and clay with a high bulk density is, on the average, extremely small. Thus, very few voids are present of large enough size to allow root penetration. Pore space for aeration is limited and water movement in the soil will be very slow.

The final soil layer in a typical soil is the C horizon, comprised of the parent material from which soil formation is occurring. The chemical composition of the regolith (the parent material of soil formation) is similar to that of the rocky mantle of the earth given in table 7.1. Little root activity occurs in the C horizon, and we will not devote much attention to this horizon. In some locations the B horizon is thin and the C horizon occurs near the surface. In some areas of the world such as the black soils of Ukraine, the B horizons extend tens of meters before the C horizon is encountered.

Soil Minerals and Plant Growth

Fertility

The nature, availability, and percent of minerals in the soil determine a soil's fertility. Fertility is distinguished from productivity. Fertility is the property that enables a soil to provide compounds in proper amounts and balance for plant growth, and this can only occur if other environmental factors are favorable. Productivity is the capability of a soil to produce a series of plants under a given management system. Often, the responsibility of the engineer is to take a fertile soil and modify the environmental factors to ensure productivity by irrigation or other means. In this section we largely restrict our attention to soil fertility and understanding those properties of the soil that contribute to soil fertility.

Soil Minerals

The chemical constituents of the soil that are vital to plant growth include carbon, hydrogen, and oxygen, and largely are obtained from the atmosphere and used in the synthesis of carbohydrates and fats in the plant. These are supplied to the plant as CO_2, O_2, and H_2O and are used by the plant in elemental form after chemical breakdown within the plant cells.

A second group of chemical constituents includes nitrogen, phosphorus, potassium, calcium, magnesium, and sulfur. This group of elements are the macro nutrients and are prefixed by *macro* because large quantities are required and are used by the growing plant.

Nitrogen is an essential constituent of proteins and enzymes in plants, and each chlorophyll molecule contains four nitrogen atoms. Thus, the absence of nitrogen affects chlorophyll formation, and, consequently, photosynthetic activity.

Phosphorus is an essential component of the energy-transfer process enabling a plant to use energy quanta from sunlight in the food production process within the plant. It is also found as a constituent of proteins and fats (lipids).

Potassium is important in the metabolic processes of the cell. Recently, it has been found that potassium is vital to the stomatal control in the green plant.

In addition to its role in controlling soil Ph, calcium is an essential part of the cell wall and middle lamella as calcium pectate. It is also essential to the formation of new cells in the meristematic regions of the plant.

Magnesium is found in the enzyme system of plants. Sulfur forms an integral part of the amino acids from which proteins are synthesized. Each chlorophyll molecule contains a magnesium atom, and, therefore, magnesium is important to food production in the plant.

A third group of elements is also needed for plant development. These are the micro-elements Fe, Mn, Bo, Cu, Zn, Mb, and Cl. Their role in plant growth is varied. Iron and manganese are necessary for chlorophyll synthesis and in the function of enzyme systems. Boron is used in the meristematic portions of the plant, and an absence of boron may cause meristematic activity to cease. Copper and zinc are thought to be a constituents of the enzyme systems of green plants. Molybdenum enters into the reduction of nitrates, thus, absence of this element

Table 7.4. Macronutrients in humid and arid soils

	Humid Soils		Arid Soils	
Nutrient	Typical Percents	Plow Layer (kg/ha)	Typical Percents	Plow Layer (kg/ha)
N	0.15	567	0.12	405
P	0.04	146	0.07	259
K	1.70	6275	2.00	7368
Ca	0.40	1457	1.00	3644
Mg	0.30	1093	0.60	2227
S	0.04	146	0.08	296
O.M	4.00	12 146	3.25	11 943

inhibits protein synthesis. Chlorine, a minor element whose role is very poorly understood, is thought to be a regulator of osmosis and enters the overall cation balance of the plant.

Several elements cannot be classified properly as microelements, though their presence in the plant when consumed by animal life is often beneficial to the consumer. Iodine, though not apparently necessary to the function of the green plant, is known to reduce incidence of goiter among humans. Fluorine in small quantities, leads to the development of strong teeth and bones. A fairly recent finding, sodium is essential to growth and the development of the plant sodium may partially replace the potassium requirement as well.

Nutrient Availability

It is one thing to realize that certain nutrients are essential to maintain fertility and another to manage a soil-plant ecosystem to ensure a high level of productivity. Although it is not possible to go into extensive detail on this subject here, certain essential facts must be pointed out and understood, so that engineering implementation can be carried out in cooperation with the recommendations of soil and plant scientists.

A plant seldom lacks carbon, hydrogen, or oxygen, since these elements are abundantly present in the atmosphere as CO_2 and H_2O. Of course, a deficiency of water that leads to water stress can be developed by any growing plant when the transpirational demands exceed the supply available through the plant root system. The engineering remedy to this is accomplished by irrigation—an engineering practice involving the application of water in amounts and at times when needed by the growing plant.

Table 7.4 gives typical amounts of macronutrients and organic matter found in humid and arid region soils of the temperate regions. Table 7.5 presents the ionic form in which the nutrient is available to the plant. Both the nutrient and the ionic form of macronutrients are important for the engineer to understand the role of soil colloids and soil biotic life in soil fertility.

Table 7.5. Ionic forms of nutrients available to plants

Nutrient	Ionic Form	
N	NO_3^-	NH_4^+
P	N_2PO^4	$HPO_4^=$
K	K^+	—
Ca	CA^{++}	—
Mg	Mg^{++}	—
S	$SO_4^=$	—

Source: Buckman and Brady (1969)

As was pointed out in a previous example, a high moisture content in the soil can cause soil pore spaces that would otherwise be available for the diffusion of gases throughout the soil to be filled with water. Thus, excessive water reduces oxygen availability to the growing region of the root zone where metabolism rates are highest and oxygen demand is greatest. This situation can be remedied by drainage, a subject that again occupies a course of study in the general area of soil and water conservation. We will point out certain aspects of drainage as they interact with irrigation and affect the balance of soil minerals essential to both soil fertility and potential productivity.

Nitrogen Availability

The first element in tables 7.4 and 7.5 is of particular importance. Although 80% of the earth's atmosphere is nitrogen, a relatively small percentage of N appears in the soil in a form available for plant growth, namely the nitrate ion (NO_3^-) and the ammonium ion (NH_4^+) The processes which lead to the formation of available ionic forms of nitrogen are complex, and briefly will be sketched out here so that the reader may more fully appreciate the effect of environmental modification environment on the presence of available forms of N for plant growth. The process of creating available N usually involves two steps. Two sources of N exist in the soil. One is N available as a component of organic matter in the form of proteins and elements of chlorophyll molecules. The second involves N present in the soil air. Nitrogen in these forms must undergo a complex set of biochemical processes involving various forms of soil bacteria before it is changed into an available form as ammonium (NH_4^+). This part of the complex N cycle is known as mineralization. When mineralization is complete, a second group of soil organisms are active in changing the ammonium with the addition of O_2 into (NO_2^-) and, by a process also requiring O_2 for its completion, into the form NO_3^- which is available for plant use (see table 5). The latter portion of the cycle is known as nitrification.

Two points are important here. First certain soil bacteria must be present before either mineralization or nitrification can occur. Second, ammonium (NH_4) and nitrite (NO_2^-), toxic to plants, must be converted to nitrate (NO_3^-) which is highly soluble and readily fixed by soil colloids. As NO_3^-, nitrogen is available for plant use. However, the conversion process yielding NO_3^- is a process requiring free oxygen. Thus, a deficiency of oxygen in the soil, such as can be caused by poor drainage, will cause the mineralization and nitrification processes to be arrested. This will result in the loss of potentially available nitrogen as well as a nitrogen deficiency due because the nitrogen will be present in a form that is unavailable for plant use. Finally, because nitrite is highly soluble, it is readily leached out of the soil profile and becomes a contaminant in the groundwater supply.

An important concept in soil nitrogen availability is the carbon-to-nitrogen ratio. The bacteria that are responsible for fixing nitrogen require energy for their existence. This energy is derived from the oxidation of carbonaceous ions in the soil. If too much carbon is present, nitrogen as nitrate disappears until the C/N ratio is reduced by the evolution of gaseous CO_2. An ideal C/N ratio for most soils is in the range of 10 or 12 to 1. In plants the C/N ratio may be as high as 20 to 30 to 1 for legumes or even 90 to 1 for straw.

Obviously arbitrarily plowing under organic matter with a high C/N ratio may result in a deficiency of N to the plant unless nitrogen is added to reduce the C/N ratio to a value beneficial to nitrate formation.

Atmospheric N is also made available to certain plants called legumes as a result of the action of bacteria that live in a symbiotic relationship with the legume host plant in nodules on the plant roots. An adequate population of these bacteria can make a legume self sufficient, since the atmospheric supply is virtually inexhaustible. It is a recommended practice to inoculate the seed of legumes such as alfalfa with nitrogen fixing bacteria before planting to ensure there is an adequate number of N-fixing bacteria present near the seed. The Nitrogen cycle is given in schematic form in figure 7.3.

Soil Acidity

It is difficult to discuss the availability of the remaining macro and micro nutrients in the soil without being aware of the influence of the pH, a measure of the acidity of the soil. pH is defined as the negative logarithm of the molar concentration per liter of hydrogen ions in a solution.

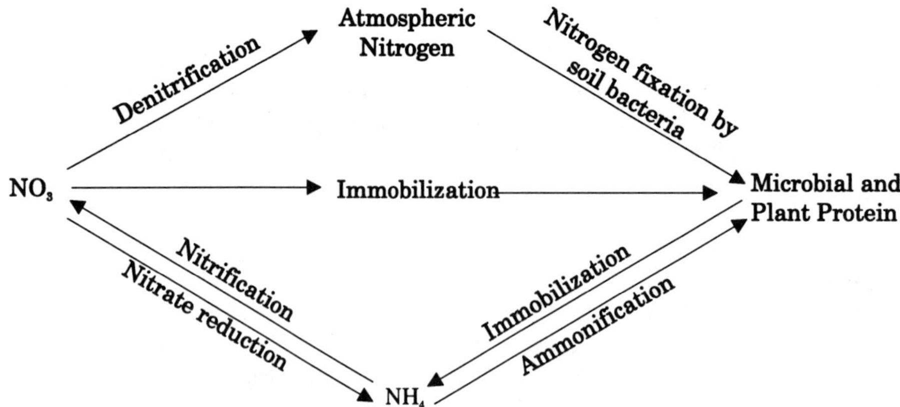

Figure 7.3. The nitrogen cycle. An excess of carbon atoms will cause the production of NO_3 to cease and will result in a nitrogen deficiency to the green plant. Poor subsurface drainage will halt the cycle at the formation of nitrite which is highly toxic to plants. In addition, nitrite is easily leached out of the soil.

Example 7.12. Hydrogen ion concentration and pH
Pure water contains H_2O molecules plus a certain number of molecules of water that have dissociated into H^+ and OH^- ions. The concentration of H^+ ions in pure water is 10^7 mol/L. What is the pH of pure water?

Solution
pH = –log(H^+ ion concentration in mol/L)

$$pH = \log \text{ of concentration} = -\log\left(10^{-7}\right) = -(-7) = 7 \qquad 7.30$$

A solution with a pH of 7 is said to be neutral. If the pH is less than 7, the solution is acidic; if greater than 7, the solution is basic.

Example 7.13. Hydrogen ion concentration and pH
A solution was found to have a pH of 5.3. What is the molar concentration of H^+ ions in the solution? How many ions/L does this represent?

Solution
pH = –log[H^+] where [H^+] denotes concentration in mol/L, so:

$$\text{concentration} = 10^{-5.3} = 5.01 \times 10^{-6}\frac{\text{mol}}{\text{L}} \qquad 7.31$$

Since 1 mole is 6.023×10^{23} ions:

$$5.01 \times 10^{-6}\frac{\text{mol}}{\text{L}} \times 6.023 \times 10^{23}\frac{\text{ion}}{\text{mol}} = 3.02 \times 10^{18}\frac{\text{ions}}{\text{L}} \qquad 7.32$$

Note: A neutral solution would contain:

$$1 \times 10^{-7}\frac{mol}{L} \times 6.023 \times 10^{23}\frac{ions}{mol} = 6.023 \times 10^{16} H^{+}\frac{ions}{L} \quad 7.33$$

As illustrated, an abundance of H^+ ions results in a lowered pH. Now we will examine the relationship between soil pH and the presence of colloidal matter in the soil.

As has been noted, colloidal material in soils (clay, organic matter) contains a large surface-to-volume ratio. Depending on the nature of the colloidal material, much of the surface is negatively charged. The negative charges readily attract positively charged ions (such as H^+) that attach themselves to the surfaces. If the ions are H^+ ions, they represent a potential source of H^+ ions for the soil solution. This can be used to increase the concentration of H^+ ions in the solution as other ions replace or exchange the H^+ ions on the colloids. Since H^+ ions are readily replaced by alkali metal ions, a soil colloidal solution containing many number H^+ ions is potentially an acidic soil that can tie up other ionic forms of plant nutrients before they can be absorbed by the plant.

Figure 7.4 illustrates the capacity of a colloidal particle or micelle to enter an exchange reaction (in this case with potassium chloride). It is obvious from figure 7.4 that colloidal particles can attract and hold ionic forms of nutrients present in the soil that are essential to plant growth. However, certain ions are more easily exchanged on the soil colloid than other. Some of the more common ions in descending order of replacing power are: H, Sr, Ba, Ca, Mg, K, NH, Na, Li.

Cation Exchange Capacity

The cation-exchange capacity of soils (CEC) is the number of cation absorption sites per unit weight of soil. It is expressed as the number of H^+ ions that could be attracted to 100 g of oven dry soil. A CEC is also expressed as a milliequivalent equal to 6.02×10^{23} ions/mol $\times$ 0.001. Thus, a soil whose cation exchange capacity is 1 milliequivalent (1 meq) would contain 6.02×10^{20} sites to absorb positively charged ions.

Example 7.14. Cation exchange capacity
A typical humid soil has a CEC of 50 meq. If the molecular weight of hydrogen is 1 g/mol and the soil is H^+ saturated, how many grams of hydrogen are there in each gram of oven-dried soil?

Solution
1 meq = 6.02×10^{20} sites, so:

$$50\text{ meq} = 50\text{ meq} \times 6.02 \times 10^{20}\frac{\text{sites}}{\text{meq}} \times \frac{1\ H^{+}\text{ion}}{\text{site}} \times \frac{1\text{ mol}}{6.02 \times 10^{23}\text{ ions}} \times \frac{1\text{ g}}{\text{mol}} \quad 7.34$$

$$= 0.05\ \frac{g_{H+}}{100\ g_{soil}}$$

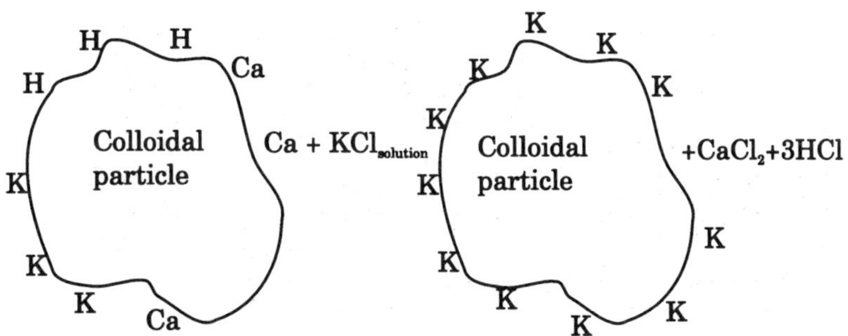

Figure 7.4. The exchange of ions which occurs on the surfaces of soil colloids may control the availability of these ions to the growing green plant.

If all the H^+ ions in the plow layer of an H^+-saturated soil were replaced by Ca^{++}, how much Ca^{++} would be required? Assume the molecular weight of Ca is 40 g/mol.

Example 7.15. Lime and soil acidity

Solution

From example 7.7 the total mass of soil in the plow layer is about 22.2×10^5 kg/ha. The total mass of H^+ would be:

$$0.05 \frac{g_{H+}}{100\, g_{soil}} \times 22.2 \times 10^5 \frac{kg_{soil}}{ha} \times 10^3 \frac{g}{kg} = 1.11 \times 10^6 g_{H+} \tag{7.35}$$

Or, expressed in moles:

$$\frac{1.11 \times 10^6}{1 \frac{g}{mol}} = 1.11 \times 10^6 mol \tag{7.36}$$

Ca^{++} has 2 positive charges, therefore, each ion of Ca^{++} would replace 2 ions of H^+ and the mass of Ca^{++} required would be:

$$1.11 \times 10^6 mol_{H+} \times 1 \frac{Ca^{++}}{2H^+} \times 40 \frac{g}{mol} = 2.2 \times 10^7 g_{Ca++} = 2.2 \times 10^4 kg \tag{7.37}$$

A mixture of ions is needed in soil and it would be foolish to attempt to saturate the plow layer of the soil given in example 7.13 and 7.15 with Ca^{++} ions. However, the magnitude of the answer in example 7.15 gives an idea of the cation exchange capacity (CEC) of a soil and the magnitudes of the masses of elements fixed by the soil colloids.

Cation exchange by soil colloids is important in the availability of phosphorus to plants. First, plants require quantities of phosphorus far more than that available in soils. The phosphorus requirement of a crop must be satisfied by addition of fertilizers. Furthermore, in acid soils, phosphorus enters into chemical reactions with iron or

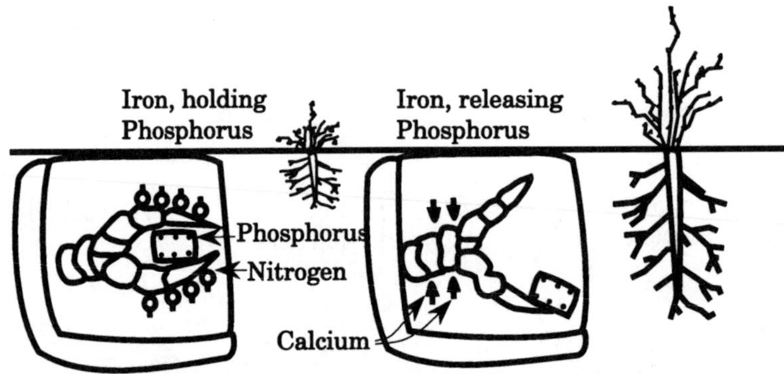

Figure 7.5. The availability of phosphorus to the growing green plant may be controlled by soil pH. Replacing a hydrogen ion with a calcium ion may release the phosphorus and make it available to the plant.

aluminum and cannot be obtained by the plant. Calcium (lime) added to a soil replaces the hydrogen ion on the soil colloids and makes the soil less acidic. Under these conditions, the iron-phosphorus or aluminum-phosphorus reaction is reversed and the phosphorus in the soil is made more available to the crop. This is depicted in figure 7.5 below. Potassium is abundant in most soil. However, in humid areas on acid soils, potassium may be leached out of the soil profile. Removal of potassium by plants varies widely and may range from 44.5 kg/ha for spinach to as much as 520 kg/ha for sugar cane. The average removal due to cropping is probably the lower of the two figures.

Calcium is necessary for soil fertility because it has the capacity to neutralize acid soils by replacing the excess H^+ ions which are potentially available to cause a low pH (acid soil) with Ca^{++} ions. This results in a more neutral soil in which certain elements such as phosphorus are more readily available to the plant. However, a pH of 7 (neutral) is not always desirable. Furthermore, the method used to determine soil pH and the nature of the calcium-containing material is very important as regards the final value attained. Since most of the calcium lost by the soil occurs as a result of leaching, it is wise to have a soil test performed periodically to be sure that suitable conditions for growing the desired crop exist.

Magnesium is made available to the plant when dolomitic limestone is applied as a fertilizer. A second and more rapid method of correcting a magnesium deficiency involves spraying the plant with a solution of magnesium sulfate. This treatment may be carried out during irrigation if a spray-type irrigation system is used. Dolomitic limestone is slower-acting but it is also the cheapest method of application.

Sulfur-deficient soils, while rare, exist in some areas in the northwestern United States. Sulfur is returned to the soil through the production of sulfur dioxide (SO_2) resulting from the burning of fossil fuels such as coal, gas, and the oil that is returned to the earth, usually by rain water, in the form $SO_4^{=}$. Plants do not possess infinite tolerance for sulfur. Serious toxic effects from SO_2 have been seen near high SO_2 producing operations. Generally the source of sulfur for plants is $SO_4^{=}$.

Micronutrients

The micronutrients are used by plants only in trace amounts. The micronutrients, the percent of the nutrients typically found in surface soils, and the ionic forms in which these minerals are usable by growing plants are given in table 7.6.

Despite the large amount of iron present in the plow layer of a typical soil, it may not be present in the form readily available to plants. This is especially true of alkaline soils or soils that are low in organic matter or sulfur.

Manganese, another trace mineral available in large amounts, may be made unavailable if excessive lime is added to the soil. We can see that the fertility of a soil is dependent not only on having the required nutrients present, but on maintenance of a delicate balance between the various constituents and on the pH of the soil solution. The productivity of a soil is strongly dependent on having all the necessary nutrients present in amounts sufficient to meet plant demand. However, productivity is limited by the least amount of a particular nutrient needed

Table 7.6. Micronutrients, their availability in a typical surface soil, and the ionic form available to the growing plant

Nutrient	Percent in Plow Layer	Plow Layer (kg/ha)	Ionic Form
Mn	0.5	1130	Mn^{++}
Fe	0.3	6670	Fe^{++}
B	0.0008	20	BO_3^{++}
Zn	0.02	45	Zn^{++}
Cu	0.02	45	Cu^{++}
Mo	0.0003	0.8	$MoO_4^{=}$
Cl	0.05	100	Cl^-

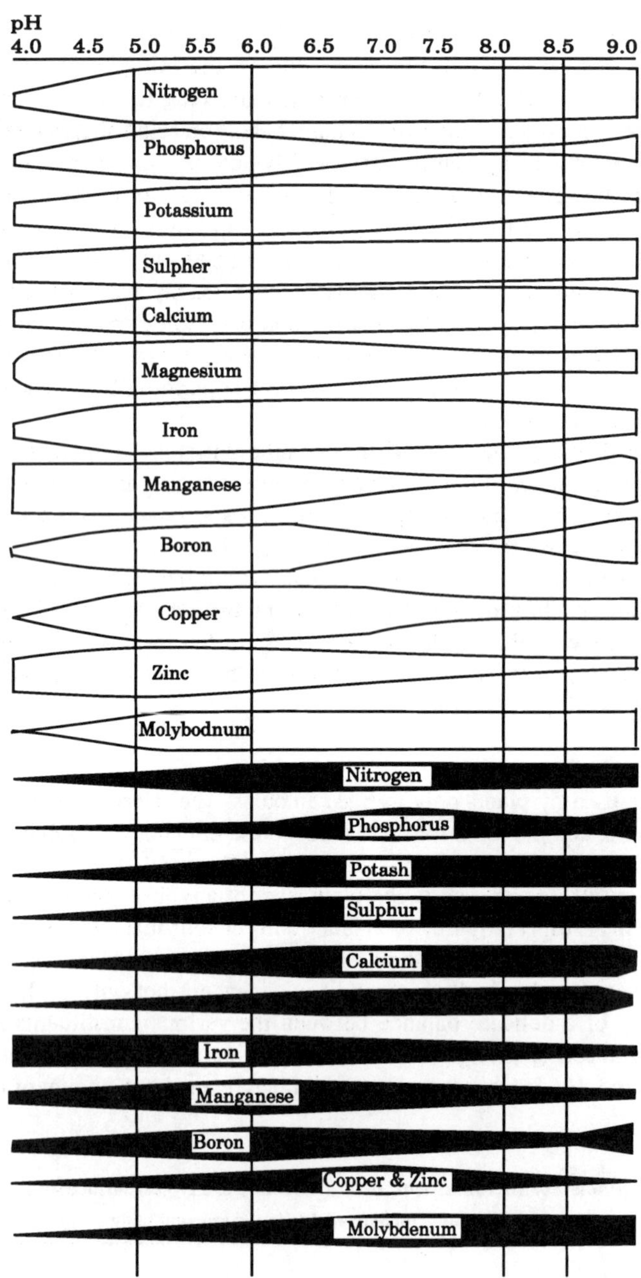

Figure 7.6. A relationship exists between the soil pH and the availability of micronutrients to the growing green plant. The width of the bar indicates the relative availability; the wider the bar, the more available the nutrient.

by a plant. Thus, a soil containing all the necessary elements but one can be no more productive than the availability of the limiting element will allow. A determination of the element necessary to restore fertility to an optimum level is the responsibility of a competent soil scientist because of the delicate interrelationships involved. Often, the addition of calcium, which replaces hydrogen on the soil colloids, will readjust the soil pH so the limiting nutrient becomes available. Figure 7.6 demonstrates schematically the relationship between soil pH and relative nutrient availability for selected elements and for mineral soils, as contrasted with soils high in organic matter (peat or muck soils).

Fertilizer and Fertility Enhancement

To remedy one or more deficiencies in the macronutrients necessary for high soil fertility, soil scientists will often recommend the use of a commercial fertilizer. Commercial fertilizers are prepared from materials rich in the essential elements and are balanced to provide for the relative usage of the essential element by the plant.

Many sources are used to supply N, P, and K, the most common ingredients of commercial fertilizers. Regardless of the source, the relative amounts of N, P, and K are expressed on fertilizer labels as percentages of N, P_2O_5, and K_2O. Efforts are presently under way to change the designation of fertilizers to actual percentages of N, P, and K present in a form readily available to plants.

Example 7.16. Fertilizer identification

A certain fertilizer is listed as 10-20-10. What amounts of N, P_2O_5, and K_2O are there in 36 kg (80 lb) of this fertilizer? What are the actual percentages of N, P, and K present? If the label were to give the values in percents of N, P, and K, what should the label read?

Solution

Percentages of substances present:

$$\text{N: } 10\% \times 36 \text{ kg} = 3.6\text{kg}_N$$

$$P_2O_5\text{: } 20\% \times 36 \text{ kg} = 7.2\text{kg}_{P_2O_5}$$

$$K_2O\text{: } 10\% \times 36 \text{ kg} = 3.6\text{kg}_N$$

$$\text{Filler: } = 21.6 \text{ kg}$$

To find the actual percentages present, we first determine the number of moles of each constituent. The gram molecular weights are:

$$\begin{array}{lll} N = 14 & N & = 14 \text{ g/mol} \\ P = 31 & & \\ O = 16 & P_2O_5 & = 142 \text{ g/mol} \\ K = 39 & K_2O & = 94 \text{ g/mol} \end{array}$$

The mole fractions of N, P, and K are:

$$\text{N: } \frac{14}{14} = 1$$

$$P_2\text{: } \frac{62\frac{g}{mol}}{142\frac{g}{mol}} = 0.44$$

$$K_2\text{: } \frac{78\frac{g}{mol}}{94\frac{g}{mol}} = 0.83$$

The actual percentages of N, P, and K are:

$$3.6kg_N \times 1 = 3.6kg_N$$

$$7.2kg_{P_2O_5} \times 0.44 = 3.17kg_{P_2O_5}$$

$$3.6kg_{K_2O} \times 0.83 = 3.0kg_{K_2O}$$

The percent of N, P, and K is:

$$N: \frac{3.6}{36} \times 100\% = 10\%$$

$$P: \frac{3.17}{36} \times 100\% = 8.8\%$$

$$K: \frac{3.0}{36} \times 100\% = 8.3\%$$

The label should read 10-8.8-8.3.

Example 7.16 demonstrates the need to understand the implications of the labeling method used for fertilizers. The latter form 10-8.8-8.3, more accurately represents the percents of N, P, and K being added to the soil even though some inaccuracy may occur due to N, P, and K dissolving in the soil solution and entering into reactions with other soil elements. This may cause some immobilization of the nutrient, but it still gives a clearer picture of the quantity of elements being added. As it stands today, all fertilizers are required by law to be labeled with the following information:

1. Name or trademark.
2. Analysis of guaranteed chemical composition expressed as elemental N, available phosphoric acid (P_2O_5) and water soluble potash (K_2O).
3. The net weight of the contents.
4. The name and address of the manufacturer.

Micronutrients, when present in the fertilizer, are required by law to be expressed as percent of the elemental available form present.

Engineering Considerations of Fertilizer Application

The method and nature of fertilizer application to soils depends upon both the fertilizer and the crop. In some cases foliar application is possible in the irrigation water. It is more convenient to apply at planting, or, later, to the soil alongside the plant. The latter is termed "side dressing".

In general, phosphorus applied to the soil remains at the point of application due to very rapid reaction with soil colloids. Therefore, to be most advantageous to the plant, fertilizers supplying phosphorus should be placed where the plant roots can intercept the phosphorus and absorb it. In addition, it is more beneficial to isolate a fertilizer containing phosphorous in a band or hilling type of application rather than broadcasting it over the soil colloids as less phosphorus becomes fixed, leaving more available to the growing plant.

Nitrogen, and to a lesser extent potassium, tend to move through the soil in the direction of water movement. Thus, placement immediately above or below the growing plant may prove harmful if water movement is downward in the first case or upward in the second. It is most logical to place K and N for row crops to the side of the seed so it remains within easy reach by the growing roots, yet does not form a concentrated soil solution that could injure the crop.

For many crops, the need for phosphorus occurs early in the plants' growth cycle. Nitrogen is more beneficial to the plant later in its development. In this case nitrogen can be applied separately as a side dressing to the growing plant. Table 7.7 can be used to guide placement once the required fertilizer application rates have been determined by soil testing. Note that calcium is not mentioned. Usually this requirement is supplied by broadcasting ground limestone to the soil surface and then mixing it with the soil. Limestone is rapidly immobilized by soil colloids and as a result does not leach out of the soil as does potassium and nitrogen.

Certain fertilizers can be applied with the irrigation water. Two conditions govern the application of fertilizers via irrigation systems. First, the fertilizer must be completely water-soluble. Second, if the application is performed via a sprinkler irrigation system, it is necessary to follow fertilizer application with an application of pure water to prevent high concentrations of fertilizer remaining on the crop leaves from damaging the leaf. Table 7.8 gives a listing that can be used to guide fertilizer application by irrigation.

Irrigation

In the design of irrigation systems the engineer must be extremely conscious of two things: (1) the water requirement and the rate at which water can be applied to a soil; (2) the chemical nature of the water being applied and the condition of the soil to which application is being made. The second is especially important in preventing salt build-up and destruction of the soil properties beneficial to plant growth. Under ordinary conditions, soils in humid areas contain 230 to 450 kg soluble salts per acre of furrow slice. Under unfavorable conditions such as are found in arid regions this value may rise tenfold to 11 000 kg/ha.

Several reasons exist for this situation. First, humid-area soils receive sufficient rainfall in addition to supplemental irrigation that excess salts present in the upper horizons of the soil profile are leached out of the root zone. In arid regions, on the other hand, excess rainfall seldom occurs. Most of the water needed for growth comes from the irrigation system. Often, the irrigation water contains quantities of soluble salts that migrate through the soil with the irrigation water. However, while the water is removed through evaporation and transpiration, the salts remain in the root zone. This results in a build-up of salt and causes saline or sodic soil. A soil is saline if it contains an excess of soluble salts, such as chlorides, sulfates, nitrates and bicarbonates. The pH of a saline soil is usually 8.5 or less and the exchangeable sodium content is less than 15%. If the soluble mineral is sodium in a concentration more than 15% of the exchange capacity of the soil, the pH rises as high as 10 and the soil becomes sodic.

Good irrigation planning can prevent either of these situations from occurring and can even reverse the trend toward build-up of an undesirable condition. In arid regions, a prerequisite of an irrigation system is a well-designed underground drainage system that can allow the excess soluble minerals to leach from the profile. Occasional heavy applications of irrigation water then will cause enough leaching that soluble salts and minerals will be carried through the soil profile.

For sodic soils containing an excess of sodium, in addition to a build-up of sodium, degradation of soil structure will occur. This takes place because sodium-saturated soil colloids lose their attraction for each other and deflocculation occurs resulting in destruction of the soil aggregates. Soils that lack aggregation have low infiltration capacity, very low hydraulic conductivity, and correspondingly low porosity. Reclamation of sodic soils to restore productivity can be hastened by applying calcium sulfate in the irrigation water, and by planting deep-rooting crops whose root systems enhance soil aggregation and encourage percolation of water through the soil profile. Prevention of either a saline or a sodic condition from occurring is the best management practice. This can be accomplished by ensuring that an adequate drainage system exists before irrigation is begun. The drainage system should be so designed that the free water surface at no time rises closer than to within 1.5 m of the soil surface. A rise of no more than 2.4 to 3 m is preferable to ensure adequate aeration for growing crops.

Table 7.7. Recommended methods of fertilizer application to selected crops

Crop	Broadcast Application	Direct Placement
Celery	1/2 to 2/3 total application.	Hill or band 1/2 to 1/3, 50 mm to the side and 50 mm below the seed.
Cucumbers, Melons Pumpkins, Squash		75 to 100 mm to the side and 50 mm deeper than the seed.
Lettuce (>780 kg/ha)	Part broadcast, remaining amount at planting.	100 mm deep, 50 mm to the side.
(<780 kg/ha)	Entire amount at planting.	
Potatoes		50 mm to the side, slightly below the seed.
Root crops		40 to 50 mm to one side and 50 to 75 mm below the seed.
Tomatoes (clay soil)	Broadcast & plow under 1100 to 2200 kg/ha with a cover crop 6 to 8 weeks before planting.	Additional amounts may be side-dressed.
(sandy soils)	Broadcast & plow under 560 to 780 kg/ha.	225 to 335 kg/ha side placed at transplanting, then side dress two more times at 3- to 5-wk intervals during
(raised from seed)		cultivation. Apply complete fertilizer 50 mm to the side and 50 to 100 mm below the seed.
Sugar beets		25 mm to the side and 50 mm below the seed at planting.
Tobacco		75 to 100 mm to each side of the row and 25 to 35 mm below crown of transplant.
Rice (Do not use Nitrate N because denitrification during flooding occurs.)		
Vegetable crops		Band 50 mm from seed.
Snap beans, Southern peas, Lima Beans		50 to 75 mm to the side and 25 to 50 mm below seed level.
Cabbage, Broccoli, Brussels sprouts, Cauliflower	Broadcast 2/3 of total	Apply remaining 50 to 75 mm to the side and 75 mm deep for transplanted crops, or 50 to 75 mm to one side and 25 to 50 mm deep for seeded crops.
Corn		50 mm to the side and 50 mm below seed drilled in continuous band. Side dress with N as anhydrous ammonia between rows 125 mm deep before corn reaches 0.3 m high.
Cotton		73 mm to one side and 50 mm below seed. On sandy soil, 100 mm to one side and 100 to 150 mm below seed.
Field beans, Soybeans		35 to 50 mm to one side and 50 mm below seed level.
Small grains		Drill with seed, 25 mm to one side and 25 mm below seed.

Table 7.8. Guide to fertilizer application through irrigation systems

Fertilizers that can be applied through irrigation systems	
Ammonium polyphosphate	Ammonium sulfate
Ammonium phosphate	Urea
Diammonium phosphate	Muriate of potash
Liquid ammonium phosphate	Potassium sulphate
Urea ammonium phosphate	Water-soluble borax or
Ammonium nitrate	Borate fertilizer
Fertilizers that should not be applied through sprinkler systems	
Water-soluble phosphoric acid	Water soluble but corrosive
Anhydrous ammonia and ammonia solutions	Water soluble but excessive N would be lost in the air.
Fertilizers that cannot be applied through sprinkler or other irrigation systems	
Superphosphates	These materials do not dissolve thoroughly in water.
Elemental sulfur	
Some mixed fertilizers	
Gypsum	
Lime	

Exercises

7.1. Assume the percentages of elements in table 7.1 are by mass. Compute a table giving the number of moles of each element in 1 kg of soil. Assume each substance is present in elemental form. What is the apparent molecular weight of the lithosphere? [20.5]

7.2. Estimate the apparent molecular weight of soil. Assume the percentages for an ideal soil are given on a volumetric basis with the mineral density of 2650 kg/m^3, air density of 1.17 kg/m^3 and molecular weight of 29 g/mol, organic matter density of 1500 kg/m^3 with a molecular formula of $C_6H_{10}O_5$, and soil water to have a very dilute solution. [20.9]

7.3. Estimate the cross-sectional area available for the movement of air or water in a square section of soil if it is packed, as in example 7.2. [3.30×10^2]

7.4. Assume the form $F = -Dlc/x$ where c is concentration of a soil gas such as oxygen in g/m^3, x is distance in m, D is the diffusion coefficient in m^2/s and f is the flux of the gas. If the diffusion coefficient is 2.4×10^{-5} m^2/s and the oxygen concentration gradient is 1.3 g/m^2, what is the rate of passage of oxygen through the soil cross-section in exercise 3 in g/s?

7.5. If all the oxygen in exercise 4 were being used by respiration of roots in the soil, what is the rate of CO_2 movement through the soil block have to be? Assume the general form of the respiration equation $C_6H_{12}O_6 + 6O_2 \Rightarrow 6CO_2 + 6H_2O + \text{Energy}$. [0.142]

7.6. What is the rate of glucose usage by the respiring roots in problem 5? [0.097]

7.7. Assume a container 20 cm tall filled with a suspension of soil and water at 20° C. If the values derived in example 7.5 apply, how long in minutes would one have to wait to ensure that all the silt particles at the upper range of diameters had settled out of the container? Above what height would you expect the solution to contain no more silt particles of the lower range of particle diameter? At what height would the solution contain no more clay particles of 1 μ diameter? [9.25, 19.8, 19.95]

7.8. A soil was found to have 15% sand, 35% clay, and 55% silt. What is the textural classification of the soil? What would you expect the "feel" of the soil to be?

7.9. Assume the movement of water through a soil can be described by Darcy's Law $F = -K\ \partial h/\partial x$, where F is the volume flux ($m^3/m^2{\cdot}s$), K is the hydraulic conductivity and $\partial h/\partial x$ is the rate of change of pressure causing flow with distance. If $\partial h/\partial x = 1$ and the hydraulic conductivity in the plow layer is 0.03 m/h while in the B horizon due to the formation of a plow pan it is 0.0001 m/h, at what maximum rate would you expect water to enter the ground surface if infiltration (entry of water through the ground surface) had been occurring for 4 h? If infiltration had been occurring for 24 h? Why?

7.10. What would the cross-sectional area available for oxygen diffusion be in a square meter of soil with a pore space of 0.5% by volume? [0.03]

7.11. Assume the model applied in exercise 4 holds and the oxygen concentration gradient is 1.9 × 10 S g/cm^4. What rate of oxygen flow would you expect through the soil in example 7.9/cm^2 soil area? -[4.56 × 10 T]

7.12. What relationship would you expect between nitrogen availability to crops and drainage?

7.13. Discuss the effect of irrigating with water containing sodium ions. Can you suggest a way of determining the presence of mineral ions in irrigation water? [Electrical conductivity]

7.14. Why is calcium important for acid soils? What effect will too much calcium have on soil productivity? Why?

7.15. Discuss the differences in the engineering implications involved in applying phosphorus as compared to potassium.

7.16. Give two reasons why nitrogen is important to plant growth. (Why does a nitrogen-rich crop appear dark green?)

7.17. What might be the concentration of soluble salts in g/L of an ideal humid area soil saturated with water? [0.67 to 1.3]

7.18. What might be the concentration (in g/L) of an arid region soil under unfavorable salt conditions (assume a pore space of 40% and a saturated soil)? [16.4]

Chapter 8

Radiant Energy in the Green Plant Environment

Energy regimes for efficient plant growth are very nearly the same as the regimes that are harmful to the growing plant. We must understand the way energy is transferred in the green plant environment to be able to appreciate the effect the environment has on agriculturally significant output. The modes of energy transfer include conduction, convection, latent energy transfer, and radiation. Conduction, convection, and latent energy transfer were discussed in chapter 3 and will be reviewed briefly here to place the importance of radiation in proper perspective.

Conduction

In conduction, the heat energy content of an entity is reflected in the vibrational activity of its molecules. Heat energy is passed on as the molecular activity of neighboring molecules increases. Energy transfer occurs through an increase in molecular motion. The medium usually assumed when considering this mode of transfer are solids. Conduction also occurs in liquids although other modes are more active in liquid media.

Within the green plant, conduction moves energy up and down the stem of the plant. It is the primary mechanism by which energy is transferred throughout the soil mass. The mathematical model governing energy transferred by conduction is the differential equation of heat transfer.

Convection

Convection is important to the atmospheric environment of the plant. Convection describes the action of a moving liquid, such as air, as the mass absorbs heat from bodies at a higher temperature and transfers the energy to bodies at a lower temperature with which the air mass comes in contact. Inasmuch as air that has been warmed is less dense than cool air, the warm air rises. Air absorbing energy from within a plant canopy is continually set in motion and rises to carry heat out of the canopy.

Convection is responsible for transferring large amounts of energy laterally over the earth's surface. Advected air masses, influenced by the forces of the earth's rotation and the topography, control the climate of entire continents. The eastern portion of the United States is warmed by advected air that has been affected by the gulf stream, an ocean current that convects water heated in the tropics northern latitudes. The midwest region, on the other hand, is influenced by advected air which moves from the Gulf of Mexico and carries both moisture and heat energy over the land mass. The southwest United States, receives air that has had to pass over mountain ranges as it moves from west to east. In doing so, it has risen to higher altitudes where the pressure is less than at sea level, and has expanded and become cooler. This has caused the moisture in the air mass to condense and precipitate. Once over the mountains, the air again drops to a lower altitude, causing it to warm up. Because it has lost its moisture in rising over the mountains, it is also dry.

Lateral transfer can be detrimental to a crop system. A crop growing downwind of a fallow field may suffer heat stress because air passing over the field absorbs energy from the heated surface. Energy that is carried into and heats the crop canopy downwind is sometimes called the *clothesline effect* and may raise the temperature within the crop canopy several degrees. This excess energy is dissipated by the latent heat of vaporization through transpiration. Lateral transfer of energy (as described above) by a moving air mass is advection.

Advective transfer of dry warm air that passes over Texas, Oklahoma, and Kansas significantly affects plant growth in the western portions of Nebraska, South Dakota, and North Dakota. Advected energy, increases evapotranspiration significantly over the amount that would be predicted as a function of solar radiation alone. The combination of dry warm air results in insufficient rainfall. Thus, advection is responsible for increasing the need for irrigation.

Example 8.1. Heat transfer by convection
Dry air at 24° C contacts the ground surface and is heated to 40° C by solar radiation. Compute the difference in densities of the 24° C and the 40° C air. Assume 1 atm pressure and M = 28.9 g/mol. What buoyancy force would act on a cubic meter of air at 40° C surrounded by 24° C air?

Solution
From chapter 3, using the perfect gas law, the density of air can be calculated as:

$$\rho = \frac{P_{air} M_{air}}{RT_{air}} = \frac{1 \text{ atm} \times 100\frac{\text{kPa}}{\text{atm}} \times 28.9\frac{g_{air}}{\text{mol}}}{\frac{8.314 \text{ kPa·L}}{\text{mol·K}}} = 347.6\frac{\text{g·K}}{\text{L}} = 347.6\frac{\text{kg·K}}{\text{m}^3} \quad 8.1$$

At 24° C = 297 K:

$$\rho_{air_{24}} = \frac{347.6\frac{\text{kg·K}}{\text{m}^3}}{297\text{K}} = 1.17\frac{\text{kg}}{\text{m}^3} \quad 8.2$$

At 40° C = 313 K:

$$\rho_{air_{40}} = \frac{347\frac{\text{kg·K}}{\text{m}^3}}{313 \text{ K}} = 1.11\frac{\text{kg}}{\text{m}^3} \quad 8.3$$

The difference in densities is (1.17 – 1.11) = 0.059 kg/m³. The buoyancy force is determined by multiplying the difference in densities by the acceleration of gravity.

$$F = 0.059\frac{\text{kg}}{\text{m}^3} \times 9.81\frac{\text{m}}{\text{s}^2} = 0.59\frac{\text{N}}{\text{m}^3} \quad 8.4$$

Example 8.2. Energy transfer by advection
Air moving over a heated surface absorbs energy that is lost as the air passes over a cooler surface. Assume that air at 45° C moving at a speed of 5 m/s enters a crop 1.5 m high and 100 m wide. The air travels a distance of 60 m through the canopy and when it exits it has a temperature of 40° C. Assume no energy is transferred through the horizontal plane level with the top of the canopy and that all moisture leaves through the 1.5 m × 100 m area. Give all fluxes in units/m²·s.

(a) What is the rate of energy loss to the canopy?

(b) What is the energy flux transferred into the canopy (for computational purposes assume the energy transfer is vertical into a plane at the crop level)?

Solution

(a) At 45° C = 318 K (using the calculation in example 8.1), the mass rate of air entering the canopy is:

$$\rho_{air_{45}} = \frac{\frac{347\ kg\cdot K}{m^3}}{318\ K} \times (100 \times 1.5)m^2 \times 5\frac{m}{s} = 818\frac{kg}{s} \quad 8.5$$

At 40° C = 313 K the mass rate of air leaving is:

$$\rho_{air_{40}} = \frac{\frac{347\ kg\cdot K}{m^3}}{313\ K} \times (100 \times 1.5)m^2 \times 5\frac{m}{s} = 832\frac{kg}{s} \quad 8.6$$

The rate of energy loss to the canopy is then:

$$Flux_{energy} = 14\frac{kg}{s} \times 1\frac{kJ}{kg\cdot C°} \times 5C° = 70\frac{kJ}{s}\ 8,25 \quad 8.7$$

(b) the energy flux into the canopy is:

$$Flux = \frac{70\frac{kJ}{s}}{(60 \times 100)\, m^2} = 1.17 \times 10^{-2}\frac{kJ}{m^2\cdot s} \quad 8.8$$

The mathematical models used in convection processes are those of fluid dynamics, since the atmosphere surrounding the plant is a low-density. An exhaustive study of this phenomenon, and of the energy regime induced by it, properly belongs to the area of micrometeorology because of the complexity of the models used, such a study is usually reserved for the advanced graduate level.

Latent Energy

Both conduction and convection involve the transfer of energy as sensible heat. However, as was shown in example 8.2, it is important to be aware energy is also transferred as latent energy. Latent heat transports energy from tropical regions to more temperate climates. All this is accomplished by a change of phase in water.

Energy is utilized within the plant environment in the production of chemical compounds that comprise the body of the plant. This energy, primarily involved in the photosynthetic process, is minuscule as compared to the amount of energy being transferred in the plant environment by conduction, convection, and latent transfer. Energy transfer in the form of latent heat can be extremely useful in the modification of the plant environment. For the most part, the models used in a basic understanding of energy transfer in the form of latent energy are those of elementary physics involving the change of phase of a substance.

Example 8.3. Energy transfer by latent heat

If moisture in example 8.2 goes to transpire water from the canopy, what is the additional ET in millimeters per day (mm/d) due to the energy absorbed by the crop?

Solution

The vertical moisture flux (ET) will be increased due to moisture flux leaving the canopy:

$$Flux_{H_2O} = \frac{65\frac{kJ}{s}}{2470\frac{kJ}{kg} \times (100 \times 60)m^2} \times 10\frac{h}{day} \times 3600\frac{s}{h} \times \frac{1\,m^3}{1000\,kg}$$

$$= 1.58 \times 10^{-4}m = 0.158\,\frac{mm}{d} \qquad 8.9$$

Electromagnetic Radiation

A very important mechanism of energy transfer involves the transfer of energy as electromagnetic radiation, or more simply, radiation. Electromagnetic radiation from the sun, particularly in the visible and near infrared ranges is the primary source of energy for the plant environment. Electromagnetic radiation literally makes possible life on earth.

The transfer of energy by electromagnetic radiation is described by a set of differential equations. These equations differ from those involved in the phenomena of heat and diffusion. Two equations are necessary, one for the electric field and the second for the magnetic field. The differential equations have the general form of the wave equation for electromagnetic radiation given as (eq. 8.1):

$$\frac{\partial^2 u_v(x,t)}{\partial^2 x} = \frac{1}{c^2}\frac{\partial^2 u_v(x,t)}{\partial^2 t} \qquad 8.10$$

The equation which describes the electric field is:

$$\frac{\partial^2 E_v}{\partial x^2} = \varepsilon_o \mu_o \frac{\partial^2 E_v}{\partial t^2} \qquad 8.11$$

That which governs the magnetic field is:

$$\frac{\partial^2 B_v}{\partial x^2} = \varepsilon_o \mu_o \frac{\partial^2 B_v}{\partial t^2} \qquad 8.12$$

The solution to the wave equation has the form:

$$u_v = A_v \sin\left[\frac{1}{\lambda}(x - ct)\right] \qquad 8.13$$

where

x = distance measured in the direction of propagation
λ = wavelength of the radiation
c = speed of light
t = time

Figure 8.1 illustrates the properties of a wave and describes parameters used in describing the wave. The amplitude A refers to the distance measured perpendicularly from the x-axis to the peak of the wave. The wavelength refers to the distance between adjacent peaks of the wave measured parallel to the x-axis. It should be clear that if the speed of the wave is known, a relationship must exist between the frequency, ν, and the wavelength, λ. This relationship is:

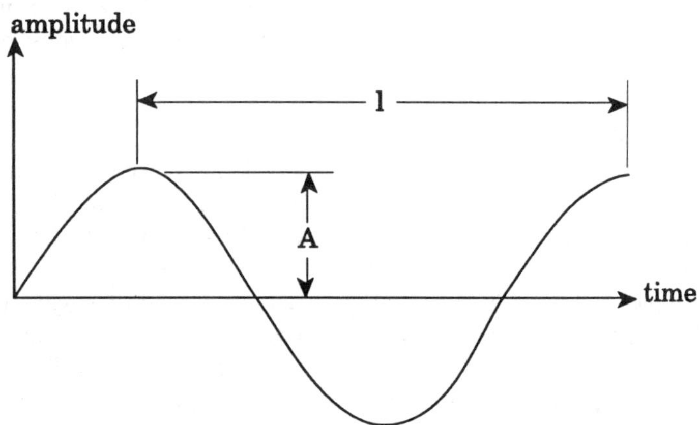

Figure 8.1. The amplitude of a wave is the distance from the midline to the crest while the wavelength is the distance the wave travels in one complete cycle.

$$\nu = \frac{c}{\lambda} \qquad 8.14$$

where ν denotes the frequency.

Concerning equation 8.13, although the solution contains a numerical value A_ν, the amplitude of the wave. The solution to 8.10 does not depend upon the numerical value of the amplitude. Likewise, equation 8.13 is a valid solution independent of the wavelength λ of the wave. The solution to the wave equation is also independent of frequency. Thus, according to the wave equation, any wavelength and frequency are transmitted, and propagation occurs at the speed of light. Electromagnetic radiation is an almost instantaneous way of transferring energy between bodies.

Example 8.4. Nature of the wave equation solution
Show that:

$$\frac{\partial u}{\partial x} = -\frac{D}{\lambda}\sin\left[\frac{1}{\lambda}(x-ct)\right] \qquad \frac{\partial u}{\partial t} = \frac{Dc}{\lambda}\sin\left[\frac{1}{\lambda}(x-ct)\right]$$

is a solution to equation 8.10, the wave equation and demonstrate that the solution in not limited to a selected frequency?

Solution
To show that the function $u_\nu(x,t)$ is a solution we must show that the second partial derivatives satisfy the equality given in equation 8.10. Accordingly:

$$\frac{\partial^2 u}{\partial x^2} = -\frac{D}{\lambda^2}\cos\left[\frac{1}{\lambda}(x-ct)\right] \qquad \frac{\partial^2 u}{\partial t^2} = -\frac{Dc^2}{\lambda^2}\cos\left[\frac{1}{\lambda}(x-ct)\right]$$

It is obvious that by multiplying $\partial^2/\partial t^2$ by $1/c^2$, the second derivative of $u_\nu(x,t)$ with respect to distance is identical to the second partial derivative of $u_\nu(x,t)$ with respect to time. Thus $u_\nu(x,t)$ satisfies equation 8.10 and is therefore a solution that in no way depends on the wavelength, λ. Since, by the model given in equation 8.14, the wavelength determines the frequency, it must be obvious that the solution above is independent of the frequency.

Electromagnetic Spectrum

The discussion about the wave equation demonstrates that energy transfer by electromagnetic radiation may take place at all frequencies and all wavelengths simultaneously. The portion of the electromagnetic spectrum most familiar is that occupied by visible light. That light is composed of a mixture of all frequencies or wavelengths is well known from experiments involving Newton's use of a prism to separate white light into distinct color bands. Visible light, however, occupies only a very brief portion of the electromagnetic spectrum. The entire spectrum includes all forms of radiation, ranging from electric power transmission (60 Hz) and radio waves down to the extremely high frequency short wavelength gamma and cosmic radiation. To aid in the discussion of electromagnetic radiation, certain portions of the continuous spectrum have been assigned names as illustrated in figure 8.2.

Spectral Classifications

In figure 8.2 the various portions of the electromagnetic spectrum are classified according to the wavelength of radiation associated with that portion of the spectrum. The visible spectrum that occupies a portion of the continuous spectrum slightly to the left of a wavelength of 1 μ is only a small portion of the complete spectrum. The visible spectrum extends from approximately 0.39 μ to 0.76 μ. To the left of visible radiation there is a region of ultraviolet radiation that is preceded by a region of x-ray, gamma, and cosmic radiation. To the right of the visible spectrum in figure 8.2 is the infrared region, extending from approximately 0.7 μ to 70 μ. To the right of this is the spectrum utilized by communication, including radio and television.

In broad terms, the continuous spectrum can be separated into two parts, a shortwave section and longwave section, the division occurring somewhere near 4 μ. The utility of this division will become more obvious when the properties of radiation emitted by bodies as a function of the body temperature are examined in more detail.

Radiancy

The energy emitted or received by a body can be expressed as a function of the radiancy, a term denoting the energy flux associated with a given wavelength of radiation. Figure 8.3 illustrates the radiancy as a function of wavelength for a body emitting radiation at the temperature of the sun. In figure 8.3 the radiancy function rises quite rapidly from zero to peak at a wavelength of approximately 0.5 μ and then decays at approximately an exponential rate. Although the general shape of the radiancy function was known, the mathematical expression that modeled the

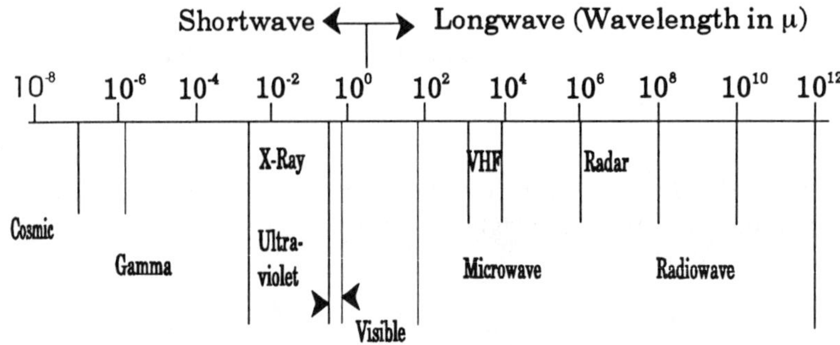

Figure 8.2. The electromagnetic spectrum. Wavelengths to the right of the visible are commonly termed longwave while radiation to the left of the visible is shortwave. The quantum value of shortwave radiation, especially at wavelengths in the ultraviolet or shorter, can cause damage to living tissue.

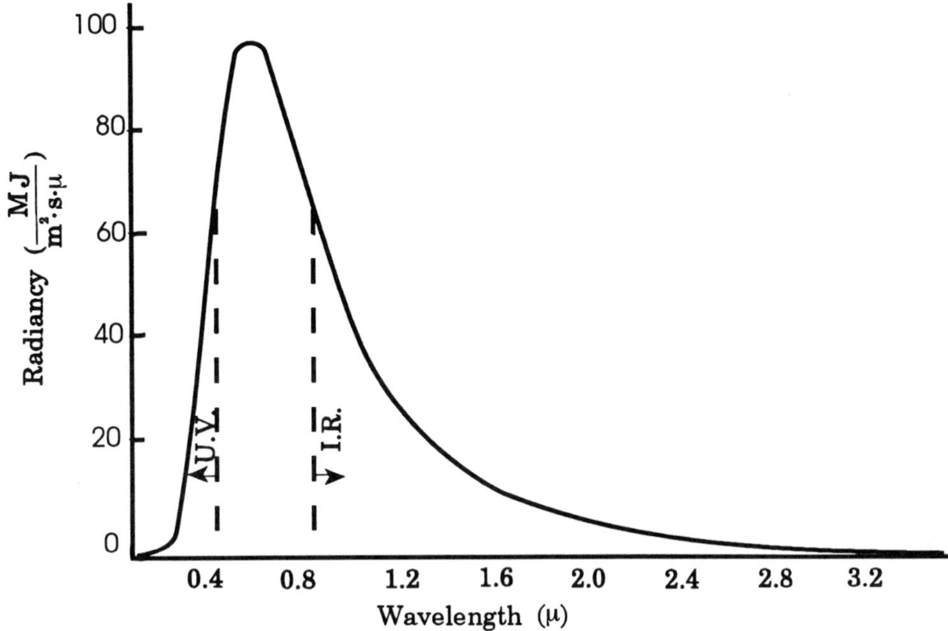

Figure 8.3. The radiancy curve for blackbody radiation at about 6000.

radiancy function and that was necessary before the function could be integrated to determine the total energy flux was determined by Max Planck. Planck experimentally determined the mathematical function that generates the radiancy function to be

$$R_\lambda = \frac{c_1}{\lambda^5 \left[\exp\left(\frac{c_2}{\lambda T}\right) - 1 \right]} \qquad 8.15$$

where

R_λ = radiancy (the energy flux for radiation with wavelength λ)
c_1, c_2 = constants
λ = wavelength
T = temperature (in absolute units)

From experimental measurements, Planck was able to determine values for the constants c_1 and c_2. The experimental values determined and the function were presented before the Berlin Physical Society on 19 October 1900. Shortly thereafter, following the formulation of the quantum concept of energy transfer, the values were verified. The quantum concept proposes that energy transfer is not continuous. Instead, energy actually is propagated in discrete units called quanta. A single quantum of energy is a function of the wavelength of radiation, and is defined by the relationship:

$$\Delta E = h\nu = \frac{hc}{\lambda} \qquad 8.16$$

where

ΔE = one quantum of energy

h = Planck's constant (6.63×10^{-34} J·s)
ν = frequency (Hz)
λ = wavelength
c = speed of light

Once the quantum concept was introduced, Planck was able to use probability theory to derive the exact form of equation 8.6 and determine the theoretical expression for the constants c_1, and c_2:

$$c_1 = 2\pi c^2 h$$
$$c_2 = \frac{hc}{k} \qquad 8.17$$

where
k = Boltzmann's constant (1.38×10^{-23} J/K)
c = speed of light (3×10^8 m/s)

Example 8.5. Magnitude of a single quanta of energy
(a) How much energy is contained in a single quantum of x-ray radiation at about 0.05 μ in the x-ray region of the spectrum?

(b) How much is a single quantum of energy for a spring mass system at a frequency of 20 Hz?

(c) Does this result have any bearing on why energy in engineering calculations is usually considered to be a continuous function?

Solution
(a) From figure 8.3 the x-ray band extends from a wavelength of about 0.0035 μ to 0.5 μ, 0.05 μ is very nearly in the center of the x-ray region. Using equation 8.7:

$$\Delta E = \frac{hc}{\lambda} = \frac{6.63 \times 10^{-34}\,\mathrm{J{\cdot}s} \times 3 \times 10^{8}\frac{m}{s}}{0.05\mu \times 10^{-6}\frac{m}{\mu}} = 3.99 \times 10^{-18}\,\mathrm{J} \qquad 8.18$$

(b) by equation 8.7:

$$\Delta E = h\nu = 6.63 \times 10^{-34}\,\mathrm{J{\cdot}s} \times 20\,\mathrm{Hz} = 1.33 \times 10^{-32}\,\mathrm{J} \qquad 8.19$$

(c) The energy levels of the spring mass oscillator contain many quantum of energy. Therefore, one individual quanta of energy in the spring mass system is so small that it is virtually undetectable. While the quanta of energy in the x-ray band is small, it is still 3×10^{14} larger than a quanta of energy in the spring mass oscillator.

Total Radiant Energy Flux

Radiancy describes the number of quanta of energy passing through, or impinging on, a unit area in a unit of time for a given wavelength. Since energy transmitted by electromagnetic radiation is independent of the amplitude or frequency of the wave, is obvious that all frequencies and amplitudes can be emitted from any body emitting electromagnetic radiation. The total energy flux being emitted by the radiating body must be found by summing the radiancy over all wavelengths. Although the energy emitted is in the form of discrete quanta, the number is so large that the radiancy can be integrated as though it were a continuous function. Integration yields the Stefan-Boltzmann Law which is named after its developers:

$$R = \sigma T^4 \qquad 8.20$$

where

R = radiant energy flux, $J/(m^2{\cdot}s)$(or W/m^2)

σ = Stefan-Boltzmann constant 5.67032×10^{-8} $J/(m^2{\cdot}s{\cdot}T^4)$ or $W/(m^2{\cdot}K^4)$

In terms of basic physical constants:

$$\sigma = \frac{2\pi^5 k}{15c^2h^3} \qquad 8.21$$

As can be determined from equation 8.15, radiancy is a function of temperature as well as wavelength. Intuitively, the wavelength at which the maximum radiancy occurs should also be a function of temperature. According to standard mathematical techniques for determining a maximum or minimum, the wavelength at which the maximum radiancy occurs can be found by taking the derivative of the radiancy function with respect to the wavelength, setting the resulting function equal to zero, and solving for the wavelength. This yields a function of λT. If we solve for λT we find:

$$\lambda T = \text{a constant} \qquad 8.22$$

As equation 8.22 indicates, if one knows the wavelength at which the maximum radiancy occurs, then the temperature at which the radiation is occurring can be determined from Wein's Law, the name by which equation 8.22 is usually known.

Example 8.6. Wein's constant determination
Determine the value of the constant in Wein's Law.

Solution
We use equation 8.15 to obtain:

$$\frac{\partial R_\lambda}{\partial \lambda} = \frac{\partial}{\partial \lambda}\left\{\frac{c_1}{\lambda^5\left[\exp\left(\frac{c_2}{\lambda\, T}\right) - 1\right]}\right\}$$

$$= -c_1\left\{\lambda^5\left[-\frac{c_2\, T}{\lambda^2\, T^2}\exp\left(\frac{c_2}{\lambda\, T}\right)\right] + \left[\exp\left(\frac{c_2}{\lambda\, T}\right) - 1\right]5\,\lambda^4\right\} = 0 \qquad 8.23$$

$$= -\frac{c_1\,\lambda^3}{T}\left[\exp\left(\frac{c_2}{\lambda\, T}\right)\right] + 5\,\lambda^4\left[\exp\left(\frac{c_2}{\lambda\, T}\right)\right] - 5\,\lambda^4$$

After multiplying both sides of the above expression by T^5 and setting $x = \lambda T$ we obtain:

$$\frac{\partial R_\lambda}{\partial \lambda} = -c_2 x^3 \left[\exp\left(\frac{c_2}{x}\right)\right] + 5x^4\left[\exp\left(\frac{c_2}{x}\right)\right] - 5x^4 = 0 \qquad 8.24$$

Now divide by x^3 to simplify the expression. This yields:

$$-c_2\left[\exp\left(\frac{c_2}{x}\right)\right] + 5x\left[\exp\left(\frac{c_2}{x}\right)\right] - 5x = 0 \qquad 8.25$$

Which can be written:

$$x = \frac{c_2\left[\exp\left(\frac{c_2}{x}\right)\right]}{5\left[\exp\left(\frac{c_2}{x}\right) - 1\right]}$$

or 8.26

$$x = \frac{c_2}{5\left\{1 - \left[\frac{1}{\exp\left(\frac{c_2}{x}\right)}\right]\right\}}$$

Since (a) contains x on both sides of the equation, a trial-and-error solution is required. Remembering that $x = \lambda T$ and by figure 8.3, for $T = 5776$ K, $\lambda \approx 0.5\ \mu$. We use these values to compute a trial value for x:

$$x = (5776\ \text{K})(0.5\ \mu) = 3000\ \text{K}\cdot\mu$$

$$c_2 = \frac{hc}{k} = \frac{6.63 \times 10^{-34}\ \text{J}\cdot\text{s} \times 3 \times 10^8 \frac{\text{m}}{\text{s}}}{1.38 \times 10^{-23} \frac{\text{J}}{\text{K}}} \times \frac{10^6 \mu}{\text{m}} = 14400\ \text{K}\cdot\mu \qquad 8.27$$

The first approximation to equation 8.26 is then:

$$x = 3000\ \text{K}\cdot\mu = \frac{1.44 \times 10^4 \text{K}\cdot\mu}{5\left\{1 - \frac{1}{\exp\left(\frac{1.44 \times 10^4 \text{K}\cdot\mu}{2903\ \text{K}\cdot\mu}\right)}\right\}} \qquad 8.28$$

or:

$$3000\ \text{K}\cdot\mu = ?\ 2904\ \text{k}\cdot\mu \qquad 8.29$$

So our choice of x was close, but still not correct. For the second approximation, use the value on the right side of the equation. This yields:

$$2904\ \mathrm{K}\cdot\mu \ = \ ?\ \frac{1.44 \times 10^4}{5\left\{1 - \left[\dfrac{1}{\exp\left(\dfrac{144 \times 14^4}{2903}\right)}\right]\right\}} \qquad 8.30$$

The approximation is numerically different only in the fourth significant digit. Therefore, Wein's constant, to three significant digits is 2900 K·μ.

Practical Considerations of Radiation

To this point we have assumed only blackbody radiation, the term that refers to an object that behaves as a perfect radiator. A blackbody radiator emits radiation according to the theoretical form of the Stefan-Boltzmann Law:

$$R = \sigma T^4 \qquad 8.31$$

As is the case with most natural phenomena, some deviation from the theoretically described behavior can be expected. For radiating bodies, if the radiation flux is not equal to the theoretically predicted blackbody radiation flux, the radiator is known as a "gray body" radiator. To account for the predicted difference in the model's ability to describe radiation from natural sources, a correction factor, the emissivity ε of the body must be introduced. The equation describing radiation from an imperfect natural radiator is:

$$R = \varepsilon\sigma T^4 \qquad 8.32$$

where

R = radiation flux
ε = ratio of actual radiation flux theoretical radiation flux
σ = Stefan-Boltzmann constant
T = kelvin temperature

In equation 8.32 the emissivity, χ, must be determined for the radiating body under consideration and is different for different bodies as well as for different wavelengths of radiation. When this occurs, the Stefan-Boltzmann law can no longer be used to predict the radiation flux emitted by such a body. Instead, detailed calculations must be carried out as a function of the emissivity associated with a given wavelength or band of wavelengths. However, for many practical engineering purposes one can assume the emissivity is constant over a large band of wavelengths.

Example 8.7. Gray body radiation calculations
A 60-W bulb has a surface area equal to a sphere 60 mm in diameter. If the average surface temperature of the bulb surface is 80° C (176° F) and the emissivity is 0.25, how much energy is the bulb radiating?

Solution
For the temperature given, using equation 8.32:

$$R = 0.25 \times 5.67 \times 10^{-8} \frac{W}{m^2 \cdot K^4} \times (353K)^4$$
$$= 220 \frac{W}{m^2} \quad 8.33$$

The area is that of a sphere or radius 0.03 m:

$$A = 4 \times \pi \times (0.03\,m)^2 = 1.13 \times 10^{-2} m^2 \quad 8.34$$

Therefore, the rate of energy loss is:

$$\text{Energy} = 220 \frac{W}{m^2} \times 1.13 \times 10^{-2} m^2$$
$$= 2.49\,W \quad 8.35$$

The above figure does not represent all the energy being radiated by the bulb, only that being radiated from the surface of the glass. As can be seen from figure 8.4, the surface is reradiating only that energy that has initially been absorbed by the glass. The remainder of the energy is passing through the surface of the bulb without being absorbed and reradiated.

Reflectivity, Transmissivity, and Absorptivity

Figure 8.4 illustrates that incident radiation arriving at a body is divided into three components. The first component is that which is reflected from the body, denoted by the symbol ρ. This fraction, since it is reflected, possesses the same wavelength as the incoming radiation. If the natural body is not opaque but exhibits some degree of translucency, then a portion of the radiation may be transmitted through the body. The fraction that is transmitted is denoted by the symbol τ. Finally, depending upon the degree of translucency, some part of the incident radiation

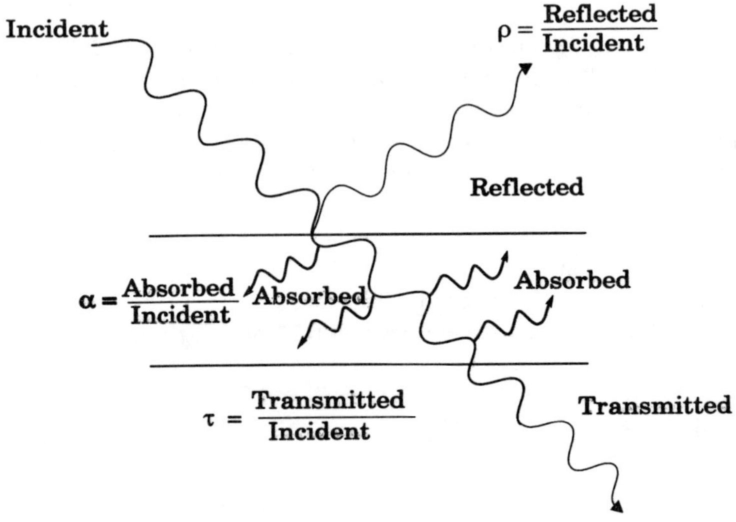

Figure 8.4. A schematic showing how a translucent medium affects electromagnetic radiation. Note that only the amplitude of the radiation reflected or transmitted is altered. The frequency remains the same.

will be absorbed and converted into sensible heat. The fraction of incident radiation absorbed is denoted by the symbol α.

The magnitudes of radiation reflected, absorbed, and transmitted are dependent on wavelength. Values for these properties may change with changing wavelength of radiation so that some natural bodies that are transparent to shortwave radiation, such as window glass, may behave as opaque to longwave radiation.

In the context of the gray body radiator described previously, one may think of a gray body absorber. Such an absorber would be one for which the reflectivity, absorptivity, and transmissivity remain the same throughout spectrum. Although few natural bodies behave throughout the spectrum as a gray body, it is common practice to treat the radiation arriving at or leaving these bodies, at least in selected temperature ranges, as though the body was behaving as a gray body. We note finally that, regardless of whether a body's transmissivity, absorptivity, and reflectivity change for different wavelengths, the sum of α, τ, and ρ is unity for a given wavelength:

$$\alpha + \tau + \rho = 1 \qquad 8.36$$

For bodies that are completely opaque to radiation, $\tau = 0$, and, therefore, the sum of the reflectivity ρ and the absorptivity α must be equal to unity.

Example 8.8. Interaction of reflectivity and absorptivity
A pane of glass used in a greenhouse has a reflectivity p = 0.05 for sunlight. If the absorptivity = 0.03, what will the radiation flux be inside the greenhouse if the solar radiation flux incident upon the glass is 488 W/m^2? If the glass is sprayed with a white film to increase the reflectivity to 0.3, what is the flux of radiation within the greenhouse?

Solution
From equation 8.36:

$$\begin{aligned} 0.03 + 0.05 + \tau &= 1 \\ \tau = 1 - 0.08 &= 0.92 \end{aligned} \qquad 8.37$$

and:

$$\tau = 0.92 = \frac{\text{radiation transmitted}}{\text{radiation incident}} \qquad 8.38$$

Since the radiation received is 488 W/mm, the radiation transmitted is:

$$R_{\text{transmitted}} = 0.92 \times 488 \frac{W}{m^2} = 449 \frac{W}{m^2} \qquad 8.39$$

If the reflectivity increases to 0.3, then the transmissivity becomes:

$$\begin{aligned} 0.03 + 0.30 + \tau &= 1 \\ \tau &= 0.67 \end{aligned} \qquad 8.40$$

Therefore:

$$\text{radiation transmitted} = 0.67 \times 488 \frac{W}{m^2} = 327 \frac{W}{m^2} \qquad 8.41$$

Spraying the greenhouse with a white film to increase the reflectivity reduces the radiant energy flux within the greenhouse by:

$$\frac{449 - 327}{449} \times 100\% = 27\% \qquad 8.42$$

Emissivity and Absorptivity

For an opaque body there exists a relationship between emissivity and absorptivity. Consider the relationship between the emissivity, ε, and absorptivity, α, of a gray body absorber. The relationship between emissivity and absorptivity can be developed by considering what will happen to a gray body absorber under idealized conditions when it is subject to radiation from a blackbody radiator at some higher temperature.

In figure. 8.5, the upper portion of the figure is an idealization of radiation being received by a gray body absorber whose temperature is T_2. The radiation is arriving at the absorber from a blackbody radiator that is at a temperature T_1, where T_1 is greater than T_2. While the gray body will absorb and reradiate from all directions, for convenience in illustration, the radiation has been idealized as arriving and leaving as a band whose width is directly proportional to the radiation flux. Since the gray body is an imperfect absorber, it will exhibit reflectivity and the arriving radiation will be only partially absorbed. This is illustrated in the upper portion of figure 8.5 where some radiation arriving is shown as reflected. Since the gray body is at some temperature $T_2 < T_1$, by the Stefan-Boltzmann Law it will be reradiating energy at a lower rate than that at which energy is received. This is shown schematically at the extreme right, where the band of radiation has the flux intensity, $\varepsilon\sigma T_2{}^4$. Thus, more energy is arriving at the gray body than is being reradiated because of the temperature difference between the blackbody radiator and the gray body absorber. Consequently, some arriving energy must be converted into heat and be used in the gray body to raise its temperature. Note that the rate at which energy is absorbed can be expressed in terms of the radiation being absorbed by the body minus the radiation being radiated, or:

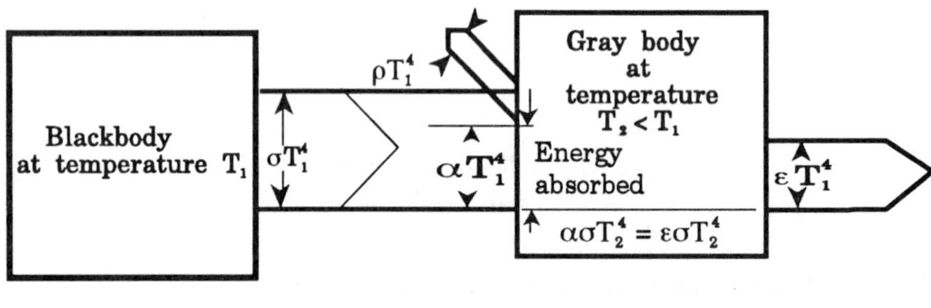

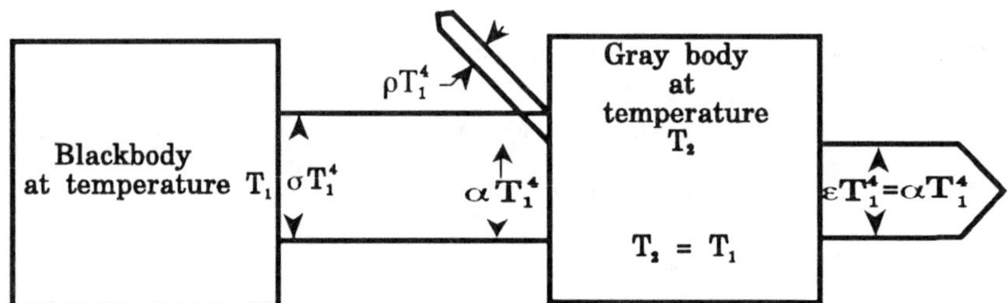

Figure 8.5. The emissivity of a gray body is the same as its absorptivity. If it were not, the absorbing body would either gain or lose continually, thus violating the second law of thermodynamics.

$$\text{rate of energy absorption} = \alpha\sigma T_1^4 - \varepsilon\sigma T_2^4 \tag{8.43}$$

or

$$\text{rate} = \sigma\left(\alpha T_1^4 - \varepsilon T_2^4\right) \tag{8.44}$$

As the temperature of the body absorbing the radiation rises, the radiation flux emitted by the gray body will increase according to the Stefan-Boltzmann Law. When the temperature of the gray body is equal to that of the blackbody emitter there can be no more absorption of energy by the gray body. If additional energy were to be absorbed it would have to be converted to heat which would result in a further increase in the temperature of the gray body. Energy would be transferred from a body at lower temperature to one at higher temperature without work being done in the system—a violation of the second law of thermodynamics. Accordingly, the rate of energy transfer when both bodies have the temperature T is as shown in the lower part of figure 8.5. Here energy is arriving at the gray body from the blackbody radiator and a portion of it is being reflected. The remainder, the energy flux which is being absorbed by the gray body is:

$$\text{Energy absorbed} = \alpha\sigma T_1^4 \tag{8.45}$$

An identical amount of energy must be leaving the body to ensure that the temperature of the gray body does not exceed that of the blackbody. The radiation flux leaving the gray body is:

$$\text{Energy emitted} = \varepsilon\sigma T_2^4 \tag{8.46}$$

Since the difference between the energy absorbed and the energy flux emitted must be zero, we can write:

$$0 = \sigma T^4 (\alpha - \varepsilon) \quad \text{or} \quad \alpha = \varepsilon \tag{8.47}$$

We have verified a fundamental relationship first derived by Kirchoff, namely, the absorptivity and emissivity of a gray body are equal.

From above, if one can determine either the reflectivity or absorptivity of an opaque body, one immediately knows the absorptivity and the emissivity of the body since the transmissivity for an opaque body is zero, and the emissivity is, therefore, equal to the absorptivity.

Example 8.9. Influence of reflectivity on energy absorption
If the overall reflectivity of an opaque leaf is 0.2, at what rate will the leaf lose energy by radiation if its temperature is 27° C? Give the answer in W/m^2.

Solution
From equation 8.36:

$$\alpha = 1 - \tau - \rho = 1 - 0.2 - 0 = 0.8 \tag{8.48}$$

By equation 8.47:

$$\alpha = \varepsilon = 0.8 \qquad 8.49$$

The radiation flux leaving the body is, by equation 8.46:

$$R = 0.8 \times 5.67 \times 10^{-8} \frac{W}{m^2 \cdot K} \times \left(300^\circ K\right)^4 = 367 \frac{W}{m^2} \qquad 8.50$$

Atmospheric Influences on Radiation

To this point we have been considering the behavior of gray body radiation and absorption for opaque bodies where the transmissivity of the body is zero. Radiation is the primary source of energy for the earth. Since the atmosphere that surrounds the earth is a translucent medium through which radiation from the sun must pass to reach the earth's surface, it is important to consider the effect a translucent medium will have on radiation that passes through the medium. Figure 8.6 illustrates what must occur as radiation of a given wavelength traverses a translucent medium.

Beer's Law

In figure 8.6, for clarity it has been assumed that the reflectivity is zero, although this need not be the case. Schematically, figure 8.6 shows that radiation entering a translucent medium experiences certain losses due to interaction between the medium and the radiation. It is desirable to be able to determine the intensity of radiation at any point within the medium in terms of the intensity of radiation incident upon the exterior surface of the medium. To this end, a model of the effect of the medium on the radiation is necessary. This can be accomplished as follows.

We have shown that the rate at which the radiation flux for a given wavelength is diminished within the medium is proportional to the magnitude of the radiation flux at that point. Mathematically, this is:

$$\frac{dR_\lambda}{dz} = -k_\lambda R_\lambda \qquad 8.51$$

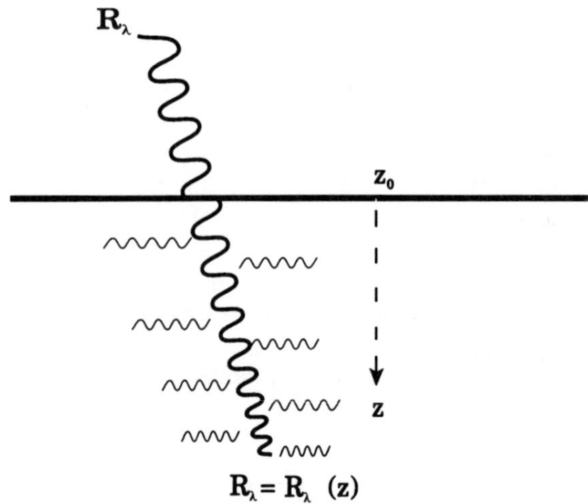

Figure 8.6. Wein's Law indicates that the amplitude of radiation of a given frequency diminishes with penetration into a translucent medium.

The solution for equation 8.51 can be determined by multiplying both sides of the equation by dz, separating variables, and integrating from some value R°_λ at z = 0 to R_λ. The value of the flux at depth z within the medium obtain:

$$\int_{R^\circ_\lambda}^{R_\lambda} \frac{d\xi}{\xi} = k_\lambda \int_0^z d\zeta$$

$$R_\lambda = R^\circ_\lambda \, e^{-k_\lambda z} \qquad 8.52$$

where

- R = radiancy
- k_λ = proportionality constant
- z = distance of penetration measured from the surface of the medium

In equations 8.51 and 8.52 the radiancy, R_λ, was being used to obtain the most general form for the energy at any point within a translucent medium. Equation 8.51 was integrated to describe the absorption of radiation of a given wavelength, λ, by a translucent body, the expression known as Beer's Law.

Example 8.10. Effect of absorptivity of visible light
In his descent in a bathysphere, Beebe found blue light at the 518-m depth in the ocean waters off Bermuda. Estimate k_λ for blue light if blue light has a wavelength of 0.47 μ. Assume R at the ocean surface is 7.45×10^7 W/(m^2·μ), and that an intensity of at least 4.19×10^6 W/m^2 is necessary before the human eye can detect light and distinguish the color blue.

Solution
Beer's Law, equation 8.52, may be used to obtain a solution. By taking logarithms of both sides of equation 8.51 and rearranging we obtain:

$$\ln(R_\lambda) - \ln(R^\circ_\lambda) = -k_\lambda z \qquad 8.53$$

So that:

$$k_\lambda = \frac{\ln(R^\circ_\lambda) - \ln(R_\lambda)}{z} = \frac{\ln\left(\frac{R^\circ_\lambda}{R_\lambda}\right)}{z}$$

$$k_\lambda = \frac{\ln\left(\frac{7.54 \times 10^7}{4.19 \times 10^6}\right)}{518 \text{ m}} = 5.56 \times 10^{-3} \text{m}^{-1} \qquad 8.54$$

If the properties of the medium through which the radiation is passing are such that the effect upon the radiation flux is the same at all wavelengths, the constant k_λ in the exponent on the right side of equation 8.51 is identical for all wavelengths. The expression may be integrated over all wavelengths of the electromagnetic spectrum to obtain:

$$R = R^\circ \, e^{-kz} \qquad 8.55$$

where

R = radiant energy flux at depth z
R° = radiant energy flux at z = 0
k = constant related to the rate of energy absorption of the medium

If k is not a function of wavelength, equation 8.55 can be used to find the radiation flux at any point within a medium providing sufficient information is given to determine k.

Example 8.11. Flux intensity variation with depth
In passing through a particular fluid, the radiant energy flux was found to vary according to $dR/dz = 2.5\ m^{-1}$. If the flux was 558 W/m^2 at the surface of the fluid, what was it at 0.20 m below the surface?

Solution
From equation 8.55, the model for radiation absorption is:

$$R = R^\circ e^{-2.5z} = \left[558\, e^{\left(-2.5\, m^{-1} \times 0.20\, m\right)}\right] \frac{W}{m^2} = 338 \frac{W}{m^2} \qquad 8.56$$

Although the behavior of radiation may not be exactly as predicted by Beer's Law, the approximation will usually be satisfactory for engineering purposes. An example of where Beer's Law can be applied to the plant environment involves predicting the level of radiation at some depth within a plant canopy. The effect of leaves and plant density on the absorption and reflection of the radiation arriving at the plant canopy is such that with increasing depth into the canopy, the radiation flux decreases. Beer's Law may be applied to this situation to calculate the solar energy flux incident upon a leaf at some level within the plant canopy. Similarly, a rough approximation of the solar radiant energy flux reaching the ground underneath the plant canopy can be obtained.

Example 8.12. Net radiation
Net radiation is the difference between the incoming radiation flux and the outgoing radiation flux at some point in space. It is considered positive if the incoming flux exceeds the outgoing flux. Within a crop canopy, the net radiation decreases with depth into the canopy as the incoming solar radiation is filtered out by the leaves. Given $-k = 0.7\ m^{-1}$, assume that net radiation can be described in a corn crop by using Beer's Law as a model. Compute the net radiation flux at 0.30 cm from the ground in a corn crop 2.60 m tall if the net radiation at the top of the crop canopy is 488 W/m^2.

Solution
Beer's Law is:

$$R_n = R^\circ e^{\left(-0.7\, m^{-1} \cdot z\right)} \qquad 8.57$$

where the subscript n implies net radiation. Under the conditions given, z = 2.60 – 0.30 = 2.30 m. Therefore:

$$R = 488 \frac{W}{m^2}\left[e^{\left(-0.7\, m^{-1} \times 2.3\, m\right)}\right]$$
$$= 98 \frac{W}{m^2} \qquad 8.58$$

Energy Transfer in the Plant Environment

Shortwave (Solar) Radiation

Figure 8.3 shows that solar radiation reaching the earth's surface has a maximum radiancy at about 0.5 μ. Although solar radiation is perceived by the human eye as white light, it is actually a mixture of colors and contains all wavelengths. Table 8.1 gives the color that is associated with a band of wavelengths centered at approximately the wavelength shown. As was shown also in figure 8.3, the maximum amount of radiant energy arrives at the earth in the visible region and has a peak near the green portion of the visible band.

Radiation from the sun must pass through the atmospheric envelope and in the process is influenced by the atmosphere. It is subjected to the properties of reflectivity, transmissivity, and absorptivity that characterize the atmospheric medium. From figures 8.2 and 8.3, since most of the radiant energy reaching the earth is shortwave radiation, it is general practice to use the terms *shortwave radiation* and *solar radiation* interchangeably and this practice will be followed in this text.

Albedo

In referring to shortwave radiation, we will also conform to general usage and define the reflectivity of an object with respect to shortwave radiation as the *albedo* of the object. The practice of differentiating between longwave reflectivity and shortwave albedo is useful because most objects exhibit completely different properties with respect to shortwave and longwave radiation. Plant leaf surfaces and different ground surfaces all possess different albedos. With respect to longwave radiation, however, these objects behave very nearly as perfect radiators, absorbing and emitting radiation as a blackbody.

The earth or, more accurately, its outer atmosphere, is assumed to have an albedo of 0.43. A considerable proportion of shortwave radiation is reflected before it reaches the plant environment. However, some radiation that is not reflected by the outer atmosphere is altered during its passage through the atmosphere. Since the plants enter the gaseous exchange cycle and since certain gases are largely responsible for modifying the radiant energy, we will become examine aspects of radiation transfer through the atmosphere that influence the plant environment.

Example 8.13. Radiant energy and albedo
A hectare of ground on the earth's surface has an albedo of 0.2. If the incoming solar radiation flux is 209 W/m^2, what is the heat load in kW/ha being reflected to outer space?

Solution
The radiation reflected is:

$$209\frac{W}{m^2} \times 0.2 = 41.8\frac{W}{m^2} \qquad 8.59$$

The heat load reflected is:

$$41.8\frac{W}{m^2} \times \frac{10\,000\ m^2}{ha} = 418 \times 10^3\frac{kW}{ha} \qquad 8.60$$

Atmospheric Effects Important to the Plant Environment

From table 8.1, the visible portion of shortwave radiation extends from a little less than 0.4 μ to about 0.7 μ. Reference to figure 8.3 shows that although most of the energy of solar radiation is concentrated in this range, some energy exists at wavelengths less than 0.4 μ and greater than 0.7 μ.

The radiation with wavelengths less than 0.4 μ is of special interest because in this wavelength range, the individual quanta of energy possess a high energy value. Radiation at wavelengths less than 0.4 μ contains x-ray radiation as well as cosmic radiation, both of which can do great damage to living tissue. It is important to understand why no appreciable damage to plants that thrive in sunlight is observed attributable to these sources. The answer lies in the interaction between certain atmospheric components and radiation in the range under discussion.

Oxygen and other gaseous constituent molecules in the atmosphere play an important role in the protection afforded by the environment against the deleterious effects of high-frequency shortwave-length radiation. They also act to trap and store energy in the lower-frequency, longwave-length infrared portions of the electromagnetic spectrum. We will briefly examine some aspects of molecular structure as the structure is affected by interactions with electromagnetic radiation.

As indicated in chapter 3, molecules are comprised of atoms of various sizes and weights. When atoms join to form a molecule, several molecular configurations may be possible depending on the number of atoms present, and the nature of the forces that bond or join them. Finally, the molecule itself is not a rigid body fixed in space. Molecules are able to absorb radiation in several ways, contributing to the modification of electromagnetic radiation in the atmosphere comprising the aboveground portion of the plant environment on both a macro and a micro scale. First, consider the energies released when two or more atoms bond to form a molecule. A certain amount, i.e., a quanta of energy of a certain magnitude, is released in the process. The magnitude of the energy released is a function of the molecular bond and it varies for different molecules because of the individuality of different molecules and atom-to-atom configurations. If a molecule absorbs a quantity of energy equal to that which was released when the atoms bonded together to form the molecule, then the molecule will dissociate and the initial atomic structure will result.

The phenomenon of dissociation occurs when molecules absorb a quantum of energy, ΔE, exactly equal to the dissociation energy of the molecule. The result is that certain molecules absorb electromagnetic energy strongly at a wavelength whose quantum value, ΔE, is equal to the dissociation energy of the molecule.

Example 8.14. Interaction of gasses and quanta of radiation
Barrow (1966) gives the dissociation energy of O_2 as 495 kJ/mol. What wavelength of electromagnetic radiation should be most strongly absorbed by O_2 in the atmosphere?

Solution
Since 1 mol of a gas contains 6.02×10^{23} molecules, the energy per molecule for dissociation can be calculated:

Table 8.1. Color associated with electromagnetic radiation in bands centered at the indicated frequency

Wavelength (μ)	Color of the Spectrum
0.37	ultraviolet
0.45	violet
0.47	dark blue
0.49	light blue
0.53	green
0.56	yellow green
0.58	yellow
0.60	orange
0.64	red
0.76+	infrared

$$\Delta E_{O_2} = \frac{495 \frac{kJ}{mol}}{6.02 \times 10^{23} \frac{molecules}{mol}} = 8.22 \times 10^{-22} \frac{kJ}{molecule} \qquad 8.61$$

This amount of energy corresponds to a single quantum of radiation with a wavelength which can be determined by the model given in equation 8.7

Note: Radiation with a wavelength of 0.24 μ is far into the ultraviolet portion of the spectrum. Dissociation of O_2 molecules by ultraviolet radiation is a factor in the removal of harmful radiation energies from the solar spectrum.

In addition to the quanta of energy required to cause dissociation of a molecule, other molecular phenomena also can cause molecules to absorb energy. One of the more important phenomena at very short wavelengths results from the vibration of the atoms within the bonded structure of the molecule. As a spring mass system possesses a characteristic resonant vibrational frequency, so do the bonds between the atoms comprising a molecule. In the case of a molecule, however, only a simple diatomic molecule could single characteristic resonant frequency. Depending on the number of atoms in a molecule, their orientation, and their relative bonding stiffness (this may be thought of in terms of the Hooke's Law spring constant that characterizes the stiffness of a spring), an individual molecule may have several resonant frequencies at which the atom can vibrate.

When a photon or quantum of electromagnetic radiation of appropriate energy strikes the molecule, it will be absorbed and the atoms of the molecule will begin to vibrate at the resonant frequency corresponding to the energy of the absorbed quantum. This concept is pictured schematically in figure 8.7(a), (b), and (c) where the possible modes of vibration of the water molecule are illustrated with the wavelength which can excite the molecules and cause them to vibrate.

In addition to vibrational motions, molecules also can undergo rotational motions such as are illustrated in figure 8.7(d). A molecule may tumble, i.e., it may experience several modes of rotation and vibration simultaneously. It should be apparent that there are many ways that atmospheric molecules can influence selective wavelengths of radiation as it passes through the mixture of gases in the comprising the atmosphere.

As indicated in example 8.14, O_2 absorbs energy very strongly at a wavelength of 0.24 μ. In this fashion, O_2 plays an important role in protecting the lower atmosphere from high-energy ultraviolet radiation. Consider O_2 with the action of several other atmospheric gases insofar as they affect the solar spectrum. The dissociation of O_2, which takes place mainly at high altitudes, is responsible for creating an abundance of elemental oxygen (O) in the outer reaches of the atmosphere. The reaction is:

$$O_2 + h\nu_{0.24\,\mu} \Rightarrow O + O \qquad 8.62$$

When elemental oxygen is present, it is chemically very active and another reaction occurs between elemental oxygen (O) and molecular oxygen (O_2). When these two collide with a third body (which may be another neutral molecule or a dust particle), they combine to form ozone (O_3).

$$O + O_2 = O_3 + \text{momentum} \qquad 8.63$$

(The collision is important because some excess momentum is present in the collusion and must be absorbed by the third body.)

Under the influence of energy with wavelengths of less than 0.24 μ, the O_3 molecule absorbs energy and breaks down again into elemental and molecular oxygen.

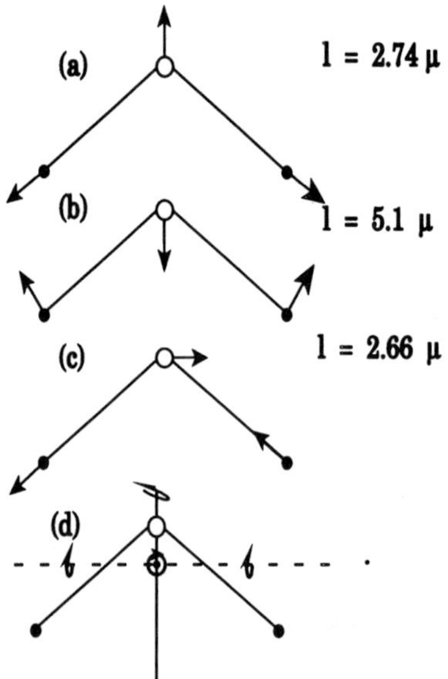

Figure 8.7. The structure of a water molecule showing the position of the two hydrogen atoms with relation to the oxygen atom. The three modes of vibration are influenced differently by the quantum energy levels of radiation of different wavelengths.

$$O_3 + h\nu_{0.24\mu} \Rightarrow O + O_2 \tag{8.64}$$

This frees elemental oxygen to create another ozone molecule. The net result of these destructive and productive processes occurring at altitudes high above the earth's surface is that x-ray and cosmic radiation are removed from the radiation as it passes through the atmosphere.

Chlorine atoms from chlorofluorocarbon compounds play an important role in the destruction of the O_3 molecule without absorbing the high energy quanta. This is because the chlorine (Cl) molecule reacts with the ozone (O_3) molecule to form chlorine monoxide (ClO) and molecular oxygen (O_2).

$$Cl + O_3 \Rightarrow ClO + O_2 \tag{8.65}$$

The chlorine monoxide molecule can then react with another oxygen atom to form a chlorine atom and O_2.

$$ClO + O \Rightarrow Cl + O_2 \tag{8.66}$$

The chlorine is then released to attack another O_3 molecule. In this fashion, the chlorine destroys the ozone. Unfortunately, in the chlorine interaction, no high energy radiation was absorbed and, thus, it is free to penetrate the atmosphere and reach the earth's surface where it can do serious damage to plant and animal life. Since the average life of a chlorofluorocarbon molecule in the atmosphere is from 60 to 100 years, a single molecule can be responsible for the removal of significant amounts of ozone.

Other components of the atmosphere are important in the radiation modification process. At selected wavelengths, a quanta of energy can interact with molecules of a gas in several ways. Electrons may be knocked off the gas molecule, creating an ion of the gas. The molecule may absorb a quantum of radiant energy that may become manifest as kinetic energy of vibration, rotation and translation. Energy transfer phenomena such as this are responsible for the modification of radiation at wavelengths greater than about 0.3 μ. The gases most important in this region of the spectrum are molecular oxygen, carbon dioxide, and water vapor. Figure 8.8 illustrates the result of the interaction of solar radiation with the atmospheric gases.

Radiation is a quantum phenomenon and the energy absorption on a molecular level is also a quantum phenomenon. Bands or windows occur in the spectrum of radiation reaching the earth, depending on the quantum value of the incoming radiation and the energy required to initiate molecular action. Figure 8.8 shows clearly the presence of certain areas of almost complete absorption of radiation by atmospheric components. Water vapor and carbon dioxide modify the infrared portion of the spectrum radically. Since infrared radiation is converted into sensible heat when it strikes a plant leaf or the surface of the earth, the filtering effect of water vapor and carbon dioxide is an important factor in the energy available for conversion into heat at any point on the earth's surface.

Although the atmosphere is very transparent to shortwave radiation, some reflection occurs from dust and pollutants near the earth's surface and from clouds at all levels. Because of the albedo of clouds, when the sun is shining through a hole in a cloud cover, radiation reflected from the cloud bank can increase the radiation flux in the plant environment significantly. This condition, however, is most likely to be temporary.

The constituents of the atmosphere affect radiation as it passes through. The atmospheric effect is proportion to the thickness of the atmosphere through which the radiation must travel. The effect of atmospheric thickness is shown in table 8.2 where the effect is tabulated in terms of optical air masses, one optical air mass being the thickness of the atmosphere measured vertically along a radius vector of the earth. As the path of solar radiation deviates from the vertical, the path length of the radiation the optical air mass value through the atmosphere increases.

In addition to increasing the optical air mass thickness, the effect of radiation striking the earth at an angle other than normal to the surface decreases the amount of energy transmitted by solar radiation to the surface. The decrease occurs because the radiation does not strike the surface perpendicularly. The radiation flux varies according to the cosine of the angle of deviation from normal to the surface. For this reason, the radiant energy flux to the earth's surface varies both with the latitude and with time. The yearly variation is shown graphically in figure 8.9 for solar radiation in the northern hemisphere.

Table 8.2 gives the radiant energy flux at sea level for various angles of incidence of the sun for the wavelengths of radiation. Table 8.2 shows that the effect of a radiation path length equivalent to the thickness of the

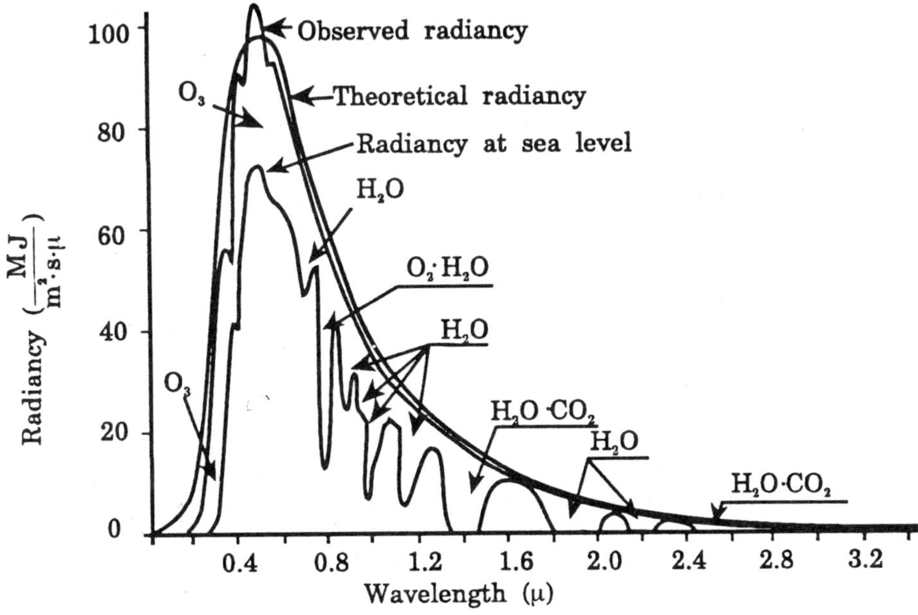

Figure 8.8. Comparison of the theoretical, observed, and actual radiancy received by the earth. The radiancy at sea level is influenced by the elements and compounds indicated. Water is the most influential in reducing the radiancy received at the earth's surface.

atmosphere normal to the earth (one optical air mass, m = 1) is to reduce the radiation reaching the ground by about 30%. Increasing the angle of incidence to the point at which five atmospheric thicknesses are traversed (m = 5) reduces the radiation by approximately 67%. The effect of the atmospheric envelope on radiation reaching the plant canopy is due to both absorption and reflection from a combination of causes. Figure 8.10 shows the overall effect of the atmosphere in modifying the incoming solar radiation and the nature of the radiation reaching the plant environment. At the left is the incoming solar radiation. Some of which is reflected from atmospheric components and by plant and soil surfaces. Depending on the relative amounts of the gases present, some radiation is absorbed, primarily by water vapor, carbon dioxide, oxygen, and ozone. These constituents also absorb some reflected radiation. Regardless of the energy source, the energy that becomes heat is reradiated as longwave radiation, some going to outer space and some returning to the earth's surface.

Ultimately a balance is reached and the incoming radiation equals the outgoing radiation. Obviously, the energy balance of the plant environment is very complex since energy is changed from the shortwave to the longwave form and possibly undergoes several recyclings before being lost to outer space.

Water vapor and carbon dioxide present in the atmosphere play an important role in the overall energy balance of the earth. Table 8.3 shows the influence of absolute humidity on the infrared portion of the solar spectrum. Figure 8.11 presents a graph of the absorptivity of water vapor per 0.1 mm of precipitatable water as a function of the wavelength of radiation out into the longwave portion of the spectrum. From figure 8.11, it can be seen that only a portion of longwave radiation leaving the earth's surface escapes directly to outer space. The remainder is absorbed by water vapor. Much of this is returned because a good absorber is also a good radiator.

Figure 8.12 illustrates the effect of this phenomenon. Radiation from the earth can be approximated by the blackbody radiation curve for a radiation temperature of about 300 K, as indicated by the outer curve in figure 8.12. The radiation actually being freely passed through the atmosphere directly from the earth is given by the inner lines. The magnitude of the passed radiation, of course, is greatly dependent on the amount of water vapor present in the

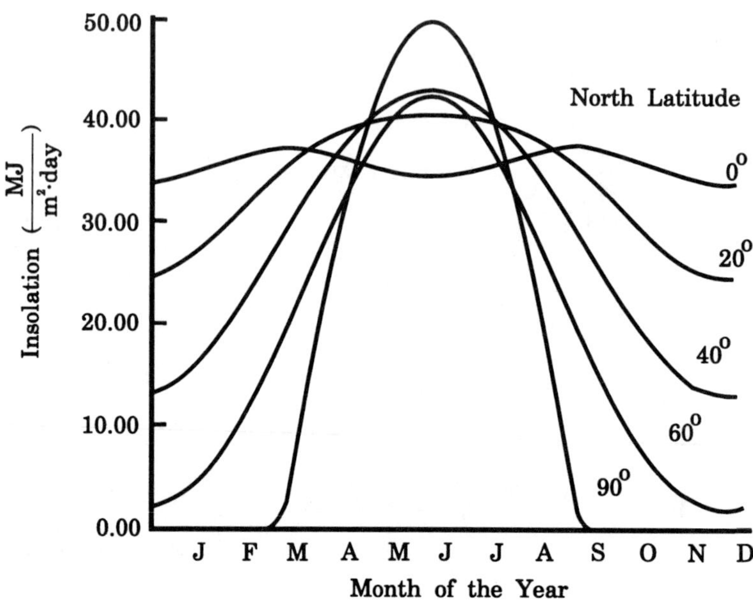

Figure 8.9. Insulation received in the earth's northern hemisphere as influenced by the time of year and the latitude. In the southern hemisphere, insulation would peak in December.

Table 8.2. Normal spectral radiancy (W/m^2) at sea level as a function of the optical air mass

Wavelength Interval	Multiple m of the Optical Air Mass					
μ	m = 0	m = 1	m = 2	m = 3	m = 4	m = 5
0.29–0.40	94.80	41.10	20.20	10.50	5.58	2.79
0.40–0.70	542.00	429.00	335.00	264.00	210.00	167.00
0.70–1.40	316.00	316.00	274.00	239.00	210.00	185.00
1.40–1.50	162.00	97.60	41.80	58.60	49.50	41.80
1.50–1.90	75.50	52.30	46.00	41.80	39.00	36.20
1.90—>	117.00	13.20	9.80	7.70	6.97	6.27
totals	1352.00	950.00	757.00	622.00	521.00	440.00
% Reduction	0.00	29.80	44.00	54.00	61.40	67.50

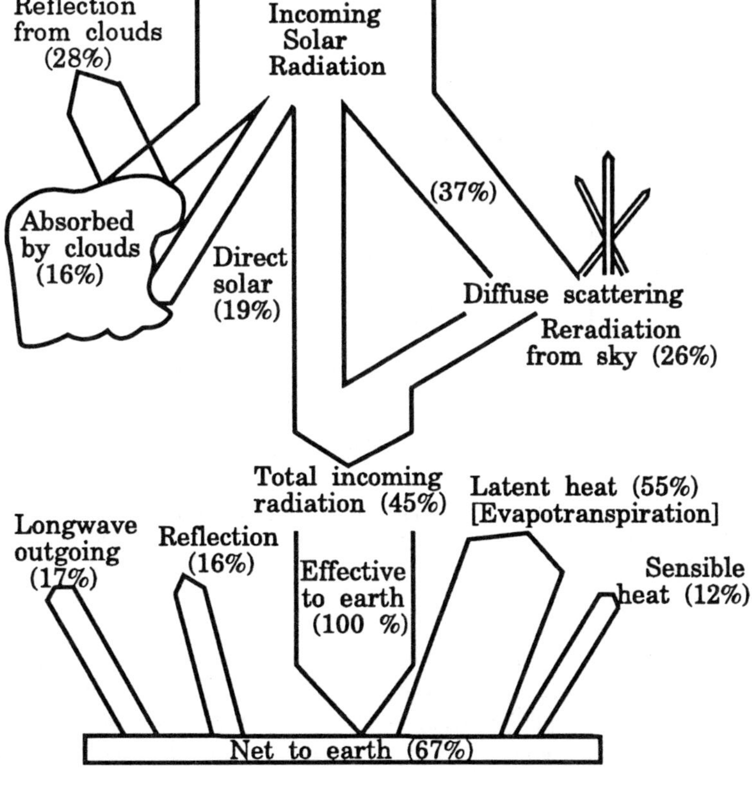

Figure 8.10. Influence of the atmosphere and biosphere on solar radiation.

Table 8.3. The radiant energy flux absorbed by water vapor in the atmosphere

		Water vapor content along the entire path the radiation traverses (kg/m^2)							
Wavelength (μ)	Radiation flux outside the atmosphere (W/m^2)	5	10	20	30	40	50	60	80
0.70–0.74	54.7	1.1	1.7	3.3	3.83	4.9	6.0	7.1	9.3
0.79–0.84	54.0	1.3	2.2	3.8	4.3	5.5	6.3	7.5	9.8
0.86–0.99	110.0	9.9	14.3	20.9	25.3	29.4	34.2	36.3	40.7
1.03–1.23	114.9	11.5	14.2	22.9	26.4	31.0	33.3	32.6	39.1
1.24–1.53	99.9	37.5	41.9	47.8	52.9	55.9	57.9	58.9	59.8
1.53–2.10	84.3	24.4	26.9	29.4	31.1	31.9	32.8	34.6	35.4
Totals	517.8	85.8	101.2	128.1	143.9	158.7	170.5	180.0	194.1

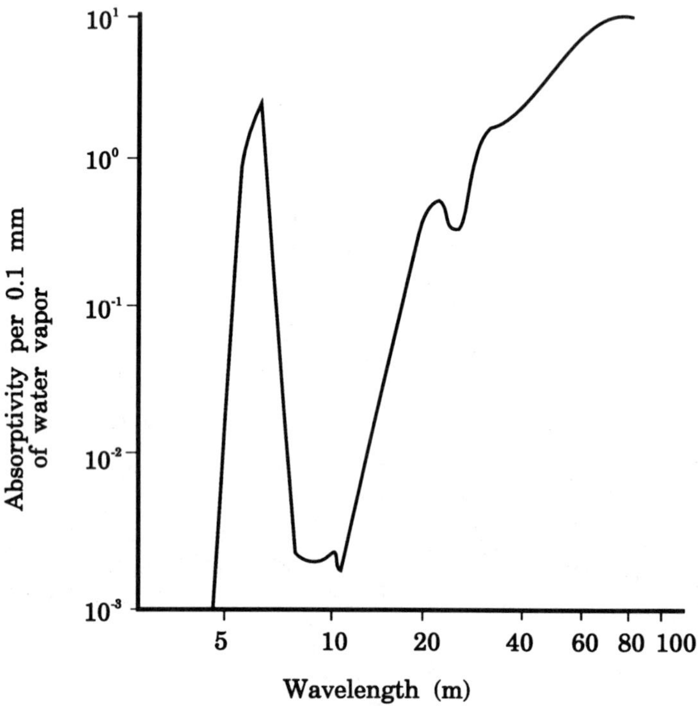

Figure 8.11. The influence of water vapor on longwave radiation. longwave radiation at less than 8 μm and greater than 15 μm is very heavily absorbed by water vapor.

atmosphere. Less radiation is lost directly through a high humidity atmosphere than through an arid, low humidity atmosphere. This explains why chances of frost are diminished when clouds are present, and why the air temperature in a desert climate drops so quickly when the sun sets. In the latter case, very rapid cooling as a result of radiation to outer space occurs because there is little water vapor in the atmosphere to intercept and re-emit the radiation back toward the earth.

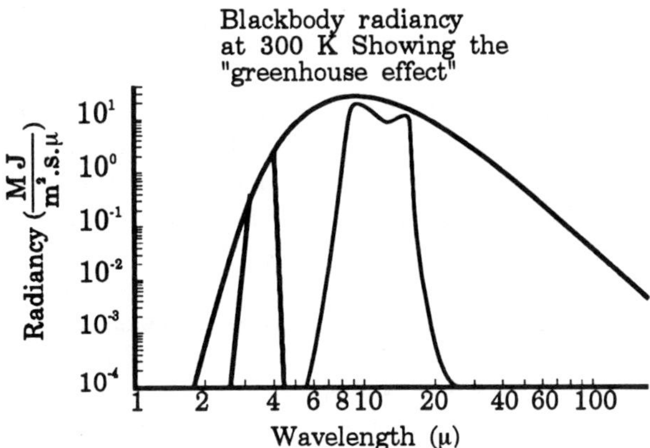

Figure 8.12. The actual radiancy curve at a wavelength of 10 μm compared to the theoretical curve. Note the effect of the earth's atmosphere on the radiancy.

Example 8.15. The affect of water vapor in the atmosphere
If the atmosphere contains 80 kg/m^2 of water vapor, calculate the total incoming radiation flux in the 0.70 to 2.1 μ band. Does the absorbed energy reach the earth? At approximately what wavelength?

Solution
From table 8.3 the incoming solar radiancy in the interval from 0.7 to 2.1 μ is about 518 W/m^2. The energy absorbed due to an absolute humidity of 80 kg/m^2 is 194.1 W/m^2, so that the energy available to reach the earth if no other influence is present is:

$$518\frac{W}{m^2} - 194.1\frac{W}{m^2} = 323.9\frac{W}{m^2} = 324\frac{W}{m^2} \qquad 8.67$$

At least part of this energy will reach the earth since a good absorber is also a good emitter. Thus, one could expect approximately 50% of the absorbed energy flux to be emitted at long wavelengths and returned to the earth.

Example 8.16. Effect of water vapor on nighttime temperatures
What effect might a clear, still, dry night have on temperatures of apple tree buds in an orchard at 2° C.

Solution
With little water vapor present in the atmosphere, and with no air movement to enhance convective heat transfer, the buds may lose energy by radiation faster than sensible heat can be supplied by conduction through the still air. Consequently, bud temperatures may drop several degrees below the ambient temperature.

Energy Balance of the Green Plant Environment

The general form of the energy balance of the plant environment will involve consideration of shortwave radiation incident on, and either absorbed or reflected by, the plant canopy, the longwave radiation from atmospheric sources, and longwave radiation from the earth, as well as that which is reradiated by the plant canopy. The energy balance must include energy transferred within the canopy in the form of latent heat and that used in chemical reactions. In addition to energy transfer as latent heat, sizable amounts of energy also may be transferred within the plant canopy as sensible heat. The energy balance equation must account for energy stored as internal energy, both in vegetation and in the soil mass.

The amount of energy stored and transferred in the form of latent heat is a direct function of the water vapor present in the plant canopy. Therefore, the amount of energy involved in latent heat transfer and storage is denoted by the symbol lE, where l denotes the latent heat of vaporization, and E is used to denote the amount of water vapor actually present. The energy present in the plant canopy in the form of sensible heat is denoted by H. The energy stored in the ground, or in the plant parts, is denoted by G. Finally, the energy utilized in photosynthetic activity is denoted by aA, where a is the chemical energy storage coefficient and has a value of approximately 15 mJ/kg of substance produced. A represents the total mass of chemical compound produced in photosynthetic activity.

Modeling the Energy Balance of the Plant Environment

In developing an equation for the energy balance of the plant environment the following symbols are used: R_s denotes shortwave radiation arriving from the sun. R_l denotes the longwave radiation. The component of radiation that is reflected from the earth's surface is then denoted by rR_s, where r is used to denote the albedo of the surface to avoid confusion with ρ which is reserved for long wave radiation. Table 8.4 presents values of albedo for certain surfaces that can be used as a guide by the student to determine magnitudes of reflected energy. From table 8.4, the albedo varies from 95% for fresh snow cover to less than 10% for water surfaces.

We define radiation toward the earth's surface as positive and express the energy balance of the plant environment mathematically as:

$$(1 - l)R_s + R_{ln} + lE + H + G + aA = 0 \qquad 8.68$$

The first two terms on the left of equation 8.68 represent the net amount of shortwave and longwave radiation at the plant. The net amount of shortwave radiation arriving at the plant is determined by measuring the incoming shortwave radiation flux and subtracting from it the flux reflected due to the albedo of the object. The net longwave radiation within the plant canopy is determined by the total amount of longwave radiation incoming to the plant from all sources, including radiation from the atmospheric constituents, other plants and, the ground surface, minus the radiation emitted by the plant. This value is denoted by R_{ln}, where the subscript is understood to imply net radiation. R_{ln} can be determined by measurement using suitable instrumentation.

In many instances it is possible to neglect certain components in equation 8.68. Over periods of several days, the total change in G is small because the inward radiation at the earth's surface during the day is compensated by outward radiation occurring during the night. For this reason, over periods of one to several days, G is often neglected. In addition, actual measurements have shown that aA seldom exceeds 2 to 3% of the total incoming shortwave radiation. Several reasons account for the low value. Although shortwave radiation has a peak at approximately 0.5 μ, i.e., in the green, the most efficient use of energy for photosynthetic activity occurs between wavelengths of 0.6 to 0.7 μ, i.e., in the red and near infrared. Throughout the remaining portion of the visible spectrum, the efficiency of radiation used for photosynthetic activity is low. Utilization approaches zero beyond wavelengths greater than 0.75 μ, as indicated in figure 8.13.

Since as little as 1.5% or less of the total energy is used in photosynthesis, it is common to ignore this component of the energy balance of the plant environment. Note, however, that an increase of from 1.5 to 3% in the total amount of incoming solar radiation used for photosynthesis could increase the total amount of carbohydrates produced by plants. No successful engineering approaches to this problem exist.

Over time the energy balance of a plant environment can change dramatically. Figure 8.14 illustrates the changes that may occur in one day in a plant environment with ample water. Both long- and shortwave radiation act as the energy source. During periods of darkness, the net effect is energy loss due to longwave radiation. In figure 8.14, the energy transfer due to sensible heat and latent heat are positive away from the plant. Figure 8.14 illustrates the effect of photosynthesis on the movement of water through the plant and use of energy by the transpiration process, since the latent heat term rises sharply as shortwave radiation becomes significant. Note that heat is added to the plant as the plant loses energy by longwave radiation and dew begins to condense.

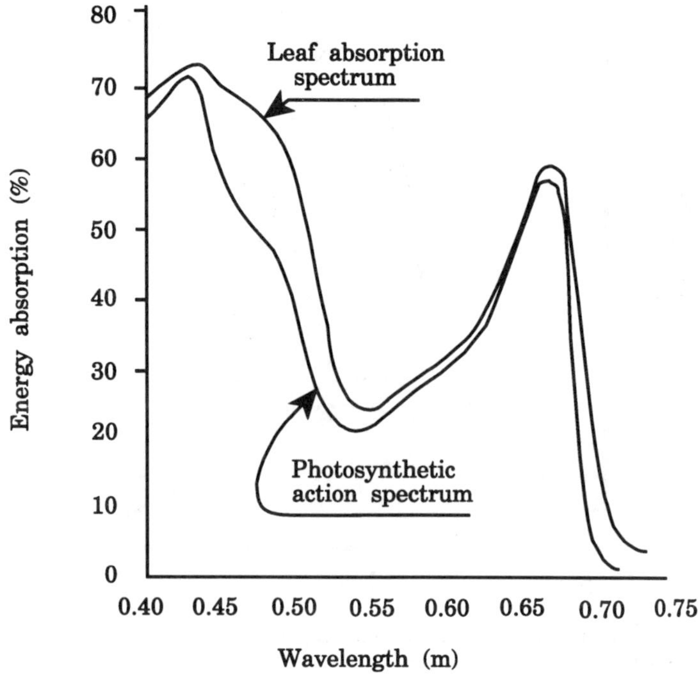

Figure 8.13. A comparison of the leaf absorption and the photosynthetic actions spectra of a leaf.

The sensible heat movement is a loss, i.e., it is in a direction away from the plant canopy as soon as energy becomes available from solar radiation to raise the temperature of the plant canopy above that of the ambient air.

The soil heat flux, G, is positive if the flux is into the ground. As figure 8.14 shows, the energy absorbed into the soil mass during the day is almost balanced out by energy lost during the night. The overall effect for one day is a negligible net transfer of energy. The behavior of the energy converted into the heat in the soil mass with depth and over periods of time is significant, however, and engineering modifications can be effected to modify this behavior. The soil heat flux is studied separately in a later chapter.

Three components of longwave energy must be considered. Energy coming from outer space as longwave is depleted by absorption and scattering as it penetrates the canopy. Radiation from surrounding leaves and from the earth under the canopy also contributes to the heat load due to longwave radiation. Finally, the leaf is radiating in the longwave due to its own temperature. Over a short period, the energy absorbed by the plant parts and the energy used for photosynthesis is very small and can be neglected.

In addition to longwave radiation by the leaf, the energy received by the leaf is dissipated by convection currents of air around the leaf, and by latent heat involved in evaporation and transpiration of water.

Example 8.17. Components of the heat load on a leaf in a canopy
What are the components of the heat load on a leaf within a crop canopy and what must be considered in estimating these components?

Solution

The shortwave energy load on the leaf is equal to shortwave radiation on the canopy, less that quantity of energy reflected by the crop albedo, less energy lost by reflection and scattering as the radiation penetrates the canopy, less that energy actually reflected by the leaf.

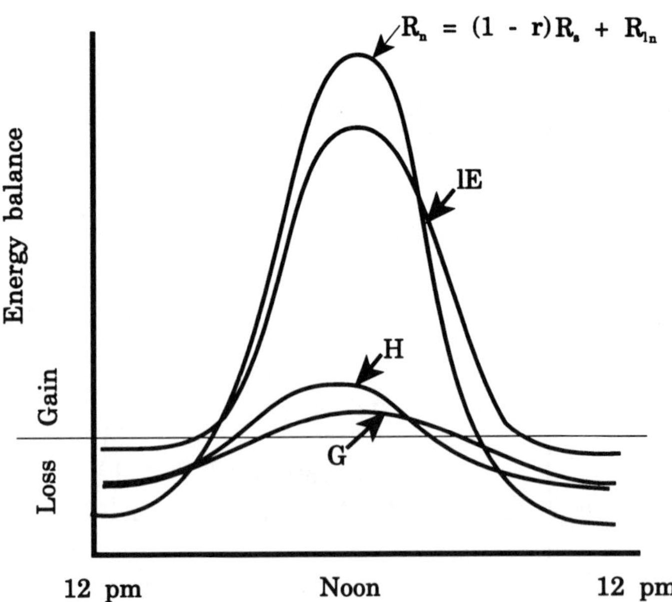

Figure 8.14. The incoming net radiation is portioned into latent heat (*l*E), sensible heat (H), and heat absorbed by the earth and vegetation (G).

Table 8.4. Albedo values for various surfaces, soils, and vegetal covers

Soil	Albedo	Vegetation	Albedo
Black soil, dry	14	Spring wheat	10–25
Black soil, moist	8	Winter wheat	16–23
Gray soil, dry	25–30	Winter rye	18–23
Gray soil,moist	10–12	High grass, densely grown	18–20
Fallow, dry	23	Green grass	26
Fallow, moist	16	Grass dried in the sun	19
Plowed field, moist	8–12	Tops of oak forest	18
Desert, loamy surface	5–7	Tops of pine forest	14
Sand, yellow	14	Tops of fir forest	10
Sand, white	29–31	Cotton	20–22
Sand, bright-fine	35	Lucerne (start and blossom)	23–32
Freshly plowed, bare black clay	34–40	Rice field	12
	Lettuce	22	
	Beets	18	
	Potatoes	19	

Influence of Surface Condition on Albedo		Influence of Snow Cover on Albedo	
Kind of Roughness	Albedo	Type of Snow	Albedo
Flat, even surface	30	Dense, dry and clean	86–95
	–	Clean, moist, fine	63–64
	31	Clean, moist, granular	61–62
Covered with pellicles, moist	27	Porous, moist, grayish	45–47
Covered with small pebbles	25	Light brown, wet	31
Covered with bigger pebbles	20	Dirty	29
Newly plowed field	17		

The nature of various effects on the sensible heat exchanges, H, within the plant environment are important. From the laws of heat transfer, exchanges of sensible heat within the plant environment occur according to heat transferred in response to a gradient of temperature. Because the green plant canopy both reflects and absorbs heat, the canopy itself affects the exchange of sensible within the canopy. The mathematical model describing the transfer of sensible heat is:

$$H = -c_p \times \rho_a \times K_H \frac{\partial T}{\partial z} \qquad 8.69$$

where

H = sensible heat flux
c_p = specific heat of air at constant pressure 1007 J/(kg·C°)
ρ_a = density of moist air
$\partial T/\partial z$ = temperature gradient
K_H = eddy transfer coefficient

From equation 8.69 it is seen that the transfer of sensible heat within the plant environment is assumed to be a convective process. The total quantity of heat transferred is related to the density of the air being moved and the specific heat of the air. The potential causing transfer is assumed to be temperature, so that the potential gradient is expressed as a temperature difference per unit of distance. The transfer coefficient, K_H, has dimensions of area per unit time.

In addition to the sensible heat in the energy balance of the plant environment, we must consider energy transfer in the form of latent heat. The equation describing the transfer of energy in the form of latent heat is:

$$lE = -l_{va} \times \rho_a \times K_w \frac{\partial q}{\partial z} \qquad 8.70$$

where

E = the flux of liquid water converted to water vapor
l_{va} = latent heat of vaporization (2.47 kJ/g)
ρ_a = density of moist air
$\partial q/\partial z$ = gradient of specific humidity
K_w = the transfer coefficient for water vapor in the atmosphere

The general form of equation 8.70 is seen to be identical with that of equation 8.69, i.e., a phenomenological expression in which a quantity, (H or lE), is proportional to the gradient of some potential (T or q). The proportionality constant for equation 8.69 and equation 8.70 is composed of products of parameters that are essential to model the process. For heat transfer, the parameters include the specific heat of air and the density of air as well as a transfer coefficient. For energy transferred in the form of latent heat, the transfer coefficient includes the latent heat of vaporization and the density of the air in addition to the transfer coefficient.

Bowen Ratio

The utility of equation 8.69 and equation 8.70 depends on determination of the transfer coefficients K_H and K_w. This determination is very complex and depends on wind speed and turbulence within the plant canopy. Under certain assumptions, however, valuable information can be obtained concerning the energy status of the plant canopy without actually determining the coefficients. Since the energy being transferred is a function of the density of the air in both instances, and because the nature of the equation indicates that the transfer depends on the air movement taking place, it is assumed that the value of K_H is identical with K_w (an assumption sufficiently accurate for engineering purposes).

If $K_H = K_w$ in equations 8.69 and 8.70, it is possible to formulate a dimensionless number whose sign and magnitude is shows the status of energy balance within the plant canopy. To do this we divide equation 8.69 by equation 8.70 to obtain:

$$\beta = \frac{H}{lE} = \frac{c_p \times \Delta T}{l \times \Delta q} \tag{8.71}$$

where β is a dimensionless number termed the Bowen ratio and the partial differentials have been replaced by finite increments. In equation 8.71 it is assumed that the values of temperature and specific humidity entering into equation 8.71 are to be determined at the same point so that Δz, the distance which separates the points at which the determinations are made, can be canceled.

The Bowen ratio can be used as an indicator of the conditions existing within the crop canopy. Consider that if the surface of the soil beneath the crop canopy is wet, evaporation will take place and the difference in specific humidity between the air at the soil surface and some point outside the crop canopy will be large. Also, the difference in temperature between the soil surface and the point outside the crop canopy may be small. Thus, most of the energy being carried away from the canopy will be transferred in the form of latent heat. Under these conditions the Bowen ratio will be a small positive number.

If, however, the surface of the soil under the plant's canopy is dry, then the difference in specific humidity between the soil surface and some point exterior to the plant canopy will be small. The temperature difference between the soil surface and the air outside the plant canopy, however, may be large, since energy entering the soil is converted into heat rather than being converted into latent energy by evaporation of water. In this case, the actual value of the Bowen ratio will be large, indicating that most of the energy is being transported as a sensible heat flux.

Under certain conditions it is possible for the Bowen ratio to be negative. A negative Bowen ratio could be expected early in the morning when the soil surface is cool and the air above the plant canopy is warm. The temperature is lower at the soil surface than outside the canopy while the specific humidity is greater at the soil surface than outside the plant canopy. With a negative Bowen ratio, the energy transfer as sensible heat will occur in a direction toward the plant canopy while latent energy will be transferred away from it.

In addition, the transfer of energy horizontally can affect the magnitude and sign of the Bowen ratio. This may be visualized by imagining air moving over a barren desert area that is at a high temperature. After passing over the barren area and absorbing heat from the surface, the air then moves into the crop canopy where it discharges its heat either in the form of latent heat of vaporization due to high transpiration from the plants (see example 8.2). Alternatively, the heat can be given off as sensible heat due to a temperature gradient. The effect may occur several times sequentially in a situation where irrigated areas are interspersed with arid, unirrigated areas. The phenomenon is termed the clothesline effect, and is not uncommon in the irrigated regions of the southwest United States. Horizontally advected heat will result in a heat load on a plant canopy greater than that predicted by radiation alone.

Figure 8.15(a) illustrates the behavior of the energy-transfer mechanism within an irrigated crop canopy. Over time, soil goes from a fully irrigated to a very dry condition as a function of the ratios of sensible heat, H, and latent heat, lE, to the net radiation, R_n. In figure 8.15(b) a value taken from the right-hand ordinate of figure 8.14(a), divided by the corresponding value taken from the left-hand ordinate yields the Bowen ratio. Immediately following irrigation, the greatest percentage of energy is transferred in the form of latent heat of vaporization due to evaporation of water from the ground surface and to evapotranspiration by the crop. Note that about 15 days following irrigation the sensible flux is into the crop canopy from the surrounding areas because of the inflow of advected warm air into the crop canopy. At about the 15th day, however, sufficient depletion of water has taken place so the primary energy transferred is in response to a temperature gradient. Because water has been depleted in the crop canopy, most of the sensible heat transfer is away from the crop and the crop canopy has become warmer than the surrounding air. At this point the water stress on the plant canopy is such that the crop could be severely harmed within several days if water is not provided by irrigation or rain. The Bowen ratio measured in the crop canopy is an indicator of the energy status of the canopy.

The Bowen ratio provides a means of estimating components of the energy balance of the plant environment. If the photosynthetic energy requirement is neglected, equation 8.68 can be written:

$$R_n \, lE + H + G = 0 \qquad 8.72$$

where R_n denotes net radiation, a value that can be measured directly. Over either a very short period (less than 1 h, for instance) or over an extended period of multiples of 24 h, the term G can be neglected. We can then make use of the Bowen ratio in the form:

$$\beta \times l_{va} \times E = 0 \qquad 8.73$$

along with our above assumptions to write:

$$R_n + lE(\beta + 1) = 0 \qquad 8.74$$

from which:

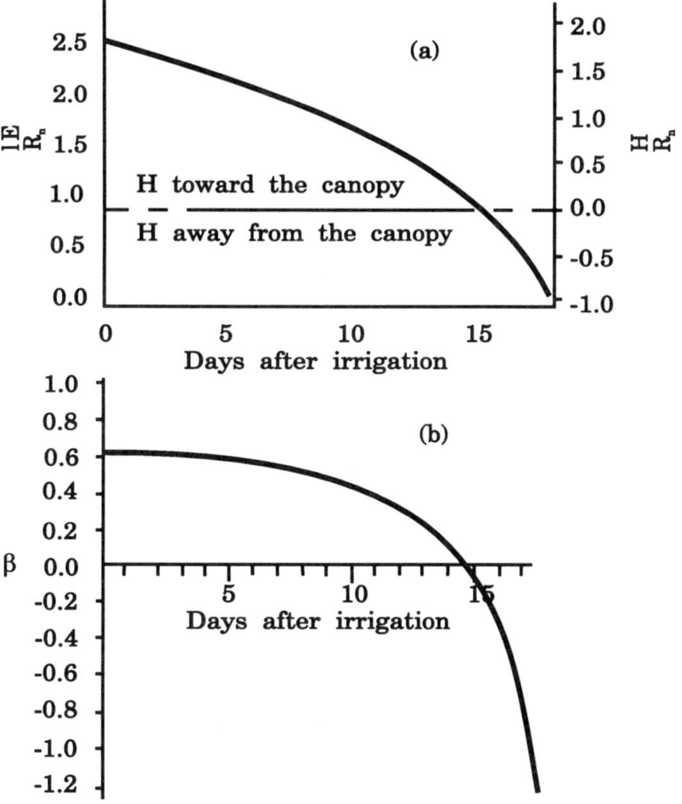

Figure 8.15. (a) Indicates the magnitude of latent energy and sensible heat over a plant canopy following an irrigation. (b) Indicates the Bowen ratio, over β for the period. The Bowen becomes negative when the plant canopy becomes warmer than the surroundings.

$$E = -\frac{R_n}{l(\beta + 1)} \quad 8.75$$

In this form, the Bowen ratio provides a means of determining water loss from the crop canopy based upon a measurement of net radiation and temperature and absolute humidity differences.

Example 8.18. Irrigation and the Bowen ratio
A farmer desires to use as little irrigation water as possible to irrigate corn. To determine when it is time to irrigate, he adopts the convention that he will irrigate when the sensible heat flux is opposite in sign from the latent heat flux, i.e., when the Bowen ratio is negative. Using a sling psychrometer he determines that the temperature and relative humidity at 0.20 m above the ground and 0. 20 m above the crop canopy are, respectively, 31° C at 57% and 29° C at 80% relative humidity. What is the Bowen ratio? Upon contacting the university, he determines the net radiation flux is 165 W/m^2 toward the crop (positive), what is the evaporative flux in kg/(m^2·h)?

Solution
It is necessary to determine q, the specific humidity. To do this the actual vapor pressure at the points of measurement must be known. We are given the relative humidity and we can make use of equation 3.2 1, the Clausius-Clapeyron equation given in example 3.10. At the top of the canopy:

$$\begin{aligned} e^\circ_{top} &= \exp\left(19.0177 - \frac{5327}{302}\right) \\ &= 3.97 \text{ kPa} \end{aligned} \quad 8.76$$

from which:

$$\begin{aligned} e_{top} &= 3.97 \text{ kPa} \times \frac{80\%}{100\%} \\ &= 3.18 \text{ kPa} \end{aligned} \quad 8.77$$

so that:

$$\begin{aligned} q_{top} &= 0.622 \frac{3.18}{100} \\ &= 0.0195 \end{aligned} \quad 8.78$$

At the ground:

$$\begin{aligned} e^\circ_{lower} &= \exp\left(19.0177 - \frac{5327}{304}\right) \\ &= 4.46 \text{ kPa} \end{aligned} \quad 8.79$$

Therefore:

$$e_{lower} = 4.46\ \text{kPa} \times 0.57 = 2.54\ \text{kPa} \qquad 8.80$$

from which:

$$\rho_{lower} = 0.622\,\frac{2.54}{100} = 0.0156 \qquad 8.81$$

It is now possible to calculate β:

$$\beta = \frac{c_p \cdot \Delta T}{l \cdot \Delta q} = \frac{1000\,\frac{\text{J}}{\text{kg}_{air} \cdot \text{C}^\circ} \times (31 - 29)\ \text{C}^\circ}{2.47 \times 10^6\,\frac{\text{J}}{\text{kg}_{water}} \times (0.0156 - 0.0195)\frac{\text{kg}_{water}}{\text{m}^3_{air}}} = -0.21 \qquad 8.82$$

The evaporative flux can now be found using equation 8.75.

$$E = \frac{R_n}{l \cdot (\beta + 1)} = \frac{165\,\frac{\text{J}}{\text{m}^2 \cdot \text{s}}}{2.47 \times 10^6\,\frac{\text{J}}{\text{kg}}(-0.21 + 1)} = 8.45 \times 10^{-7}\,\frac{\text{kg}}{\text{m}^2 \cdot \text{s}} = 0.30 \times \frac{\text{kg}}{\text{m}^2 \cdot \text{h}} \qquad 8.83$$

Note: If the flux calculated above is assumed to be the average, it can be used to estimate the daily rate of water loss for the crop in mm/day. This is about 3 mm in a 12-h day. In severe cases, given ample water and a positive Bowen ratio, rates as high as 7 to 8 mm/d may occur.

Exercises

8.1. Compute the constant c_1 in equation 8.15, Planck's Law that models the radiancy for a body emitting electromagnetic radiation in according to the theoretically possible amount.
Give your answer in $W\mu^4/m^2$.

8.2. Compute the radiancy R_λ associated with radiation from the sun at 5776 K at $\lambda = 0.8\ \mu$ (in the near infrared) and at $\lambda = 8.0\ \mu$ (in the far infrared). What is the ratio $R_{0.8}/R_{8.0}$?
[You will have to have c_2, the second radiation constant].

8.3. (a) Compute the radiancy ratio associated with radiation from a body such as the earth at 300 K for $\lambda = 0.8\ \mu$ and $\lambda = 8\ \mu$. Compare your answers with those of exercise 2.

(b) What must be concluded about the wavelengths at which electromagnetic radiation is transmitted by bodies at the two temperatures?

8.4. N_2 is one of the most abundant gases in the atmosphere. Barrow (1966) gives the dissociation energy of N_2 as 942.4 kJ/mol.

(a) What wavelength of radiation would be necessary to furnish a quantum of energy to dissociate N_2?

(b) What portion of the spectrum would this lie in?

(c) What is the magnitude of the radiancy from the sun at 5776 K at this wavelength?

8.5. Figure 8.7a, b, and c illustrate the three possible modes of vibration of a water molecule.

(a) In what part of the spectrum (longwave or shortwave) would you expect to find principal absorption bands due to quanta of electromagnetic radiation that have been absorbed and used to excite molecular vibration?

8.6. The sun is approximately 1.4×10^6 km in diameter with a surface area of 6.093×10^{12} km^2. Assume the temperature of the sun is 5776 K and that the mean distance from the sun to the outer layer of the earth's atmosphere is 1.494×10^8 km. Estimate the solar constant, i.e., the solar energy flux that would impinge on 1 m^2 area perpendicular to the earth's surface.

(Hint: Calculate the total amount of energy per second emitted by the entire surface of the sun. If no losses occur through outer space, then the same rate of energy transmission would occur through a spherical surface centered at the sun and passing through the outer reaches of the earth's atmosphere. Dividing the rate by the surface area of the imaginary sphere should yield an estimate of the solar constant.)

8.7. The solar constant is about 1394 W/m^2 (83.74 kJ/(m^2·s). The earth absorbs radiation on its projected surface. However, since it rotates, it heats more or less uniformly and radiates from its spherical surface. Assume the earth's mean radius is 6 370 km. If the earth-atmosphere system has an albedo of 0.43, (a) what is the radiation temperature of the earth, and (b) where does it occur?

8.8. Radiative heating has been proposed for frost protection. Assume an apple bud is 0.5 cm in diameter and must be maintained at 0.5° C to protect it from freezing.

(a) If the farthermost bud is 8 m from a radiant heater, what will the radiant energy flux from the bud be at 0.5° C?

(b) At what rate must the radiant heater supply energy to maintain the bud at 0.5° C, assuming the bud absorbs as a blackbody?

(c) If the source is a sphere of radius R, what energy flux must the heater supply assuming that heat transfer through the bud is virtually instantaneous?

8.9. Assuming sufficient energy is available, is the scheme in problem 8 feasible as a means of frost protection? Why?

8.10. Goody (1964) offered the following expression to model the absorptivity of the atmosphere to solar radiation as a function of the water contained in a column of column of air 1 m^2.

$$\log(\alpha) = -0.74 + 0.347\left[\log(\text{w.c.})\right] - 0.056\left[\log(\text{w.c.})^2\right] - 0.006\left[\log(\text{w.c.})^3\right]$$

Assume the relative humidity is 0.6, the temperature 25° C, and the air is mixed so that the water vapor content is uniform throughout the depth of 1 km. Calculate the absorptivity of the atmosphere.

Chapter 9

Heat Flow in the Soil

An important component of the plant environment is the soil mass. The soil provides support for the plant and serves as a medium through which water and nutrients are stored and transported to the plant root system. In addition to these two very important functions, the soil mass absorbs, stores, and releases heat energy.

The incoming radiant energy fluctuations at the ground surface are quickly damped out by the soil so the temperature extremes that influence the portion of the plant above the ground level are not present below the ground level. We analyze the movement of heat in the soil to gain an appreciation of the thermal regime of the plant root system as influenced by the soil. This is done by studying the solution of the differential equation modeling heat flow.

Heat Flow Equation

An analysis of the heat-flow equation in the soil mass and of its solution is essential for describing the thermal regime of the soil. The differential equation modeling heat flow can be developed as follows.

A basic assumption for the derivation of this equation is that the flow of heat in a given direction in a medium can be described by a first-order phenomenological model:

$$Q_s = -k_x(x,y,z)\frac{\partial T}{\partial x} \qquad 9.1$$

This model is usually written assuming that the dependence of k_x on position in the medium is understood:

$$Q_x = -k_x\frac{\partial T}{\partial x} \qquad 9.2$$

where

Q_x = heat flux (energy/area-time)
k_x = thermal conductivity
T = temperature
x = distance in the direction of movement of heat energy

The phenomenological model given in equation 9.1 states that the transport of heat energy is in the direction of decreasing potential T. It further states that the rate of transport is proportional to the gradient of temperature with the proportionality constant identified as the thermal conductivity.

The first-order phenomenological form is analogous to the form of Ohm's Law: the current is the electron flux, the driving potential is the voltage drop, and the thermal conductivity is analogous to 1/R, where R is the resistance to current flow. It is the same form used in the flow of water through a porous media where the water flux is proportional to the gradient of pressure measured in the direction of flow and the proportionality constant is the hydraulic conductivity. It is the same form encountered in the description of diffusion phenomena, where the mass flux of diffusing material is proportional to the gradient of the concentration. Although a second- or higher-order dependence of flux on the driving potential may occur, the first-order dependence is satisfactory for engineering purposes

Example 9.1. Heat energy flux
Equation 9.1 can be used to determine the energy loss through a wall. If the temperatures on the inside and outside wall are expressed in °C, the thickness of the wall in m, and the flux in W/m², what are the dimensions of k?

Solution
We can determine the dimensions required for k by expressing the dimensions for each parameter in the expression in brackets beside the symbol used to delineate the parameter.

$$-k_x = \frac{Q_x \left[\frac{W}{m^2}\right]}{\frac{\partial T}{\partial x}\left[\frac{°C}{m}\right]} \qquad 9.3$$

The dimensions of k are seen to be:

$$-k_x \left[\frac{W}{m \cdot °C}\right] \qquad 9.4$$

It is most important that the proper set of dimensions be used. Example 9.1 provides a method by which the appropriate dimensions can be determined. Note that in conventional units, k is dimensioned as:

$$-k_x \left[\frac{BTU}{f \cdot h \cdot \frac{°F}{f}}\right] \qquad 9.5$$

The negative sign preceding k_x arises because the temperature gradient in the direction of energy flow is negative. Heat energy flows down the gradient from higher to lower temperature. Customarily, the direction x is taken as positive in the direction of flow. When this is done, the partial differential of temperature with respect to distance x can be expressed as:

$$\frac{\partial T}{\partial x} = \lim_{\Delta x \to 0} \frac{\Delta T}{\Delta x} \approx \frac{T_1 - T_2}{x_1 - x_2} \qquad 9.6$$

But:

$$T_1 > T_2 \text{ if } x_1 < x_2 \qquad 9.7$$

Therefore:

$$\frac{\partial T}{\partial x} < 0 \qquad 9.8$$

The differential equation describing heat flow can be obtained by applying the conservation of energy to an elemental volume of soil as depicted in figure 9.1. Figure 9.1 illustrates an elemental volume of soil within which the individual components of the energy-balance equation are to be computed to arrive at the differential equation describing heat energy flow in the soil mass.

In figure 9.1 the heat flux, Q, is shown only for the x-coordinate direction. It is assumed, however, that similar components apply to the remaining coordinates. Considering only that changes in the flux may occur within the elemental volume as the heat flux passes through, the heat flux leaving the element must be equal to the flux entering any changes that occur within the volume. Multiplying by the cross-sectional area through which the heat flux passes and summing in the x-coordinate direction yields:

$$\left[Q_x - \left(Q_x + \frac{\partial Q_x}{\partial x}\,dx\right)\right] dy\,dz = -\frac{\partial Q_x}{\partial x}\,dx\,dy\,dz \qquad 9.9$$

Similar expressions hold for the y and z coordinate directions so that the rate of change of heat energy within the elemental volume is:

$$\left(-\frac{\partial Q_x}{\partial x} - \frac{\partial Q_y}{\partial y} - \frac{\partial Q_z}{\partial z}\right)\Delta V = \dot{Q} \qquad 9.10$$

where

Q_i = soil heat flux (i = x, y, z)
ΔV = elemental soil volume (dx dy dz)
$\dot{Q}$ = rate of change of heat energy within the elemental volume

Example 9.2. Rate of energy change within an elemental volume
An elemental volume of soil with the z-axis oriented positive downward has dimensions of 3 mm per side. Given that the elemental volume is being heated by solar energy from the upper side there are no fluxes in the x- and y-directions, if the z-component heat flux has the form:

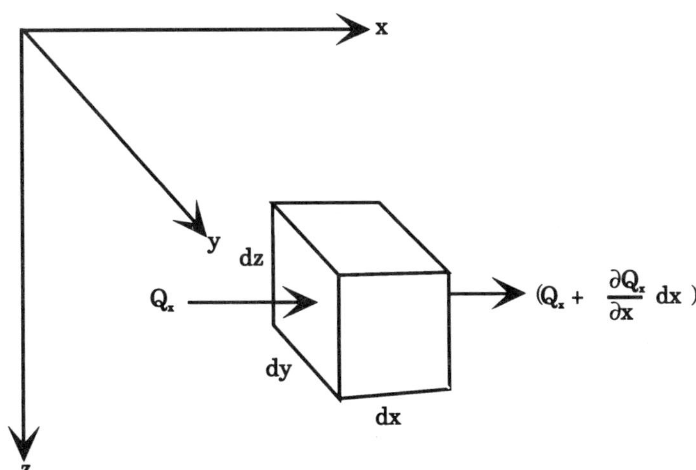

Figure 9.1. A schematic of the effect an element unit of soil has on the flux of heat passing through it. The same effect could be seen on the other two faces corresponding to the z- and y-directions.

$$\left\{Q_z - \left[Q_z + \frac{\partial Q_z}{\partial z} dz\right]\right\} = 0.002 \frac{J}{mm^2 \cdot h} \qquad 9.11$$

compute the rate at which energy is being stored within the elemental volume.

Solution

The cross-sectional area through which the flux moves is 9 mm^2. Therefore, the rate of change of energy within the elemental volume is:

$$0.002 \frac{J}{h \cdot mm^2} \times \frac{h}{3600\ s} \times 9\ mm^2 = 5 \times 10^{-6}\ W \qquad 9.12$$

The changes that may take place within the elemental volume to affect the heat energy flux as modeled by equation 9.2 are restricted to those changes that are peculiar to the entire soil mass including the soil particles and the water contained within the soil. The heat energy flux will be affected by the specific heat of the water and the soil particles and will cause the temperature of the mass to change. Let ρ denote the mass density and c the specific heat of the volume so that the rate of energy change per unit volume causing a change in temperature ΔT of the volume is modeled by:

$$\Delta E = \rho c\, dx\, dy\, dz\, \Delta T \qquad 9.13$$

Since the change occurs in some finite interval of time, dividing the increment of energy by $\Delta\tau$, the interval over which the change occurs yields:

$$\frac{\Delta E}{\Delta \tau} = \rho c\, dx\, dy\, dz \frac{\Delta T}{\Delta \tau} \qquad 9.14$$

After taking limits as $\Delta\tau$ goes to zero an expression for the change in energy within the volume due to the interaction of the energy flux with the specific heat and mass density of the soil is obtained:

$$\dot{Q} = \rho c \frac{\partial T}{\partial \tau} \Delta V \qquad 9.15$$

where

$\dot{Q}$ = rate of change of energy within the elemental volume
ρ = mass density
c = specific heat per unit mass
T = temperature
ΔV = volume element (dx dy dz)

Equation 9.14 and 9.15 can now be equated to obtain:

$$\left(-\frac{\partial Q_x}{\partial x} - \frac{\partial Q_y}{\partial y} - \frac{\partial Q_z}{\partial z}\right) \Delta V = \rho c \frac{\partial T}{\partial \tau} \Delta V \qquad 9.16$$

Example 9.3. Temperature changes in a soil

Compute the rate at which the temperature of the elemental volume of soil in example 9.2 would change if the bulk density and specific heat of the soil are 1470 kg/m^3, and 812 J/(kg·° C), respectively. These values are for a clay soil. Similar values for a sand would be 1440 kg/m^3 and 1089 J/(kg·° C). Compare the rate at which a temperature change could be expected for the two soils.

Solution

From example 9.2, the rate of change of energy is 5×10^{-6} W or 1.8×10^{-2} J/h. Therefore, from equation 9.15:

$$\frac{\partial T}{\partial \tau} = \frac{\dot{Q}}{\rho\, c\, dx\, dy\, dz} = \frac{0.018\,\frac{J}{h} \times 10^9\,\frac{mm^3}{m^3}}{1470\,\frac{kg}{m^3} \times \frac{812\,J}{kg\cdot{}^\circ C} \times 27\,mm^3} = 0.56\,\frac{{}^\circ C}{h} \qquad 9.17$$

For the sand, a similar computation gives a rate of 0.43° C/h. The clay exhibits a faster temperature rise because is has a lower specific heat. The values are for a Yolo clay with 29% volumetric water, and for a Kalkaska loamy sand with a water content of 5.7%. The behavior of these soils under field conditions will depend on the thermal conductivity of the soils, and the rate at which incoming energy is being supplied to the elemental volume.

In addition to energy changes resulting from the interaction of the heat energy flux with the specific heat and mass density of the bulk soil, additional contributions to the rate of energy change may occur as a result of biological, chemical, and biochemical changes that may be occurring at localized points within the soil mass. At some locations within the soil system decaying organic matter may be generating energy. Biochemical reactions of metabolism associated with root growth at other locations could be occurring that might result in the production of energy. These changes may be viewed as localized sources or sinks for heat energy that are independent of the effect of the soil properties.

While knowledge of the specific nature of the source/sink term may be necessary for a complete solution of the heat equation, at this point, it is sufficient to denote symbolically that such production or utilization occurs. The source/sink can be incorporated into the equation through including an additional term $\dot{q}\Delta V$ that has dimensions of energy production or utilization. By including the source/sink term and cancelling the element of volume ΔV, equation 9.16 becomes:

$$\left(-\frac{\partial Q_x}{\partial x} - \frac{\partial Q_y}{\partial y} - \frac{\partial Q_z}{\partial z}\right) + \dot{q} = \rho\, c\,\frac{\partial T}{\partial \tau} \qquad 9.18$$

which is solvable if the fluxes Q_x, Q_y, and Q_z are expressed in a recognizable form by making use of equation 9.1, the first order phenomenological form for heat flow. Substitution of equation 9.1 into 9.16 yields:

$$\frac{\partial}{\partial x}(k_x)\frac{\partial T}{\partial x} + \frac{\partial}{\partial y}(k_y)\frac{\partial T}{\partial y} + \frac{\partial}{\partial z}(k_z)\frac{\partial T}{\partial z} + \dot{q} = \rho\, c\,\frac{\partial T}{\partial \tau} \qquad 9.19$$

Homogeneity and Isotropy

A solution to the heat-flow equation as given above is difficult to obtain since the nature of the dependence of thermal conductivities on the coordinates must be known. However, certain characteristics of the soil medium can simplify the mechanics of the solution. Although the thermal conductivities, k_x, k_y, and k_z, are unique for the directions of heat flow indicated by the subscript, except for special cases, isotropy and homogeneity apply.

The property of isotropy states that the thermal conductivities are equal in two or more directions. This means that the *magnitude* of the thermal conductivity is independent of the direction of the flow of heat energy. For a medium that is isotropic in all three coordinate directions:

$$k_x(x,y,z) \equiv k_y(x,y,z) \equiv k_z(x,y,z) \qquad 9.20$$

If the thermal conductivity in a given direction is independent of *location* within the medium, the medium is homogeneous. This means that the thermal conductivity is not a function of the coordinate position. It simply means that the thermal conductivity at any point along the coordinate axis is identical to the conductivity at any other point along the axis. Another way of looking at homogeneity is that the rate of change of thermal conductivity with respect to the coordinate dimension is zero. For example, homogeneity in the x-direction implies:

$$\frac{\partial k_x(x,y,z)}{\partial x} \equiv \frac{\partial k_x(x,y,z)}{\partial y} \equiv \frac{\partial k_x(x,y,z)}{\partial z} \equiv 0 \qquad 9.21$$

with a similar implication for k_y and k_z if homogeneity holds in these coordinate directions as well. If the thermal conductivity is both homogeneous and isotropic in all directions, then:

$$k_x(x,y,z) \equiv k_y(x,y,z) \equiv k_z(x,y,z) \equiv k \equiv \text{a constant} \qquad 9.22$$

Under the assumption as given in equation 9.22, k can be factored out so equation 9.19 can be written:

$$\frac{\partial^2 T}{\partial x^2} + \frac{\partial^2 T}{\partial y^2} + \frac{\partial^2 T}{\partial z^2} + \dot{q} = \frac{\rho c}{k}\frac{\partial T}{\partial \tau} \qquad 9.23$$

Equation 9.23 is the differential equation governing heat flow in a homogeneous, isotropic medium in which heat is being produced or used at the rate, $\dot{q}$.

Example 9.4. Isotropic and/or homogeneous thermal conductivity
Assume a medium is isotropic throughout, but homogeneous only along the x- and y-axes, if no heat is being generated or used within the medium and depth is measured in the z-direction. Write the general equation governing heat flow in the medium.

Solution
By virtue of equation 9.20, isotropy implies:

$$k_x(x,y,z) \equiv k_y(x,y,z) \equiv k_z(x,y,z) \equiv k_z(z) \qquad 9.24$$

The above expression states that the magnitude of the thermal conductivity is the same regardless of direction at any vertical position in the medium. However, the magnitude of the thermal conductivity still varies with depth because of the lack of homogeneity. Because of inhomogeneity in the z-direction:

$$k_z(x,y,z) = k_z(z) \qquad 9.25$$

So that:

$$\frac{\partial}{\partial z}\left(k_z \frac{\partial T}{\partial z}\right) = \frac{\partial k_z}{\partial z}\frac{\partial T}{\partial z} + k_z \frac{\partial^2 T}{\partial z^2} \qquad 9.26$$

Since there is no heat generated or used within the medium:

$$\dot{q} = 0$$

Equation 9.19 can now be written:

$$\left(k_z(z)\right)\left(\frac{\partial^2 T}{\partial x^2} + \frac{\partial^2 T}{\partial y^2}\right) + \frac{\partial k_z(z)}{\partial z}\frac{\partial T}{\partial z} + \left[k_z(z)\right]\frac{\partial^2 T}{\partial z^2} = \rho c \frac{\partial T}{\partial \tau} \qquad 9.27$$

The condition of isotropy implies that the magnitudes of k_x and k_y are identically equal to the magnitude of k_z. Since homogeneity exists in the x and y-directions, the partial differentials $\partial k_x/\partial x \equiv \partial k_y/\partial y \equiv 0$ and only the second partial derivative of temperature remains from in the x- and y-coordinate directions. The lack of homogeneity with respect to the z coordinate direction requires that the partial of k_z with respect to z be preserved. Reflection on the part of the reader will show that homogeneity is a scalar property while isotropy is a vector property.

Example 9.5. Isotropic versus homogeneous behavior
It is usually the case that *within* a soil horizon homogeneity prevails in all three coordinate directions, however, isotropy may hold only in a plane parallel to the surface. In terms of equation 9.20 and 9.21, how should equation 9.19, the partial differential equation governing heat flow in a soil that is isotropic in x and y and homogeneous in x, y, and z be expressed?

Solution
Since homogeneity prevails throughout the soil horizon, the thermal conductivities are constant for each direction and:

$$\frac{\partial k_x}{\partial x} \equiv 0$$

$$\frac{\partial k_y}{\partial y} \equiv 0$$

$$\frac{\partial k_z}{\partial z} \equiv 0$$

However, because the medium is not isotropic in the z-direction, equation 9.19 must be written:

$$k_x\left(\frac{\partial^2 T}{\partial x^2} + \frac{\partial^2 T}{\partial y^2}\right) + k_z \frac{\partial^2 T}{\partial z^2} = \frac{\partial T}{\partial \tau} \qquad 9.28$$

where k_x and k_y have the same magnitude, while k_z has a different magnitude. However, k_x and k_z are constant throughout the soil horizon though $k_x \neq k_z$ have different magnitudes.

Application of the Heat Flow Equation to the Soil Mass

Equation 9.19 must be simplified before it can be applied to a soil mass. Since a soil is comprised of various layers or horizons, each with characteristic physical properties, the thermal properties of each layer may be different. It would appear that the soil is not homogeneous. However, In most cases within a horizon, the thermal conductivity in the vertical direction is constant. Furthermore, adjacent horizons may have thermal conductivities that differ by only a small amount so an average thermal conductivity can often be assumed for several adjacent soil horizons. In such cases, a solution which uses a single value of thermal conductivity for the entire medium may be sufficiently accurate for engineering purposes.

In other cases, if the thermal conductivities of several adjacent horizons are known, a representative average thermal conductivity may be computed. The procedure for estimating a representative average thermal conductivity is demonstrated in example 9.6 below.

Example 9.6. Thermal conductivity of inhomogeneous media
Two adjacent horizons of soil have thermal conductivities of k_1 and k_2, respectively. Assume the temperatures are T_0 at z_0, T_1 at z_1, and T_2 at z_2 so that $\Delta T_1 = T_1 - T_0$, and $\Delta T_2 = T_2 - T_1$, where $T_2 < T_1 < T_0$, and $z_2 > z_1 > z_0$. If a constant heat flux, Q, is flowing parallel to the z-direction through the composite medium, compute the average thermal conductivity, k_{avg}, of the composite medium.

Solution
By virtue of the first-order phenomenological model, equation 9.1, the heat flux through each layer and through the composite of two layers is:

$$Q = -k_1 \frac{T_1 - T_o}{z_1 - z_o} = -k_1 \frac{\Delta T_1}{\Delta z_1} \qquad 9.29$$

$$Q = -k_2 \frac{T_2 - T_1}{z_2 - z_1} = -k_2 \frac{\Delta T_2}{\Delta z_2} \qquad 9.30$$

and if a steady-state flow of heat energy exists (i.e., no sources or sinks within the medium):

$$\begin{aligned} Q &= -k_{avg} \frac{T_2 - T_0}{z_2 - z_0} \\ &= -k_{avg} \frac{\Delta T_1 + \Delta T_2}{\Delta z_2 + \Delta z_1} \qquad 9.31 \\ &= -k_{avg} \frac{\Delta T}{\Delta z} \end{aligned}$$

To determine an expression for k_{avg} we note as indicated above, that $\Delta T = \Delta T_1 + \Delta T_2$. Therefore, we can solve 9.29 and 9.30 for ΔT_1 and ΔT_2.

$$Q \frac{\Delta z_1}{-k_1} = \Delta T_1 \tag{9.32}$$

$$Q \frac{\Delta z_2}{-k_2} = \Delta T_2 \tag{9.33}$$

Since $\Delta z = \Delta z_1 + \Delta z_2$, we can form an expression using the above and equate the expression to equation 9.31, namely:

$$Q = -k_{avg} \frac{\dfrac{Q\Delta z_1}{-k_1} + \dfrac{Q\Delta z_2}{-k_2}}{\Delta z_1 + \Delta z_2} = -k_{avg} \frac{\Delta T}{\Delta z} \tag{9.34}$$

The left equality can be solved for k_{avg} since the flux Q is the same in both layers:

$$k_{avg} = \frac{\Delta z_1 + \Delta z_2}{\dfrac{\Delta z_1}{k_1} + \dfrac{\Delta z_2}{k_2}} = \frac{\Delta z}{\dfrac{\Delta z_1}{k_1} + \dfrac{\Delta z_2}{k_2}} \tag{9.35}$$

We can see that a knowledge of the thermal conductivities and thicknesses of the layers is sufficient to enable one to determine the average thermal conductivity of the composite layer.

Using a computer makes it possible to solve complex problems involving layered soils possessing markedly different thermal conductivities. In solving such a problem, the thermal condition at the upper boundary of the topmost layer is defined and a solution is obtained for that layer. The predicted thermal condition at the lower boundary of the layer then yields the boundary condition for the upper boundary of the next layer. In this fashion, the complete solution may be obtained for the entire composite medium.

An assumption that is most likely to be valid in a soil is that of isotropy. Experience indicates that, since many engineering problems require only a knowledge of temperature fluctuations with depth, a lack of isotropy between the vertical and horizontal directions is of little significance. There is virtually no horizontal heat flow in the plant environment because the soil surface is uniformly supplied with energy by solar and long wave atmospheric radiation. Furthermore, the rate of energy produced or used within the soil mass by sources or sinks is small compared to the magnitude of the overall daily energy fluctuation reaching the plant-soil-root system. As mentioned previously, the production or use of heat energy occurs as a result of moisture changes, chemical reactions, and biological action due to respiration or other biochemical processes. Although a source or sink term may improve the accuracy of solutions of the differential equation modeling heat flow in the soil, the error incurred by ignoring the term is small. Thus, for engineering simplification, can be assumed to be zero.

Example 9.7. Energy production by organic decay
A particular soil contains 3% organic matter. Estimate the rate of heat production for the soil if all the organic matter is converted into H_2O and CO_2 over a period of six months.

Solution

The chemical energy storage coefficient, a, is about 15 000 kJ/kg of organic matter. Since the organic matter is to be completely decomposed, this amount of heat must be liberated. In the following calculations, we use the subscript om to denote organic matter, and the subscript s to denote soil.

The organic matter in 1 kg of soil is 0.03 kg_{om}/kg_s. The heat energy released per kg of soil is then:

$$\text{Heat} = 0.03\,\frac{kg_{om}}{kg_s} \times 15\,000\,\frac{kJ}{kg_{om}} = 450\,\frac{kJ}{kg_s} \qquad 9.36$$

The rate of heat energy production is:

$$\dot{q} = \frac{\dfrac{450\ kJ}{kg_s}}{6\ mo \times 30\,\dfrac{d}{mo} \times 24\,\dfrac{h}{d}} = 0.104\,\frac{kJ}{h} = 104\,\frac{J}{h} \qquad 9.37$$

It is highly unlikely that a soil would lose 3% organic matter over one summer. Indeed, the organic matter content of most soil is constant. In addition, since organic matter decay is a biochemical process, its rate is governed by the Q_{10} rule so the rate of decomposition is closely related to the soil temperature. The result is that the magnitude of the term is very small and, for engineering purposes, can usually be neglected.

Boundary and Initial Conditions for an Infinite Soil

A solution of the differential equation of heat flow for the soil yields considerable information regarding the temperature regime of the soil portion of the plant environment. Before a solution can be obtained, we must explicitly define the boundary and initial conditions to which the solution will apply. In view of the previous discussion we assume the soil is a homogenous, isotropic medium with no sources or sinks. Under these assumptions equation 9.23 can be rewritten,

$$\frac{\partial^2 T}{\partial x^2} + \frac{\partial^2 T}{\partial y^2} + \frac{\partial^2 T}{\partial z^2} = \frac{1}{\alpha}\frac{\partial T}{\partial \tau} \qquad 9.38$$

where

T = temperature
x,y,z = cartesian coordinates (z positive downward)
α = thermal diffusivity = $k/\rho c$
τ = time

For convenience in the homogeneous isotropic case the parameters of thermal conductivity, specific heat, and density that characterize the soil mass have been combined into a single parameter, the thermal diffusivity, α.

To solve equation 9.38, we must specify the initial and boundary conditions that apply to the soil mass. The solution will then describe the thermal regime which exists in the soil component of the plant environment.

Consider the x-y plane in figure 9.1 to be coplanar with the soil surface, with z positive downward into the soil mass. If the source of energy is solar radiation, and if a uniform radiation flux unimpaired by clouds or other obstructions is assumed, then the temperature at the soil surface will be everywhere equal for a given surface condition. Under this assumptions, no gradient of temperature in the x- or y-directions will exist. This condition at the upper boundary is expressed as:

$$\frac{\partial^2 T}{\partial x^2} = \frac{\partial^2 T}{\partial y^2} = 0 \qquad 9.39$$

So the three-dimensional heat flow problem reduces to a one-dimensional problem:

$$\frac{\partial^2 T}{\partial z^2} = \frac{1}{\alpha}\frac{\partial T}{\partial \tau} \qquad 9.40$$

The solution we obtain will describe the thermal behavior of the soil mass. Equation 9.40 is second order in z and first order in τ, and will require three initial or boundary conditions for its solution. The boundary conditions exist at the soil surface and deep within the soil mass while the initial condition is the temperature profile within the soil. At the soil surface, the boundary condition is the energy input due to solar and long wave radiation impinging on the soil surface. Since the radiation pattern is periodic, the temperature at the surface must fluctuate. Thus, the soil surface condition for the plant environment requires a sinusoidal variation of temperature with time. We can express this as:

$$T(0,\tau) = T(\text{sinisoidal } f[\tau]) \qquad 9.41$$

The exact function used in equation 9.41 will depend on the nature of the solution desired. For example, a solution intended to define daily fluctuations in soil temperature would require an input function at the surface that is periodic on a daily basis. Conversely, a solution to describe yearly fluctuations would require an input with a periodic component of 1 year. In general, the input function will be a combination of fluctuations from all sources, while the solution will be a summation of solutions corresponding to the frequencies that correspond to the periods present at the boundary. In practice, available data can be used to describe a Fourier series that can then be used as the input function, (eq. 9.41).

The condition that must be satisfied at great depth into the soil is that the temperature fluctuations introduced at the surface completely disappear leaving only a temperature that is constant and equal to the mean yearly average temperature. This is a result of the damping effect that the soil mass has on the temperature fluctuations that will become evident when we obtain and examine the solution more closely. The boundary condition for this case is expressed as though the depth were infinite, even though a finite depth will suffice. The effective depth, z, for which this will occur will vary with soil type, but the condition will be satisfied for values of z much less than infinity. The controlling factor is the thermal diffusivity of the soil mass defined previously. This condition is:

$$T(\infty,\tau) = T_{avg} \qquad 9.42$$

The third condition is the initial condition for the soil mass. For this condition, any known temperature variation with depth will suffice. Therefore we assume that the variation with depth is some temperature function. In the solution process this variation is defined by the parameters used in the process and will depend on the diffusivity of the soil. If a finite difference technique is used, it is sufficient to assume that a constant temperature equal to the average yearly temperature exists. The finite difference solution then can be carried out to determine at what depth this condition is satisfied. Mathematically, the third condition can be expressed:

$$T(z,0) = \text{Existing temperature} \qquad 9.43$$

Example 9.8. Modeling the surface boundary condition
Assume that the temperature at the surface of a given soil has an average value of 5° C and fluctuates yearly as a cosine function with an amplitude of 5° C. Determine a mathematical expression that can be used as boundary condition, equation 9.41, for these conditions if the maximum temperature occurs near 1 July. Assume that time is measured from 1 January.

Solution
The period of the fluctuation will be 365 days, so the argument of the cosine function must undergo one complete cycle of 2π radians in 365 days. If the variable is time, τ, the argument must be:

$$\omega(\tau) = \frac{2\pi}{365\,d}\,\tau \qquad 9.44$$

The magnitude of fluctuation is 5° C, so the general form of the cosine term (with the argument in radians) is:

$$T(\tau) = 5\cos\left(\frac{2\pi}{365}\,\tau\right) \qquad 9.45$$

where it is understood that temperature is in ° C and time is in days.

To have a maximum occur near 1 July, it is necessary that a minimum occur near 1 January, This condition will be fulfilled if the argument of the function has the value 0 when τ = 0. To obtain this condition, the argument must be shifted in phase. Therefore, the argument must become:

$$T(\tau) = 5\cos\left(\frac{2\pi}{365}\,\tau - \pi\right) \qquad 9.46$$

Finally, since the average temperature is 5° C:

$$T(0) = 5 + 5\cos\left(\frac{2\pi}{365}\,\tau - \pi\right) \qquad 9.47$$

Solution of the Differential Equation

We seek the solution for equation 9.40 that will yield the temperature distribution pattern in the soil from which information of engineering interest for the green plant environment can be obtained. The approach presented here will enable the student to develop those concepts that are pertinent to temperature fluctuations in the soil. The transient solution for T(z,τ) which would account for abrupt changes at T(0,τ), such as those which occur when the sun is suddenly obscured by a cloud, will be ignored This solution is rapidly damped and does not penetrate very deeply into the soil surface. Note, however, that a transient solution added to the solution we are seeking must provide an accurate description of temperature in the soil very near the soil surface such as might be caused by a nonperiodic source caused by drifting clouds that obscure the sun. Since a periodic source is adequate for most engineering applications, and because the error in estimating the thermal diffusivity for a given soil often exceeds the error introduced by neglecting the transient term, the transient term will be ignored.

To obtain a steady-state solution for equation 9.40, we use the technique of separation of variables. We begin by assuming a solution that can be written as the product of two functions, G(z), depending only on depth, and F(τ) depending only on time. The solution must have the form:

$$T(z,\tau) = G(z)F(\tau) \qquad 9.48$$

Assume the dependence of the functions on the independent variable is understood:

$$T = GF \qquad 9.49$$

We differentiate equation 9.49 and substitute the result into equation 9.40 to obtain:

$$\frac{d^2 G(z)}{dz^2} = \frac{1}{\alpha}\frac{dF(\tau)}{d\tau} \qquad 9.50$$

Since these are now ordinary differential equations, we can use primes to designate differentiation:

$$G'' F = \frac{1}{\alpha} GF' \qquad 9.51$$

Separating variables gives:

$$\frac{G''}{G} = \frac{1}{\alpha}\frac{F'}{F} \qquad 9.52$$

Equation 9.52 can be correct only as an equality if it is true for all z and τ. This can be true if both sides are equal to some constant that we arbitrarily choose to represent as Ω. Equation 9.52 then can be written:

$$\frac{G''}{G} = \frac{1}{\alpha}\frac{F'}{F} = \Omega \qquad 9.53$$

From which the pair of ordinary differential equations:

$$G'' - \Omega G = 0 \qquad 9.54$$

and:

$$F' - \Omega \alpha F = 0 \qquad 9.55$$

can be obtained.

A solution to equation 9.55, a first-order, ordinary differential equation can be obtained by inspection as:

$$F(\tau) = C_1 e^{\alpha\Omega\tau} \qquad 9.56$$

We have already indicated that we need a solution that will be periodic in time and one that neither increases or decreases in magnitude away as time progresses. For equation 9.56 to be periodic in time, it is sufficient that Ω

be an imaginary number so the solution can be written in terms of sine and cosine functions. This will be accomplished if we define:

$$\Omega = \pm i\lambda^2 \qquad 9.57$$

Both the positive and negative signs are included because both yield valid solutions, and, as yet, there is not sufficient information to determine which sign is needed. After substituting equation 9.57 into equation 9.56 we can write:

$$F = C_1 e^{\pm i \lambda^2 \alpha \tau} \qquad 9.58$$

Since equation 9.58 actually yields two solutions depending on which sign is used, we write equation 9.58 as:

$$F_1 = C_1 e^{+ i \lambda^2 \alpha \tau} \qquad 9.59$$

and:

$$F_2 = C_2 e^{- i \lambda^2 \alpha \tau} \qquad 9.60$$

Example 9.9. Existence of dual solutions to the equation
Show that either the plus or the minus sign will determine the solution to equation 9.55.

Solution
Write equation 9.58 as:

$$F_? = C_? e^{* i \lambda^2 \tau} \qquad 9.61$$

where the asterisk denotes either the + or the – sign, and "?" can stand for either 1 or 2. We now differentiate the equation with respect to time.

$$\frac{d F_?}{d \tau} = * i \lambda^2 C_? e^{* \lambda^2 \alpha \tau} \qquad 9.62$$

Since, by our assumption, $\pm i\lambda^2 = *i\lambda^2 = \Omega$ this enables equation 9.62 to be written:

$$\frac{d F_?}{d \tau} = \Omega \alpha C_? e^{\Omega \alpha \tau} \qquad 9.63$$

while equation 9.61 is:

$$F_? = C_? e^{\Omega \alpha \tau} \qquad 9.64$$

Expressions 9.63 and 9.64 can be substituted into equation 9.55 to obtain:

$$\Omega\,\alpha\,C_? e^{\Omega\alpha\tau} - \Omega\,\alpha\,C_? e^{\Omega\alpha\tau} = 0 \qquad 9.65$$

This is true since either equation 9.59 or 9.60 is a solution independent of the constant $C_?$.

A solution to equation 9.54 can be found by a table of solutions to ordinary differential equations or by standard mathematical techniques. The solution is a set of two solutions depending on the sign adopted for the term $\Omega = \pm i\lambda^2$, a decision dependent on the boundary conditions. The solutions are:

$$G_1 = C_3\, e^{+\sqrt{i}\,\lambda z} + C_4 e^{-\sqrt{i}\,\lambda z} \qquad 9.66$$

and:

$$G_2 = C_5\, e^{+\sqrt{-i}\,\lambda z} + C_6 e^{-\sqrt{-i}\,\lambda z} \qquad 9.67$$

The above expressions can be simplified. A college algebra text shows that:

$$\sqrt{i} = \frac{1 + i}{\sqrt{2}}$$

$$\sqrt{-i} = \frac{1 - i}{\sqrt{2}} \qquad 9.68$$

If we apply the conditions shown in equation 9.68 to the solutions for G given in equation 9.66 and 9.67, we find:

$$G_1 = C_3 e^{+(1+i)\left(\frac{\lambda z}{\sqrt{2}}\right)} + C_4 e^{-(1+i)\left(\frac{\lambda z}{\sqrt{2}}\right)} \qquad 9.69$$

and:

$$G_2 = C_5 e^{+(1-i)\left(\frac{\lambda z}{\sqrt{2}}\right)} + C_6 e^{-(1-i)\left(\frac{\lambda z}{\sqrt{2}}\right)} \qquad 9.70$$

We note that G_1 arose from the $+i\lambda^2$ portion and G_2 from the $-i\lambda^2$ portion of the solution.

The constants C_3 and C_5 must be identically zero since the real part of the exponential which accompanies these two is positive. With increasing z this part of the equation would grow larger, approaching infinity with increasing depth. After setting C_3 and C_5 to zero and expanding the remaining portions of the solutions, the solution to the equation for penetration into the soil mass becomes:

$$G_1 = C_4 e^{\left[-\left(\frac{\lambda z}{\sqrt{2}}\right) - i\left(\frac{\lambda z}{\sqrt{2}}\right)\right]} \qquad 9.71$$

and:

$$G_2 = C_6 e^{\left[-\left(\frac{\lambda z}{\sqrt{2}}\right) + i\left(\frac{\lambda z}{\sqrt{2}}\right)\right]} \qquad 9.72$$

To form the complete solution, we now must multiply G_1 by F_1, both of which were derived from the $+i\lambda^2$ portion of the assumption for Ω, and, correspondingly, multiply G_2 by F_2, both derived from the $-i\lambda^2$ portion to obtain:

$$T_1 = F_1 G_1 = C_1 C_4 e^{-\left(\frac{\lambda z}{\sqrt{2}}\right) + i\left[\lambda^2 \alpha \tau - \left(\frac{\lambda z}{\sqrt{2}}\right)\right]} \tag{9.73}$$

and:

$$T_2 = F_2 G_2 = C_2 C_6 e^{-\left(\frac{\lambda z}{\sqrt{2}}\right) - i\left[\lambda^2 \alpha \tau - \left(\frac{\lambda z}{\sqrt{2}}\right)\right]} \tag{9.74}$$

By the theory of solutions to differential equations, the sum of two solutions to a differential equation is itself a solution. Therefore, we add the two solutions above to obtain a solution that encompasses the attributes of both solutions:

$$T(z,\tau) = e^{-\left(\frac{\lambda z}{\sqrt{2}}\right)} \left\{ a \cdot e^{+i\left[\lambda^2 \alpha \tau - \left(\frac{\lambda z}{\sqrt{2}}\right)\right]} + b \cdot e^{-i\left[\lambda^2 \alpha \tau - \left(\frac{\lambda z}{\sqrt{2}}\right)\right]} \right\} \tag{9.75}$$

where $a = C_1C_4$ and $b = C_2C_6$. Now we can use Euler's Theorem from college algebra* to express equation 9.75 as a sum of sine and cosine term:

$$T(z,\tau) = e^{-\left(\frac{\lambda z}{\sqrt{2}}\right)} \left[A \cos\left(\lambda^2 \alpha \tau - \left(\frac{\lambda z}{\sqrt{2}}\right)\right) + B \sin\left(\lambda^2 \alpha \tau - \left(\frac{\lambda z}{\sqrt{2}}\right)\right) \right] \tag{9.76}$$

where A denotes a real coefficient multiplying the cosine function and B is an imaginary coefficient multiplying the sine function.

Equation 9.76 is a solution to the differential equation describing heat flow in a semi-infinite soil mass. However, it is not the most general form for the solution. Equation 9.76 contains three constants, A, B, and λ that must be specified for the soil, and it is clear that each specification of A, B, and λ will be a specific solution. To obtain the general form we denote by the subscript n the *nth* specific solution for which the constants are A_n, B_n, and λ_n, and form the most general solution by summing all specific solutions. This yields:

$$T(z,\tau) = \sum_{i=0}^{\infty} e^{-\left(\frac{\lambda z}{\sqrt{2}}\right)} \left[A_n \cos\left(\lambda_n^2 \alpha \tau - \left(\frac{\lambda_n z}{\sqrt{2}}\right)\right) + B_n \sin\left(\lambda_n^2 \alpha \tau - \left(\frac{\lambda_n z}{\sqrt{2}}\right)\right) \right] \tag{9.77}$$

The general solution will be complete if we can choose values of A_n, B_n, and λ_n in such a manner that the boundary conditions, equation 9.41 a, b, c, are satisfied. The temperatures at the ground surface can be expressed in a Fourier Series:

$$T(0,\tau) = a_o + \sum_{n=1}^{\infty} a_n \cos\left(\frac{2n\pi}{t}\right)\tau + b_n \sin\left(\frac{2n\pi}{t}\right)\tau \tag{9.78}$$

where

- T = temperature
- a_i = coefficients determining the amplitude of each frequency (i = 1, 2,...)
- t = the maximum length of record from which the series was derived
- τ = specific point for which a value of temperature is desired

---

* $ae^{i\theta} + bi^{-i\theta} = a\cos(\theta) + ia\sin(\theta) + b\cos(\theta) - ib\sin(\theta) = (a+b)\cos(\theta) + i(a-b)\sin(\theta) = A\cos(\theta) + B\sin(\theta)$

Note that τ in equation 9.78 has the same dimensions as t. If daily mean temperatures for an entire year provide data to formulate the constants a_n and b_n, then τ will be expressed in days.

To particularize equation 9.77 and solve for temperatures within the soil mass, we set z equal to zero and equate equation 9.77 and 9.78:

$$A_o \cos(\lambda_o^2 \alpha \tau) + B_o \sin(\lambda_o^2 \alpha \tau) + \sum_{i=1}^{\infty} [A_n \cos(\lambda_n^2 \alpha \tau) + B_n \sin(\lambda_n^2 \alpha \tau)]$$
$$= a_o + \sum_{n=1}^{\infty} \left[a_n \cos\left(\frac{2n\pi}{t}\tau\right) + b_n \sin\left(\frac{2n\pi}{t}\tau\right)\right] \qquad 9.79$$

where we have rewritten equation 9.79 so the summation begins with n equal to unity.

A fundamental theorem of calculus states that two infinite series are equal if and only if the coefficients of like terms are equal. In equation 9.79, if the arguments of the sine and cosine terms are made equal, then the two series would be comprised of identical terms. This is true if:

$$\lambda_n^2 \alpha \tau = \frac{2n\pi}{t}\tau$$

So that:

$$\lambda_n = \sqrt{\frac{2n\pi}{\alpha t}} \qquad 9.80$$

If equation 9.80 is substituted into equation 9.79, the requirement that like terms be equal in the two series implies that:

$$A_o = \alpha_o$$
$$A_n = \alpha_n$$
$$B_n = b_n \qquad 9.81$$
$$\text{while}$$
$$B_o = 0$$

After substituting equations 9.81 back into equation 9.79, we find the particular solution to equation 9.40 for the case where the energy input to the soil surface is expressed as a Fourier series†.

$$T(z,\tau) = \alpha_o + \sum_{n=1}^{\infty} \left(\exp\left[-\sqrt{\frac{n\pi}{\alpha\tau}}\,z\right) \times \left(\alpha_n \cos\left[\frac{2n\pi}{t}\tau - \left(\sqrt{\frac{n\pi}{\alpha\tau}}\right)z\right]\right.\right.$$
$$\left.\left. + b_n \sin\left[\frac{2n\pi}{t}\tau - \left(\sqrt{\frac{n\pi}{\alpha\tau}}\right)z\right]\right) \qquad 9.82$$

† For readability we express e^x as exp(x).

The requirement that the temperature must remain finite at great depths requires that only the positive square root in equation 9.82 be used. Although the form of the particular solution implies an infinite number of terms for the series, in practice it is possible to obtain accuracy sufficient for engineering purposes with as few as two or three terms.

Example 9.10. One solution to the soil heat flow equation
What is the particular solution in equation 9.40 if the boundary condition, 9.41a, is given by:

$$T(0,\tau) = 5 + 5\cos\left(\frac{2\pi}{365}\tau - \pi\right) \quad 9.83$$

Solution
Using the formula for the cosine of the sum of two angles, one can write this expression as:

$$T(0,\tau) = 5 + 5\left[\cos\left(\frac{2\pi\tau}{365}\right)\cos(-\pi) + \sin\left(\frac{2\pi\tau}{365}\right)\sin(-\pi)\right]$$

Which reduces to:

$$T(0,\tau) = 5 - 5\cos\left(\frac{2\pi}{365}\tau\right) \quad 9.84$$

Expression 9.84 is a Fourier series containing only one periodic term. Following the procedure used to obtain equation 9.78:

$$\begin{aligned} & A_o\cos(\lambda_o^2\alpha\tau) + B_o\sin(\lambda_o^2\alpha\tau) \\ & + \sum_{i=1}^{\infty}\left(A_n\cos\left[\lambda_n^2\alpha\tau - \left(\frac{\lambda_n z}{\sqrt{2}}\right)\right] + B_n\sin\left[\lambda_n^2\alpha\tau - \left(\frac{\lambda_n z}{\sqrt{2}}\right)\right]\right) \\ & = 5 - 5\cos\left[\left(\frac{2n\pi}{365}\right)\tau\right] \end{aligned} \quad 9.85$$

and choose:

$$\lambda_o = 0$$

$$\lambda_1 = \sqrt{\frac{2\pi}{365\,\alpha}}$$

$$a_o = 5$$

$$a_1 = -5$$

Note that all other coefficients are zero. The particular solution from equation 9.76 is:

$$T(z,\tau) = 5 - 5\left[\exp -\left(\sqrt{\frac{\pi}{365\alpha}}\right)z\right] \cos\left[\left(\frac{2\pi}{365}\right)\tau - \left(\sqrt{\frac{\pi}{365\alpha}}\right)z\right]$$

Properties of the Soil Influencing Heat Flow

Much information about the temperature regimes of soil can be obtained from close examination of equation 9.82. The general behavior of the temperature fluctuations in the soil will be the same for all homogeneous isotropic soils, but the magnitude of the fluctuations will depend on the thermal diffusivity of the soil.

Mean Yearly Temperature

Equation 9.82 contains two parts, a constant contribution determined by the magnitude of a_0, and a contribution whose magnitude and sign depend on the variables z and α, along with the values of a_n and b_n for the soil under consideration. For any soil, a_0 can be taken as the average temperature of the soil surface over a time equal to one complete cycle of the longest periodic component of the Fourier Series solution. In practice a_0 would be determined by averaging temperatures over a duration equal to the length of record used to obtain the Fourier Series representation. Information on the surface temperature of different soils is available; the most probable sources of such information for any given region of the United States are the climatological records of the U.S. Weather Bureau (formerly the National Oceanic and Atmospheric Administration), available from the National Climatic Data Center, Ashville, North Carolina.

The variable contribution to equation 9.82 has a product form in which the magnitude of the fluctuating portion is modified by multiplication with an exponential function whose value is determined by the thermal diffusivity and the depth z at which the fluctuation is desired. Since at great depths the value of the exponential will approach zero, it is seen that a_0, in addition to being the average value of temperature at the soil surface, is also the average value at the depth beyond which the fluctuations are damped out. The exponential multiplier, therefore indicates to what depth a fluctuation with a frequency of n/t, [cycles/(unit of time, t)] will penetrate. It is apparent that the lowest-frequency components of the Fourier series expansion influence temperatures at greater depths than do high-frequency components. Thus, one would expect yearly temperature fluctuation to be detectable at far greater depths than the daily fluctuations. If measurements taken at a short enough interval, say hourly, are graphed against depth, only very low frequency variations appear below depths of several meters.

Damping Depth

The depth to which a fluctuation will penetrate the soil mass is related to a parameter called the damping depth that we will denote by D. The damping depth is defined as the depth at which the amplitude of a temperature fluctuation of a given frequency is diminished by a factor of 1/e. In terms of the solution to the differential equation governing heat flow in the soil, the amplitude of a fluctuation going with a particular frequency (or equivalent, period of fluctuation) is:

$$\exp -\left(\sqrt{\frac{n\pi}{\alpha\tau}}\right)D = e^{-1} \qquad 9.86$$

From which the damping depth D is found to be:

$$D = \sqrt{\frac{\alpha\tau}{n\pi}} \quad \text{or} \quad \sqrt{\frac{kt}{\rho c n\pi}} \qquad 9.87$$

The damping depth has been expressed also in terms of the thermal conductivity, soil specific heat, and soil mass density. Equation 9.87 demonstrates that the damping depth is different for each frequency of the input energy. For an annual variation, n=1 and:

Table 9.1. Thermal properties of selected soils and soil materials

Soil	Density Content kg/m^3	Water Heat %	Specific Capacity J/(kg·°C)	Heat Conductivity J/(m^3·°C)	Thermal Diffusivity J/(h·m·°C)	m^2/s	Source
Yolo clay	1180	0.00	812	9.87×10^{5}	2.11×10^{3}	2.23×10^{-3}	Johnston, 1939
	1470	29.00	812	3.01×10^{6}	1.25×10^{4}	4.18×10^{-3}	Johnston, 1939
Granitic	1290	0.00	946	1.22×10^{6}	2.56×10^{3}	2.12×10^{-3}	Johnston, 1939
Sandy l.	1560	22.70	946	2.95×10^{6}	1.07×10^{4}	3.64×10^{-3}	Johnston, 1939
Calc.	1140	0.00	644	7.32×10^{5}	1.19×10^{3}	1.62×10^{-3}	Johnston, 1939
Fine l.	1080	24.40	644	1.80×10^{6}	7.23×10^{3}	3.96×10^{-2}	Johnston, 1939
Granaitic	1320	0.00	854	1.13×10^{6}	2.06×10^{3}	1.87×10^{-3}	Johnston, 1939
s.	1900	13.10	854	2.66×10^{6}	1.63×10^{4}	6.11×10^{-3}	Johnston, 1939
Barnes	1160	5.10	837	1.21×10^{6}	6.18×10^{3}	5.04×10^{-4}	Smith, 1939
l.	1150	8.90		1.38×10^{6}	6.78×10^{2}	5.04×10^{-4}	Smith, 1939
	1070	13.00		1.46×10^{6}	7.38×10^{2}	5.04×10^{-4}	Smith, 1939
	1180	26.60		2.30×10^{6}	1.30×10^{3}	5.76×10^{-4}	Smith, 1939
Chester	1440	2.00		1.34×10^{6}	6.78×10^{2}	5.04×10^{-4}	Smith, 1939
l.	1190	4.70		1.21×10^{6}	7.68×10^{2}	6.12×10^{-4}	Smith, 1939
	1120	13.40		1.55×10^{6}	1.31×10^{3}	8.64×10^{-4}	Smith, 1939
Herman	1420	1.30		1.26×10^{6}	7.38×10^{2}	5.76×10^{-4}	Smith, 1939
Sandy l.	1240	4.40		1.26×10^{6}	9.04×10^{2}	7.20×10^{-4}	Smith, 1939
	1230	8.70		1.46×10^{6}	1.30×10^{3}	9.00×10^{-4}	Smith, 1939
Kalkaska	1550	.80		1.34×10^{6}	9.04×10^{2}	6.84×10^{-4}	Smith, 1939
loamy s.	1430	2.60		1.34×10^{6}	1.04×10^{3}	7.92×10^{-4}	Smith, 1939
	1440	5.70		1.55×10^{6}	1.87×10^{3}	1.19×10^{-3}	Smith, 1939
N'thway	1560	6.60	753	1.61×10^{6}	1.96×10^{3}	1.22×10^{-3}	Kersten, 1940
s. l.	1540	16.40		2.22×10^{6}	3.16×10^{3}	1.44×10^{-3}	Kersten, 1940
	1570	22.50		2.66×10^{6}	3.77×10^{3}	1.40×10^{-3}	Kersten, 1940
	1430	16.50		2.06×10^{6}	2.71×10^{3}	1.30×10^{-3}	Kersten, 1940
	1820	14.50		2.48×10^{6}	4.52×10^{3}	1.84×10^{-3}	Kersten, 1940
Fair-	1440	12.30	753	1.82×10^{6}	3.01×10^{3}	1.66×10^{-3}	Kersten, 1940
banks	1440	18.00		2.17×10^{6}	3.77×10^{3}	1.73×10^{-3}	Kersten, 1940
sl. cl. l.	1440	25.40		2.62×10^{6}	4.22×10^{3}	1.62×10^{-3}	Kersten, 1940
	1450	30.00		2.91×10^{6}	4.37×10^{3}	1.51×10^{-3}	Kersten, 1940
	1280	12.30		1.62×10^{6}	2.26×10^{3}	1.40×10^{-3}	Kersten, 1940
	1620	12.50		2.07×10^{6}	3.92×10^{3}	1.91×10^{-3}	Kersten, 1940
Dakota	1350	1.90	753	1.13×10^{6}	8.89×10^{3}	7.92×10^{-4}	Kersten, 1940
s. l.	2110	4.90		2.02×10^{6}	8.13×10^{3}	4.03×10^{-3}	Kersten, 1940
Black	1450	1920		1.41×10^{6}	5.57×10^{2}	3.96×10^{-4}	David, 1940
cotton	2000			1.87×10^{6}	1.01×10^{3}	5.40×10^{-4}	David, 1940
soil	2200			2.09×10^{6}	9.04×10^{2}	4.32×10^{-4}	David, 1940
	2300			2.18×10^{6}	7.83×10^{2}	3.60×10^{-4}	David, 1940
Quartz	2700		800	2.00×10^{6}	3.17×10^{4}	1.58×10^{-2}	Marshall, 1979
Clay	800			2.00×10^{6}	1.04×10^{4}	5.20×10^{-3}	Marshall, 1979
Org. M.	1100	2500		2.70×10^{6}	9.00×10^{2}	3.33×10^{-4}	Marshall, 1979
Water	1000	4200		4.20×10^{6}	2.16×10^{3}	5.14×10^{-4}	Marshall, 1979
Ice (0° C)	900	2100		1.90×10^{6}	7.92×10^{3}	4.17×10^{-3}	Marshall, 1979
Air (20° C		1.2	1000	1.20×10^{3}	9.00×10^{1}	7.50×10^{-2}	Marshall, 1979

$$D = \sqrt{\frac{\alpha \tau}{\pi}} \qquad 9.88$$

Example 9.11. Calculation of the annual damping depth
What is the damping depth of the soil in example 9.10 if the thermal diffusivity is 0.0018 m^2/h?

Solution

From equation 9.88, the model for the damping depth is the inverse of the constant portion of the exponent of the exponential multiplier of the solution. In example 9.10, this term is:

$$D = \sqrt{\frac{365 d \alpha}{\pi}}$$

$$= \left(\frac{365\,d}{\pi} \times \frac{0.0018\ m^2}{h} \times \frac{24\ h}{d}\right)^{\frac{1}{2}} \qquad 9.89$$

$$= 2.24\ m$$

From equation 9.88, the damping depth is directly dependent on the square root of the thermal conductivity, while it is inversely dependent on the soil mass density, the soil specific heat, and the frequency of the fluctuations. Based on equation 9.88, soils with a high thermal conductivity can be expected to have a greater damping depth than soils with lower thermal conductivity values.

The influence of mass density shows that denser soils have a lesser damping depth. Since the specific heat of water is significantly larger than most natural soil materials, wetter soils also will damp the temperature fluctuations faster than the same soil when it is dryer. Table 9.1 contains a compilation of thermal properties of various soils and soil materials. A wide range of variation in thermal properties exists in soil, depending mainly on water content, and influenced by soil texture.

Van Wijk (1963) has developed a model for the product ρc for a soil provided the fractions of soil minerals, soil organic matter, and soil water content are known. This expression is:

$$\rho c = \left[1.9\chi_m + 2.5\chi_{om} + 4.19\theta\right]\left(\frac{mJ}{m^2 \cdot {}^\circ C}\right) \qquad 9.90$$

where

x_m = volume fraction of soil minerals (usually $\rho_m = 2650\ kg/m^3$)
x_{om} = volume fraction of soil organic matter (usually $\rho_{om} = 1300\ kg/m^3$)
θ = volume fraction of soil water ($\rho_w = 1000\ kg/m^3$)

Example 9.12. Estimating thermal parameters of soil

Estimate the term ρc for a soil with a bulk density of 1500 kg/m³, comprised of 50% minerals, 7% organic matter, and 18% water. All percent values are based on the air-dry soil mass. Compute the thermal diffusivity if k = 12.48 W/(m·° C).

Solution

We can use the model given in equation 9.88. First, however, we must compute the volume fractions required as parameters for the model. On a per kilogram basis:

$$Vol_m = \frac{1\,m^3_{mineral}}{2650 kg_{soil}} \times 0.5 = 1.89 \times 10^{-4} \frac{m^3_m}{kg_{soil}}$$

$$V_{om} = \frac{1}{1300} \times 0.07 = 5.38 \times 10^{-5} \frac{m^3_{om}}{kg_{soil}}$$

and

$$V_w = \frac{1}{1000} \times 0.18 = 1.8 \times 10^{-4} \frac{m^3_w}{kg_{soil}} \qquad 9.91$$

Since the bulk density of the soil is 1500 kg/m^3:

$$Vol_{soil} = \frac{1\,m^3}{1500\,kg} = 6.67 \times 10^{-4}\,\frac{m^3_{soil}}{kg_{soil}} \qquad 9.92$$

$$\chi_m = \frac{1.89 \times 10^{-4}\,\frac{m^3_m}{kg_{soil}}}{6.67 \times 10^{-4}\,\frac{m^3_{soil}}{kg_{soil}}} = 0.28\,\frac{m^3_m}{m^3_{soil}} \qquad 9.93$$

Similarly:

$$\chi_{om} = \frac{5.38 \times 10^{-5}\,\frac{m^3_{om}}{kg_{soil}}}{6.67 \times 10^{-4}\,\frac{m^3_{soil}}{kg_{soil}}} = 0.081\,\frac{m^3_{om}}{m^3_{soil}}$$

$$\theta = \frac{1.8 \times 10^{-4}\,\frac{m^3_w}{kg_w}}{6.67 \times 10^{-4}\,\frac{m^3_{soil}}{kg_{soil}}} = 0.27\,\frac{m^3_w}{m^3_{soil}} \qquad 9.94$$

from which:

$$\rho c = 1.9 \times 0.28 + 2.5 \times 0.081 + 4.19 \times 0.27 = 1.87\,\frac{mJ}{m^3\,{}^\circ C} \qquad 9.95$$

Table 9.2 gives damping depths for several values of thermal diffusivity and for temperature fluctuations with periods of 1 year, 1 day, and 1 hour.

From table 9.2, it is obvious that high-frequency temperature fluctuations are rapidly damped out, especially if the thermal diffusivity is low. However, daily fluctuations will penetrate about 0.05 m even in a soil with a very low thermal diffusivity. Table 9.2 shows that since the energy source driving the soil temperature fluctuations is periodic with a daily fluctuation due to solar radiation and a rotating earth system, the rooting zone of many shallow rooted plants is subjected to significant temperature extremes. In particular, since most crops are planted at depths

Table 9.2. Damping depths for annual, daily, and hourly temperature fluctuations

Period of Fluctuation	Damping Depth in Meters for Thermal Diffusivities Given		
	$\alpha = 0.0004\ m^2/h$	$\alpha = 0.004\ m^2/h$	$\alpha = 0.002\ m^2/h$
Yearly	1.05	3.34	2.36
Daily	0.055	0.175	0.120
Hourly	0.011	0.036	0.025

of 5 cm or less, the seed undergoes wide variations in temperature during the period of germination. Fluctuating temperatures for germinating seeds are generally beneficial and result in increased germinations

The effect of soil water and organic matter on thermal diffusivity as might be inferred by examination of the definition of ρc as given in Van Wijk's (1963) model, equation 9.90 is deceptive. A cursory examination makes it appears that for a relatively constant volume fraction of organic and mineral matter, the thermal diffusivity might be:

$$\alpha = \frac{k}{C_1 + 4.19\,\theta} \qquad 9.96$$

where θ is the volumetric fraction of water. With increasing water content, the thermal diffusivity appears as though it should decrease. In fact, the opposite occurs as saturation approaches. This is because of the very rapid increase in thermal conductivity and volumetric thermal capacity with increasing water content. The explanation is related to how energy is transferred, and to the thermal property of water. In a relatively dry soil, energy transfer is from soil particle to soil particle at the points of contact. For a wetting soil, however, water begins to fill the entire soil pore and the relatively high thermal conductivity and specific heat of water begins to play an important role. Table 9.1 indicates a value of 2.16×10^3 J/(m·h·K) for still water. For many soils, an increase in water content increases the thermal conductivity as rapidly or more rapidly than the term χ is increased. For most soils, the result is an increase in water content increasing the thermal diffusivity and, from table 9.2, also the depth to which the fluctuating thermal wave will penetrate. However, although the thermal fluctuations may penetrate further as soil water increases, the very high specific heat of water [4.186 kJ/(kg·K)] will cause a wet soil to take a longer time to reach temperatures favorable for germination than a soil that is well-drained.

There is an interplay between soil water content, penetration of temperature fluctuations, soil temperature, and conditions favorable for seed germination. The engineer modifying the soil environment by drainage, mulching, or irrigation must be aware of this interplay and the effect it may have on the germination of the seed and the growth and development of the plant. Mulching with material having a very low thermal diffusivity may so damp the temperature fluctuations that plants requiring some variation in temperature to germinate and grow may be adversely affected.

The temperature of the root environment is not completely described by knowledge of the thermal diffusivity and the damping depth. From the model given by equation 9.82, the argument of the sine and cosine terms contains an angular frequency that is a function of time, and a phase shift that is a function of depth. The result of the phase shift is that the time of occurrence of the maximum or minimum temperature is shifted with depth. Figure 9.2 shows the effect of depth on a soil whose fluctuation at the surface is a pure cosine function. Over a 24-h period, the effect of the phase shift can be seen. Although the maximum temperature at the surface may occur at noon, the maximum temperature at 0.10 m for the soil illustrated does not occur until shortly after 3 P.M. At that time, the surface temperature is almost the same as the temperature at 0.10 m. As can be seen from figure 9.2, the soil mass is an important factor in altering the extremes of temperature to which the plant is exposed. While the fluctuation in temperature at the surface is 20° C, the fluctuation at 10 cm is only about 4° C, and by 0.2 m the temperature is virtually constant at the mean temperature for the day.

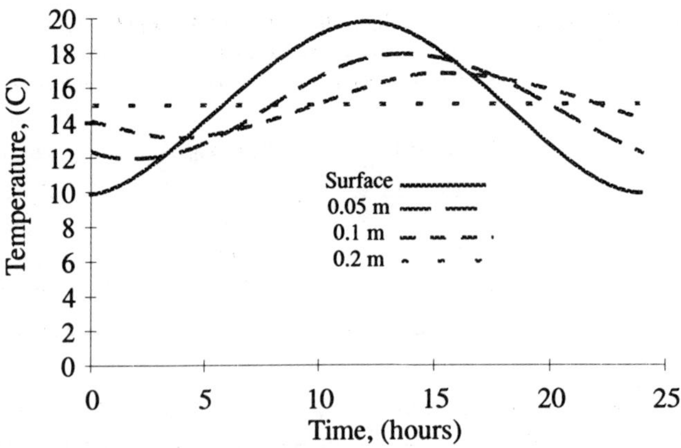

Figure 9.2. Higher frequency fluctuations of heat flux are rapidly damped out by the soil mass. In this illustration, the variation disappears by the time the heat has penetrated 0.2 m.

An analogous situation exists over the course of a year as can be seen from figure 9.3. Here a thermal diffusivity of 0.0018 $m^2\ h^{-1}$, acting over 365 days tempers the long-term temperature fluctuation. The mean daily surface temperature reaches a maximum about 28 June, and fluctuates about 10° C. However, at a depth of 2 m, the maximum is not reached until about 5 October, and the fluctuation is reduced to only about 4° C. In addition, if it were possible to include the daily fluctuations, it could be seen that the daily fluctuations are totally absent, only the long-term yearly fluctuation persists at this depth.

As was indicated earlier, the mathematics presented can be extended to analyze the effects of layered soils. The approach requires that the condition existing at the lower boundary become the boundary conditions for the next layer. A different solution, however, must be found for a totally isolated soil layer since the boundary condition with increasing depth can no longer be used to simplify the solution.

Table 9.3. Thermal conductivities of selected substances

Substance	Thermal Conductivity kJ/(m·h·K)	Reference
Silver	1520.0	Geiger, 1950
Iron	241.0	Geiger, 1950
Granite	8.28	Beskow, 1947
Basalt	7.52	Beskow, 1947
Limestone	6.77	Beskow, 1947
Concrete	8.75	Beskow, 1947
Brick	1.94	Ingerson, 1954
Peat	0.271	Beskow, 1947
Sawdust	2.11	Beskow, 1947
Snow (139 kg/m^3)	0.437	Crawford, 1952
Snow (193 kg/m^3)	0.558	Crawford, 1952
Snow (279 kg/m^3)	0.859	Crawford, 1952

Example 9.13. Temperature variation with soil depth
What is the temperature at the surface and at depths of 0.1 m and 0.4 m at 12.00 h in a soil with a thermal diffusivity of 2.16×10^{-3} m²/h where the temperature, in degrees Celsius, is modeled by:

$$T(z,\tau) = \left\{15 - 5\exp - \left(\sqrt{\frac{\pi}{24\alpha}}\right) z \cos\left[\frac{\pi}{12}\tau - \left(\sqrt{\frac{\pi}{24\alpha}}\right) z\right]\right\} \qquad 9.97$$

Solution

$$\sqrt{\frac{\pi}{24\text{ h} \times 0.002\,16 \frac{\text{m}^2}{\text{h}}}} = 7.78\text{ m}^{-1} \qquad 9.98$$

So that:

$$T(z,\tau) = \left[15 - 5e^{-\left(\frac{7.78}{\text{m}}\right) z} \cos\left(\frac{\pi}{12\text{ h}}\tau - \frac{7.78}{\text{m}} z\right)\right] {}^\circ\text{C} \qquad 9.99$$

At 12:00 h on the surface, z = 0.10 and the equation becomes:

$$\begin{aligned} T(0,12) &= \left\{15 - 5e^{-\frac{7.78}{\text{m}} \times 0\text{ m}}\left[\cos\left(\frac{\pi}{12\text{ h}} 12\text{ h} - \frac{7.78}{\text{m}} \times 0\text{ m}\right)\right]\right\} {}^\circ\text{C} \\ &= \{15 - 5 \times 1 \times \cos(\pi - 0)\} {}^\circ\text{C} = 20^\circ\text{ C} \end{aligned} \qquad 9.100$$

At 0.10 m depth:

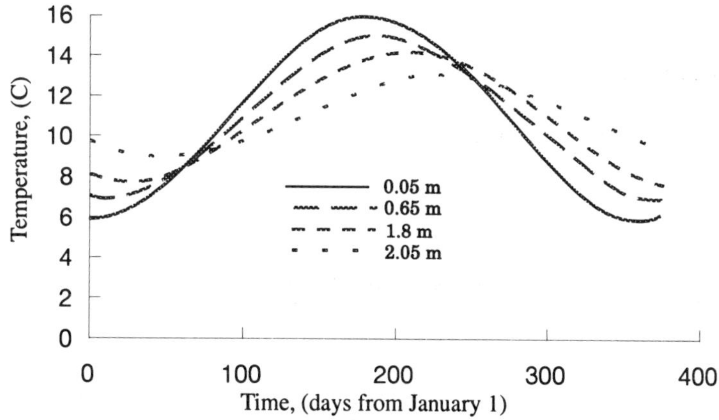

Figure 9.3. Lower frequency fluctuations in the order of one year penetrate much deeper into the soil than the daily fluctuations. The appreciable amplitude remains at a depth of 2 m.

$$T(0.10,12) = \left[15 - 5e^{-\frac{7.78}{m} \times 0.10\,m} \cos\left(\pi - \frac{7.78}{m} \times 0.10\ m\right)\right] {}^\circ C$$

$$= \left[15 - 5 \times 0.459 \times \cos(\pi - 0.778)\right] {}^\circ C = 16.6^\circ C \quad 9.101$$

and at 0.40 m

$$T(40,12) = \{15 - 5 \times 0.0445 \times \cos[\pi - (3.112)]\} {}^\circ C = 14.8^\circ C \quad 9.102$$

While the temperatures at 0.1 m and 0.4 m are nearly the same, the exponential decay term has a much larger magnitude at the 0.1 m depth, hence, considerably larger temperature fluctuations should be expected.

Determination of Thermal Diffusivity

As indicated earlier, homogeneity and isotropy are assumptions made to simplify the solution of the differential equation describing heat flow in the soil. Example 9.6 showed that, under certain assumptions, an average thermal conductivity may be calculated for soils that are comprised of distinct layers or horizons. The thermal diffusivity, α, has an integrating effect on the soil heat regime. This can be used in an inverse manner to obtain information on soil thermal parameters, providing measurements of soil temperature are available at several depths.

Equation 9.36, the general model for temperature in a homogeneous, isotropic medium, can be recast in a slightly different form involving only a single trigonometric component using trigonometric addition formulas. The general technique is illustrated in example 9.14.

Example 9.14. Use of trigonometric identities
Reduce the expression:

$$A\cos(\theta) + B\sin(\theta) \quad 9.103(a)$$

to one involving only a cosine function.

Solution
From elementary trigonometry:

$$A\cos(\theta - \phi) = A\cos(\theta)\cos(\phi) + A\sin(\theta)\sin(\phi) \quad 9.103(b)$$

If we define:

$$a = A\cos(\phi) \quad \text{and} \quad b = A\sin(\phi) \quad 9.103(c)$$

Then:

$$\phi = \arctan\left(\frac{b}{a}\right) \quad 9.103(d)$$

Furthermore, by squaring and adding both expressions above:

$$a^2 + b^2 = A^2\cos^2(\phi) + A^2\cos^2(\phi)$$
$$= A^2\left[\cos^2(\phi) + \sin^2(\phi)\right] \qquad 9.103(e)$$
$$= A^2$$

Therefore, expressions (d) and (e) provide a relationship between the phase angle, ϕ, and the amplitude, A, of the equality:

$$a_n\cos\left[\frac{2n\pi}{t}\tau - \left(\sqrt{\frac{n\pi}{\alpha t}}\right)z\right] + b_n\sin\left[\frac{2n\pi}{t}\tau - \left(\sqrt{\frac{n\pi}{\alpha t}}\right)z\right]$$
$$= A_n\cos\left[\frac{2n\pi}{t}\tau - \left(\sqrt{\frac{n\pi}{\alpha t}}\right)z - \phi\right] \qquad 9.104$$

By virtue of example 9.14, equation 9.81 becomes:

$$T(z,\tau) = A_o + \sum_{n=1}^{\infty}\left[\exp - \left(\sqrt{\frac{n\pi}{\alpha}t}\right)z\right]A_n\cos\left\{\frac{2n\pi}{t}\tau - \left[\left(\sqrt{\frac{n\pi}{\alpha t}}\right)z + \phi_n\right]\right\} \qquad 9.105$$

The amplitude A_n is now the absolute amplitude of the temperature fluctuation. The expression:

$$\left(\frac{\sqrt{n\pi}}{\alpha t}\right)z + \phi_n$$

is the lag angle that is evidenced as the thermal fluctuation penetrates the soil mass. Either of the expressions, when used with suitable measurements of temperature fluctuations in a soil, can provide average values for the thermal diffusivity. In particular, if the maximum amplitudes A are measured at two depths, z_1 and z_2, where $z_1 < z_2$ and the mean temperature is subtracted so that the difference in maximum amplitude $\Delta A = A_{n1} - A_{n2}$ is found, from the model for heat flow in the soil, it can be inferred that:

$$A_{n_2} = A_{n_1}\exp\left[\left(-\sqrt{\frac{n\pi}{\alpha t}}\right)\Delta z\right] \qquad 9.106$$

Since the only part of the model affecting the maximum amplitude, A_n of the *nth* harmonic (as the angular frequency $2n\pi/t$ is usually termed) is the exponential damping term, based upon equation 9.106, the thermal diffusivity can be found from:

$$\frac{A_{n_2}}{A_{n_1}} = \exp\left[-\left(\sqrt{\frac{n\pi}{\alpha t}}\right)\Delta z\right] \qquad 9.107$$

or inverting to eliminate the negative sign in the exponential:

$$\frac{A_{n_1}}{A_{n_2}} = \exp\left[\left(\sqrt{\frac{n\pi}{\alpha t}}\right)\Delta z\right] \qquad 9.108$$

and after taking the logarithms of both sides:

$$\ln \frac{A_{n_1}}{A_{n_2}} = \left[\left(\sqrt{\frac{n\pi}{\alpha t}}\right)\Delta z\right] \qquad 9.109$$

which can be squared and rearranged to solve for α. Performing these operations yields

$$\alpha = \left(\frac{n\pi(\Delta z)^2}{t \cdot \left(\ln \frac{A_{n_1}}{A_{n_2}}\right)^2}\right) \qquad 9.110$$

where

n = the harmonic going with the angular frequency $2n\pi/t$
t = the maximum length of record from which the Fourier series solution was derived
$\Delta z = z_2 - z_1, z_1$

Example 9.15. Estimating thermal diffusivity from actual data
A soil temperature record shows that the maximum temperature at a depth of 0.05 m was 30.5° C. At 2.00 m the maximum temperature was 27.4° C. The average temperature at this location was 21.2° C. Estimate the thermal diffusivity from these data.

Solution
By the model in equation 9.110, assuming t = 365 days and n = 1 (the first harmonic term in a Fourier Series model covering one year's daily temperature data):

$$\alpha = \frac{\pi}{365\ \text{d}} \times \frac{\text{d}}{24\ \text{h}} \times \left(\frac{2.0\ \text{m} - 0.6\ \text{m}}{\ln\left[\frac{(30.5 - 21.2)}{27.4 - 21.2}\right]}\right)$$

$$= 0.0042 \frac{\text{m}^2}{\text{h}} \qquad 9.111(a)$$

Note: This value is an average value for what are probably several horizons of soil. Nevertheless, the value is in reasonable agreement with values found by other investigators for loam soils.

The lag term in equation 9.105 also can be used to determine the thermal diffusivity of several horizons. The difference in time of occurrence of the peak temperature at two depths is determined by the difference in the phase lag term $[(n\pi/\alpha t) - \Delta z + \phi_n]$ at the two depths. The phase lag $[(n\pi/\alpha t)^{1/2}\ \Delta z + \phi_n]$ may be compensated for by a decrease in the angular frequency that may be expressed as $(2n\pi/t)\Delta\tau$, where $\Delta\tau$ is the time difference between the occurrence of the maximum temperature of the *nth* harmonic at depth z_1 and the maximum temperature at depth $z_2 > z_1$.

To estimate thermal diffusivity from this approach, proceed as follows. Consider at time τ_1 that the maximum temperature occurs at depth z_1. The argument of the cosine term in equation 9.105 must be:

$$\left[\frac{2n\pi}{t}\tau_1 - \left(\sqrt{\frac{n\pi}{\alpha t}}\right)z_1 - \phi_n\right] = 0\ (\text{or } 2\pi) \qquad 9.111(b)$$

At some later time $\tau_1 + \Delta\tau$ the maximum temperature diminished in magnitude and occurs at depth z_2. The argument of the cosine term is now:

$$\left[\frac{2n\pi}{t}(\tau_1 + \Delta\tau) - \left(\sqrt{\frac{n\pi}{\alpha t}}\right) z_2 - \phi_n\right] = 0(\text{or } 2\pi) \qquad 9.111(c)$$

Since the arguments at the two times corresponding to the two depths are depths are equal, we can equate them, expand, and cancel like terms, to get:

$$\frac{2n\pi}{t}\tau_1 - \left(\sqrt{\frac{n\pi}{\alpha t}}\right) z_1 - \phi_n = \frac{2n\pi}{t}\tau_1 + \frac{2n\pi}{t}\Delta\tau_1 - \left(\sqrt{\frac{n\pi}{\alpha t}}\right) z_2 - \phi_n \qquad 9.111(d)$$

Which reduces to:

$$\left(\sqrt{\frac{n\pi}{\alpha t}}\right)(z_2 - z_1) = \frac{2n\pi}{t}\Delta\tau \qquad 9.111(e)$$

Substituting $\Delta z = z_2 - z_1$, and squaring both sides yields:

$$\frac{n\pi}{\alpha t}\Delta z^2 = \left(\frac{2n\pi}{t}\right)^2 (\Delta\tau)^2 \qquad 9.111(f)$$

After solving for α:

$$\alpha = \frac{\frac{n\pi}{t}}{\frac{4n^2\pi^2}{t^2}} \frac{(\Delta z)^2}{(\Delta\tau)^2} \qquad 9.111(g)$$

Example 9.16. Estimation of thermal diffusivity from actual data
From the same climatological data as used in example 9.15, the maximum temperature at 0.6 m occurred on 12 July. At 2.0 m, the maximum was delayed until 6 August. Since the records are given only to the nearest day, some error is to be expected. However, assume the dates given are representative and estimate the thermal diffusivity, α, from the available information.

Solution
We use the model derived in equation 9.111(a). Again assume that t = 365 days with n = 1 (the first harmonic). Under these assumptions and given the conditions stated:

$$\Delta z = (2.0 \text{ m} - 0.6 \text{ m}) = 1.4 \text{ m}$$

$$\Delta\tau = (31 - 12 + 6) = 25\text{d}$$

Therefore:

$$\alpha = \frac{t}{4\pi}\left(\frac{\Delta z}{\Delta\tau}\right)^2 = \frac{365\text{d}}{4\pi}\left(\frac{1.4}{25}\right)^2 \frac{\text{m}^2}{\text{d}^2} \times \frac{\text{d}}{24\text{h}} = 0.0038\frac{\text{m}^2}{\text{h}} \qquad 9.112$$

Although the two are not exactly equal, the agreement is reasonably good. In using either of the two techniques, it is desirable to have hourly temperature data to ensure that the temperature observed is actually a daily maximum and not a value that is influenced by a higher harmonic such as would be produced by the daily temperature cycle.

We note that higher harmonics, i.e., larger values of n, would not give better accuracy to the determination of α since, with increasing n, the penetration of the temperature fluctuation into the soil rapidly diminishes. At shallow depths given sufficient data, the daily (t = 24 h) temperature curve may be used to estimate α for depths less than the damping depth. In any case, care is advised in the application of equations 9.106 and 9.110.

Exercises

9.1. A particular soil has a thermal conductivity that varies with depth and can be modeled with $k_z = k_1 - k_2 e^{cz}$. Particularize the differential equation of heat flow [9.5] if $k_x = k_y$, $\partial T/\partial y = 0$ and $\dot{q} = 0$. Assume that k_1, k_2, and c are constants.

9.2. Describe a physical medium that might be homogeneous with respect to thermal conductivity but is definitely anisotropic. Could you conceive of a soil mass that might possess such properties?

9.3. Assume that the specific heat and bulk density of a layered soil can be expressed as a volumetric average of the soil and obtain an expression for the thermal diffusivity of a two-layered soil with depths Δz_1 and Δz_2. Use the expression for thermal conductivity derived in example 9.6.

9.4. Show that:

$$T_o \left[\exp\left(\sqrt{\frac{n\pi}{\alpha t}}\right) z\right] \cos\left[\left(\frac{n\pi}{\alpha t}\right)\tau - \left(\sqrt{\frac{n\pi}{\alpha t}}\right) z\right]$$

is a solution by differentiating and substituting into equation 9.23.

9.5. Using either equation 9.58 or equation 9.59, prove the validity of equation 9.54.

9.6. Assume that instead of a cosine function in example 9.7, it was desired to use a sine function. For the same conditions, derive the expression that will model the temperature fluctuations at the ground surface.

9.7. Assume time will be specified in hours from midnight on 1 January and express the ground surface temperature if the mean temperature is 5° C, the magnitude of the average yearly fluctuation is 5° C, and an average daily fluctuation of 7° C is superimposed on the yearly fluctuation. Use a sinusoidal function and assume the maximum temperature occurs at 12:00 M.

9.8. If the solution in exercise 7 represents a Fourier Series description of temperature fluctuations at the earth's surface, using the general solution to temperature fluctuations in the soil, find the particular solution for $\alpha = 5.76 \times 10^{-4}$ m²/s.

9.9. Assume clouds passing across the sky obscure the sun periodically at about 15-min intervals so a sine wave of period 15 min is introduced into the series representing temperature at the soil surface. What is the damping depth of such a fluctuation in energy input if the soil is a clay loam with $\alpha = 1.693 \times 10^{-7}$ m²/h.

9.10. From figure 9.2, estimate the thermal diffusivity from the damping depth formula using the amplitude at 0.05 m and at 0.65 m. To what do you attribute any deviations that may exist between your value and the given value?

9.11. From figure 9.2, estimate the thermal diffusivity using the lag formula, choosing your own depths.

9.12. Assume the daily temperature fluctuations in a soil can be described by the model in example 9.8. What are the maximum and minimum temperatures that one could expect at a depth of 4 cm near a germinating seed about 20 May. Assume 365 days per year.

9.13. Geiger (1965) indicates that the thermal conductivity of very dry clay may fall as low as 1.8×10^{-4} m²/h. What is the daily damping depth for very dry clay with this value of thermal diffusivity? [3.7 cm]

9.14. What range of temperature fluctuations could be expected at 0.04 m if the surface temperature varied according to:

$$T = 5°\text{C} + 7°\text{C}\exp\left[-\left(\frac{\pi}{24\alpha}\right)^{\frac{1}{2}} z\right] \sin\left[\left(\frac{2\pi}{24}\right)\tau - \left(\sqrt{\frac{\pi}{24\alpha}}\right) z - \frac{\pi}{2}\right]$$

and $\alpha = 1.8 \times 10^{-4}$ m²/h.

Chapter 10

Water Movement in the Soil

The movement of water in soil is important to the agricultural and biological systems engineer. This chapter will develop models for water movement from basic concepts and equations. Our development will enable the engineer to modify the soil environment to complement the physiological needs of the green plant.

From an engineering point of view, many important problems can be handled with knowledge of the physics of water movement. The water loss by seepage through irrigation canal walls is a consideration in areas where water needs are critical. The rate at which irrigation water can be applied and the depth to which it will penetrate are both problems requiring knowledge of water flow in soils. In arid regions where furrow irrigation is practiced, the lateral water movement of water away from the irrigation furrow must be known so optimum plant placement can be effected. This will ensure an adequate supply of water to the plant without waterlogging the root system.

In humid regions water must be removed from the soil so soil pore space can be made available for the diffusion of gases into and out of the soil-root system. Removing excess water from the soil mass is a problem of subsurface drainage. The design of drainage systems requires an understanding of water flow.

The movement of water in soil occurs in two ways, each caused by different forces. Seepage from canals and ditches and the removal of excess soil water by drainage occurs due to saturated flow. In saturated flow, all soil pore space is filled with water and the potential driving water movement is due to gravity.

Water movement continues after saturated flow has ceased due to unsaturated flow. The movement of water through the soil mass toward plant roots is a problem of unsaturated flow and is important in the understanding of the overall plant-soil-water relationships.

Model Development

The differential equation governing specific water flow situations can be derived in a manner completely analogous to that used to develop the heat flow equation. The development is based on the conservation of mass in the form:

rate at which water enters a volume element	+	rate of change of water content in volume element	=	rate at which water leaves a volume element	10.1

The second term in equation 10.1 commands particular attention. Two things can affect the rate of change of water content within a volume of soil. Water may appear or disappear from the soil volume due to a source (or sink) within the volume element such as a leak from a pipe or the withdrawal of water by plant roots. In addition, the soil pore space may retain some of the water that passes through, or it may lose some of the water already present. In the development of the equation governing water movement, both possibilities must be considered. We will proceed as follows.

The change in the flux of water that will occur if a source (or sink), $\dot{Q}$, is described as a volume rate of addition (or withdrawal) per unit of volume of soil multiplied by the size of the volume element, dx x dy x dz.

$$\text{Effect of a source (or sink)} = \dot{Q}(dx \times dy \times cz) \qquad 10.2$$

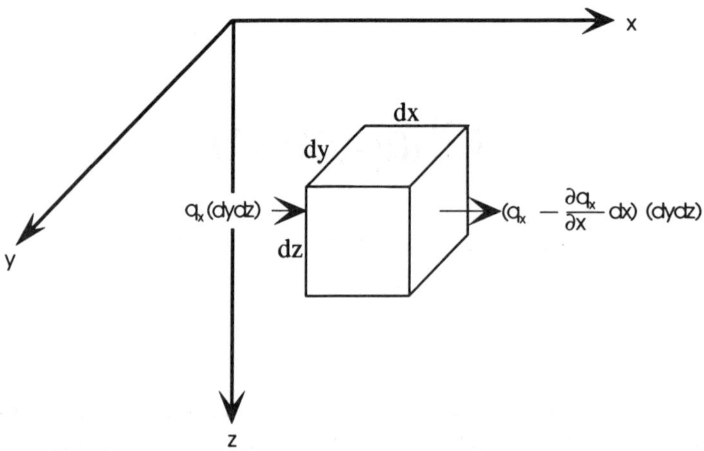

Figure 10.1. A schematic of the effect of an elemental unit of soil on the movement of water. The figure could be duplicated in each of the three cartesian coordinates.

The change in the flux that will occur if water is being stored or withdrawn from the soil volume can be modeled as a change in the volume of water ΔV_{water} in the soil volume occurring over some duration of time $\Delta\tau$, or:

$$\text{Change due to storage or withdrawal} = \frac{\Delta V_{water}}{\Delta \tau} \qquad 10.3$$

A model for water movement must be obtained from a solution of the partial differential equation governing water flow. To obtain the differential equation for the flow of water in the soil we refer to figure 10.1. In figure 10.1, water is shown entering a volume of soil, V_{soil}, dx × dy × dz. This is modeled as a flux of water, q_x, multiplied by the cross section of the volume element dy × dz.

$$q_x \times (dy \times dz) = \text{rate at which water enters the volume element} \qquad 10.4$$

Because of what may be happening to the flow within the volume element, the rate of flow exiting the volume element is written

$$\left(q_x + \frac{\partial q_x}{\partial x} dx\right) dy \times dz = \text{rate at flux exits the volume element} \qquad 10.5$$

The difference between the entering flux and the exiting flux, i.e., the difference between equation 10.4 and 10.5 must be equal to the sum of equations 10.2 and 10.3 or:

$$\left[q_x - \left(q_x + \frac{\partial q_x}{\partial x} dx\right)\right] dy \times dz = \dot{Q}\,(dx \times dy \times dz) + \frac{\Delta V_{water}}{\Delta \tau} \qquad 10.6$$

After cancelling and dividing through by V_{soil} and adopting the convention:

$$\frac{V_{water}}{V_{soil}} = \Delta\,\theta \qquad 10.7$$

where θ is the volumetric water content of the soil, equation 10.6 can be written:

$$-\frac{\partial q_x}{\partial x} = \dot{Q} + \frac{\Delta\,V_{water}}{\Delta\,V_{soil} \times \Delta\,\tau} = \dot{Q} + \frac{\partial\theta}{\partial\tau} \qquad 10.8$$

where the limit as $\Delta\tau$ approaches zero has been taken to obtain the partial differential form.

$$-\frac{\partial q_x}{\partial x} = \dot{Q} + \frac{\partial\theta}{\partial\tau} \qquad 10.9$$

While flow in only one dimension is used to obtain equation 10.9, it should be obvious that if flow were occurring simultaneously in all three cartesian coordinate directions, the three dimensional form:

$$-\frac{\partial q_x}{\partial x} - \frac{\partial q_y}{\partial y} - \frac{\partial q_z}{\partial z} = \dot{Q} + \frac{\partial\theta}{\partial\tau} \qquad 10.10$$

would be obtained. This is left as an exercise.

As in most engineering modeling, we must adopt the phenomenological equation which models the flux. Then this can be used as the basis for further development. For this purpose, a first-order phenomenological form (Darcy's Law) similar to that encountered in studies of other phenomena is assumed to hold for the movement of water:

$$q_s = -K_s\frac{\partial\psi}{\partial s} \qquad 10.11$$

where

q_s = water flux in the "s" direction
K_s = phenomenological coefficient, hydraulic conductivity
ψ = total soil water potential
s = coordinate in the direction of flow, i.e., s = x, y, or z

To complete the derivation and obtain the differential equation for water flow in soil, we make use of the phenomenological form given in equation 10.11. Substituting this expression with appropriate subscripts into equation 10.10 yields:

$$-\frac{\partial}{\partial x}\left(-K_x\frac{\partial\psi}{\partial x}\right) - \frac{\partial}{\partial y}\left(-K_y\frac{\partial\psi}{\partial y}\right) - \frac{\partial}{\partial z}\left(-K_z\frac{\partial\psi}{\partial z}\right) = \dot{Q} + \frac{\partial\theta}{\partial\tau} \qquad 10.12$$

Example 10.1. Adoption of the model for water flow
Write equation 10.12 as though it were applied to flow only in the z-direction.

Solution
First, note that the effect of a source (or sink) and the effect of changes due to storage or withdrawal are unaffected by the choice of flow direction, therefore, equation 10.2 and 10.3 can be used as they are shown above. With respect to the volume rate of flow, the only portion of equation 10.5 that changes is the subscript indicating direction, and the coordinate direction with which the partial derivative is taken. Therefore:

$$\left(q_x + \frac{\partial q_x}{\partial x} dx\right) dy \times dz = \text{rate at flux exits the volume element}$$

Thus, for flow in the x-direction, we have:

$$-\frac{\partial q_z}{\partial z} = \dot{Q} + \frac{\partial \theta}{\partial \tau}$$

Saturated Flow

We first consider the application of equation 8.8 to a soil completely saturated with water. We make the following assumptions to simplify equation 10.12.

Consider first the right side of equation 10.12. The volumetric water content of a soil is a function of the volume of pore space within the soil that is filled with water. By definition, in a saturated condition all the pores are filled with water. Since water is incompressible, the volumetric water content, θ, is constant. The partial derivative of θ with respect to time vanishes. Therefore, the last term on the right of equation 10.12 vanishes for a saturated soil.

For saturated flow we may be interested in the path water follows in moving through a soil mass en route from or to a source or a sink. It is often possible to choose the boundaries of the system in such a manner that no sources or sinks exist within the system but instead occur at the boundaries and become incorporated into the boundary conditions. In such a case the term also vanishes:

$$\dot{Q} = 0 \qquad 10.13$$

Homogeneity and Isotropy

The hydraulic conductivity of a saturated soil is not a function of the water content. However, it may be that the soil is not homogeneous or isotropic with respect to the hydraulic conductivity. The homogeneity of a medium will relate to a given parameter such as the hydraulic conductivity, and the term is used with respect to a coordinate direction. As explained in chapter 9, homogeneity implies that a parameter is independent of position within the medium. If a medium is homogeneous with respect to a parameter such as hydraulic conductivity without specifying a coordinate direction, the implication is that the medium is homogeneous in all directions.

Isotropy is a vector property and refers to a property with regard to at least two coordinate directions. In referring to hydraulic conductivity, isotropy implies that the parameter has the same magnitude in two or more flow directions. If the term is used without specification of the directions, the implication is that the medium is isotropic with regard to all coordinate directions.

It is shown in advanced texts that by suitable transformations applied to the coordinate variables, a nonhomogeneous soil can be treated as though it were homogeneous, providing the soil is isotropic. Likewise, if a soil is anisotropic but homogeneous, a coordinate transformation can be made that will enable one to treat the soil as isotropic. Although no theoretical method of handling a soil which is both nonhomogeneous and anisotropic is known, there are methods that can be used in many situations. Therefore, for practical purposes, it is sufficient to assume that a saturated soil is homogeneous with regard to hydraulic conductivity. This assumption implies that K_x, K_y, and K_z are not a function of position, so that:

$$\frac{\partial K_x}{\partial x} = \frac{\partial K_y}{\partial x} = \frac{\partial K_z}{\partial x} = 0 \qquad 10.14$$

The assumption that the soil is isotropic implies that the value of hydraulic conductivity does not change with the direction of flow, therefore:

$$K_x = K_y = K_z \qquad 10.15$$

We can apply the results expressed in equations 10.14, 10.15, and 10.16 to equation 10.12 to obtain, for saturated flow in a homogeneous, isotropic medium:

$$\frac{\partial^2 \psi_{tot}}{\partial x^2} + \frac{\partial^2 \psi_{tot}}{\partial y^2} + \frac{\partial^2 \psi_{tot}}{\partial z^2} = 0 \qquad 10.16$$

Equation 10.16 is Laplace's equation named after the mathematician who first derived it.

Soil Water Potential

In addition to the simplifications above, we must examine the nature of the total soil water potential function, ψ_{tot}, for saturated flow. From the development in chapter 4, the total soil water potential is:

$$\psi_{tot} = \psi_v + \psi_s + \psi_p + \psi_m + \psi_g \qquad 10.17$$

where

ψ_{tot} = total soil water potential
ψ_v = atmospheric component due to vapor
ψ_s = component due to the presence of solutes
ψ_p = component due to pressure of confinement in a cell
ψ_m = soil matric component (includes the component due to surface tension)
ψ_g = gravitational component (includes both hydrostatic and positional potential)

As in any situation involving the water potential function, we must determine which components of the water potential apply. For the case of saturated soil, no vapor is present, therefore, ψ_v is zero. Usually the soil solution for a saturated soil is extremely dilute and ψ_s approaches zero. There is no confining membrane for flow through soil and, therefore the ψ_p term is zero. Further, no curved air interfaces exist because the soil is completely saturated so that the soil matric component is absent. For saturated flow, the only term which remains is the gravitational component, ψ_g so the water potential function is:

$$\psi_{tot} = \kappa \rho g h \qquad 10.18$$

where

ρ = density of liquid water (kg/m^3)
g = acceleration of gravity
h = distance to the free water surface measured from some arbitrary reference plane
κ = conversion factor to express the water potential in appropriate units

For the range of temperatures encountered in the plant environment, the density of water, ρ, can be considered constant. Under the assumptions discussed previously we can write:

$$\kappa \rho g \left(\frac{\partial^2 h}{\partial x^2} + \frac{\partial^2 h}{\partial y^2} + \frac{\partial^2 h}{\partial z^2} \right) = 0 \qquad 10.19$$

After dividing by ρgh and rewriting the head to include the coordinates as a reminder that the water potential is a function of position:

$$\frac{\partial^2 h(w,y,z)}{\partial x^2} + \frac{\partial^2 h(w,y,z)}{\partial y^2} + \frac{\partial^2 h(w,y,z)}{\partial z^2} = 0 \qquad 10.20$$

Because of the simplifications which were made, the differential equation governing saturated flow through a homogeneous, isotropic medium reduces to an equation for which the only dependent variable is the hydrostatic pressure at some point h(x,y,z) expressed in terms of the distance from the datum to the free water surface. This value, h(x,y,z), is commonly called the head.

Within a saturated soil, hydrostatic pressure is the potential causing flow to occur. For saturated conditions, the hydraulic conductivity does not enter into the solution of equation 10.17. Therefore, the solution to equation 10.20 is a function of only the boundary conditions which relate to: (1) the geometry of the medium under consideration, and (2) the value of hydrostatic pressure or head, h, at the boundaries. The solution to equation 10.20 defines the head, h, as a function of position, i.e., the solution is a function giving the potential at every point within the system. Therefore, a function that satisfies equation 10.20 is called a potential function.

One-dimensional Saturated Flow

To examine the implication of equation 10.20 as it relates to flow through a saturated medium we refer to the simple flow situation given in figure 10.2. In figure 10.2, the coordinate axes are are chosen so x measures distance in the direction of flow and y is oriented vertically. The y distance at any point along the x-axis from the datum to the free water surface is the head or potential and the hydrostatic pressure at any point along the vertical within the medium is the distance from that point to the free water surface. Note, however, that the potential, h, is the distance from the datum to the free water surface, thus, at any point along the vertical, the potential is the same. We will show this by solving the Laplace's equation for the conditions on figure 10.2.

In figure 10.2, flow in a direction perpendicular to the plane of the paper is assumed to be uniform. Furthermore, since flow is defined to occur only parallel to the x-axis, equation 10.20 becomes:

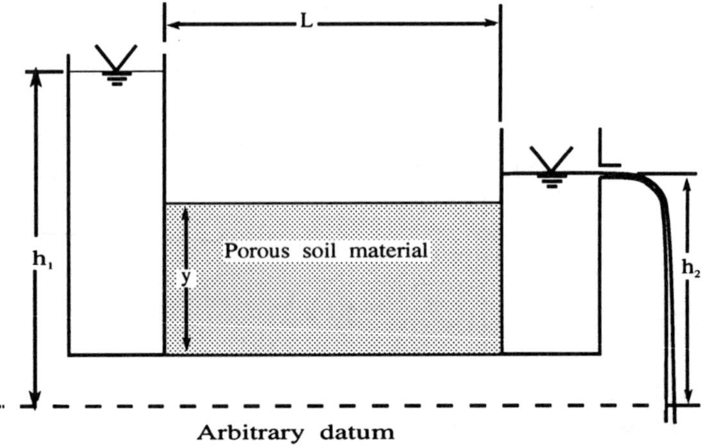

Figure 10.2. Water flowing horizontally through a porous soil media under the influence of a constant boundary condition at the inlet and outlet.

$$\frac{\partial^2 h(x)}{\partial x^2} = 0 \qquad 10.21$$

At the boundaries:

$$h(0) = h_1$$
$$h(L) = h_2 \qquad 10.22$$

A solution to equation 10.21 is straightforward. Integrating equation 10.21 once yields:

$$\frac{\partial h}{\partial x} = C_1 \qquad 10.23$$

A second integration yields:

$$h = C_1 x + C_2 \qquad 10.24$$

From the first of the boundary conditions (eq. 10.22):

$$h_1 = C_1 \times 0 + C_2 = C_2 \qquad 10.25$$

By the second boundary condition:

$$h_2 = C_1 \times L + h_1$$
$$\text{or} \qquad 10.26$$
$$C_1 = \frac{h_2 - h_1}{L}$$

The solution is an expression for the potential, h, as a function of position within the medium:

$$h = \left(\frac{h_2 - h_1}{L}\right) x + h_1 \qquad 10.27$$

Now the solution can be used to determine the lines of constant value of potential, h, as illustrated in figure 10.3. The solution (eq. 10.27), clearly shows that the head is a function only of the distance measured along the flow path, x. In figure 10.3, it is seen that for a given x_n, the y distance gives a value of potential or head $h(x_n)$ measured from the datum.

If the hydraulic conductivity, K, of the medium in figure 10.2 is known, the water flux can be found by differentiating the potential function given in equation 10.27 and multiplying the result by the negative of the hydraulic conductivity to obtain:

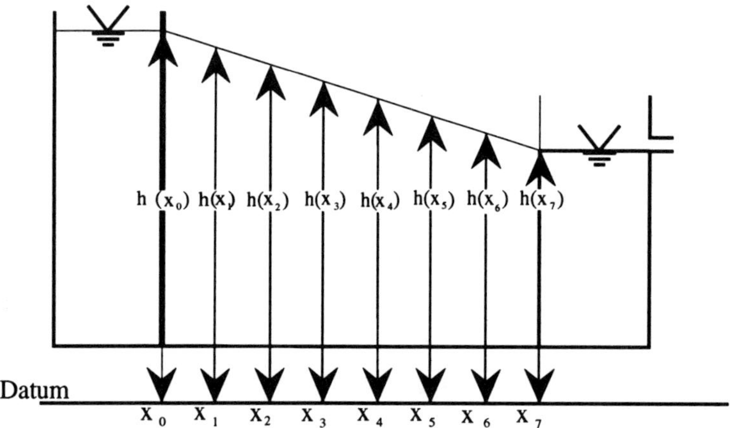

Figure 10.3. A schematic showing the lines of constant potential. Differences in adjacent potentials cause the water to flow. Conversely, the dissipation of potential as water moves through the media is responsible for the potential decrease in the direction of flow.

$$q_x = -K\left(\frac{h_2 - h_1}{L}\right) \qquad 10.28$$

Flow Net

We indicated above that the flow in figure 10.2 is parallel to the x coordinate. If lines in the direction of flow are drawn and superimposed on the lines of equal potential, the resulting diagram is called a flow net. In a homogeneous and isotropic medium, the flow lines are always perpendicular to the lines of equal potential. This is because flow occurs in the direction of the steepest gradient and is determined by the shortest distance between the equipotential lines.

A flow net is important for several reasons. In a flow net, individual flow paths are bounded by the flow lines and no liquid can escape from one flow path to another. The result is that the velocity of the moving fluid is proportional to the distance between flow lines. Since the amount of potential lost is proportional to the speed with which the liquid is flowing, the potential must decrease most quickly where the flow lines are close together.

The potential within a medium is independent of the hydraulic conductivity as shown in the development leading to equation 10.21 and is a function only of the geometry and boundary conditions. With some practice, it is often possible to sketch a flow net for a known geometry without actually solving equation 10.20. Flow lines and equipotential lines in a homogeneous, isotropic medium will always intersect at right angles, since the direction of flow is in response to the maximum gradient of potential and this always occurs where adjacent potential lines are closest together. If the flow net is drawn so the sides of the figure formed by intersecting flow and equipotential lines are very nearly equal, given an estimate of the hydraulic conductivity of the medium, the rate of flow through the medium can be estimated. The method can be illustrated by reference to figure 10.4. The gradient of potential can be expressed as:

$$\frac{h_i - h_{i+1}}{\Delta s} = \frac{\Delta h}{\Delta s} \qquad 10.29$$

Therefore, the flux of water for a unit depth into the paper, using the first-order phenomenological form that was assumed to apply, is:

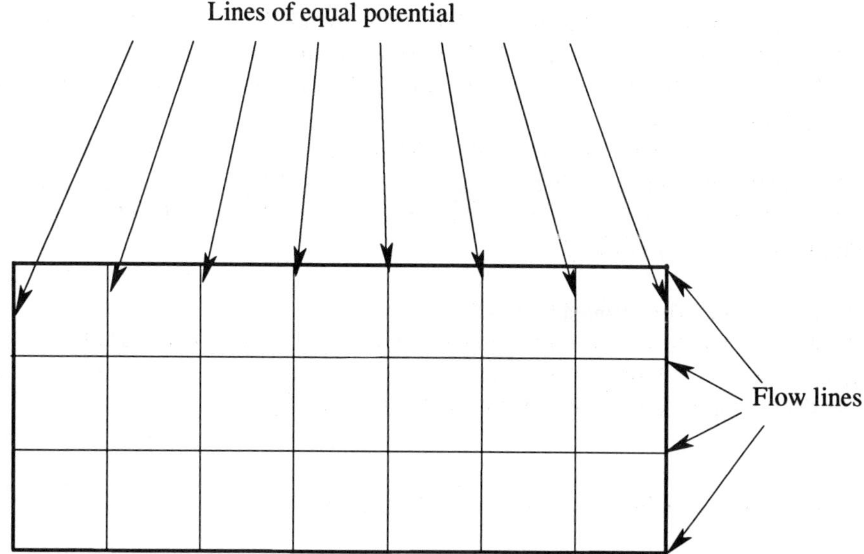

Figure 10.4. The flow net for the situation depicted in figure 10.2. In a homogeneous, isotropic medium, flow, and equipotential lines always meet at right angles.

$$q = K \frac{\Delta h}{\Delta s} \qquad 10.30$$

To get the total amount of water seeping through the cross-section, we must multiply the flux by the flow area. For the geometry shown in figure 10.4, assuming a unit of depth into the paper:

$$A = \Delta n \times N_f \qquad 10.31$$

where Δn is the distance between lines of flow and N_f represents the number of flow spaces. After multiplying equation 10.29 by equation 10.30:

$$Q_w = K \times N_f \times \frac{\Delta n}{\Delta s} \times \Delta h \qquad 10.32$$

where Q_w denotes the total seepage rate and the negative sign has been disregarded because the ratio is greater than zero. If, as we suggested above:

$$\Delta n = \Delta s \qquad 10.33$$

then the fraction $\Delta n / \Delta s$ is unity. Since Δh is the potential difference between equipotential lines and the equipotential lines are approximately equal distances apart:

$$\Delta h = \frac{h_1 - h_2}{N_e} \qquad 10.34$$

where $(h_1 - h_2)$ is the total difference in head between the flow boundaries, and N_e is the number of equipotential spaces. After substituting equation 10.33 into equation 10.31 and rearranging slightly, the flow rate is a function of the ratio of the number of flow paths to the number of equipotential spaces multiplied by the total head difference and by the hydraulic conductivity.

Often, a flow net can be sketched for a two-dimensional geometry so the sides of the figures formed by intersecting flow lines and equipotential lines are approximately equal. By counting flow paths and equipotential spaces, an estimate of the total amount of water seeping through the medium can be obtained from equation 10.33 providing the hydraulic conductivity is known.

Example 10.2. Estimating flow from a flow net
Assume the datum for the flow situation illustrated in figure 10.2 is such that h = 2 m while h = 1.8 m. If y(0) – y(L) = 1 m, L = 1.75 m, and K = 0.000 01 m/s, estimate the water flux using (a) the analytic approach, and (b) the approximation obtained from the flow net (figure 10.4).

Solution
(a) We can apply the model obtained in equation 10.22 to obtain:

$$q = -K\frac{h_2 - h_1}{L}$$

$$= 0.000\,01\ \frac{m}{s} \times \frac{-0.2\ m}{1.75\ m} \qquad 10.35$$

$$= 1.14 \times 10^{-6}\ \frac{m^3}{m^2 \cdot s}$$

(b) According to the approximate approach described by equation 10.27:

$$Q_w = K \times \frac{N_f}{N_e} \times (h_1 - h_2)$$

$$= 0.000\,01\ \frac{m}{s} \times \frac{4}{7} \times 0.2\ m$$

$$= \frac{1.0 \times 10^{-6}\ m^3}{s \cdot m_{depth}}$$

For a thickness $y_1 - y_2 = 1$ m:

$$q = \frac{1.0 \times 10^{-6}\ m^3}{1\ m \times 1\ m_{depth} \times s}$$

$$= \frac{1.0 \times 10^{-6}\ m^3}{m^2 \cdot s}$$

Several properties of flow are useful as a guide for sketching a flow net. If the inlet and outlet are submerged, the boundaries of the porous medium at these locations are equipotential lines. Furthermore, impermeable boundaries are flow lines since no water can flow across an impermeable boundary. No water will flow across a geometry that possesses a line of symmetry. Therefore, the line of symmetry also must act as a flow line and the intersection of all flow lines with equipotential lines must form right angles.

The above method cannot be used where the flow lines would intersect a boundary at an angle different from 90° such as would occur in a well where the water level in the well due to pumping is at a lower elevation than the free water surface in the soil exterior to the well. In such a case a surface of seepage occurs, and such a surface is neither a flow line nor an equipotential line. Complex mathematical techniques must be used to solve this flow problems and the reader is referred to advanced texts for more information on such techniques.

Two-dimensional Saturated Flow into an Unlined Drain

The movement of water through the homogeneous, isotropic saturated soil into an unlined drain can be modeled analytically. The physical situation is shown in figure 10.5.

A solution to fit the conditions as shown in figure 10.5 by using a mathematical model involves particularizing the general solution of a differential equation to accommodate the boundary conditions. In figure 10.5, with the coordinate axes oriented so the z is positive vertically with the x-axis oriented down the centerline of the drain, there is no variation with respect to the y-coordinate. Therefore, equation 10.17 can be simplified by using:

$$\frac{\partial Q_h}{\partial y} = 0$$

Thus equation 10.17 reduces to the two-dimensional LaPlace's equation:

$$\frac{\partial^2 \psi}{\partial x^2} + \frac{\partial^2 \psi}{\partial z^2} = 0 \qquad 10.36$$

The boundary conditions necessary to solve equation 10.36 for the physical situation given in figure 10.5 can be determined by assuming the datum from which the water potential for the saturated case is to be measured is taken at z = 0. Then the water potential at the boundaries of our system are:

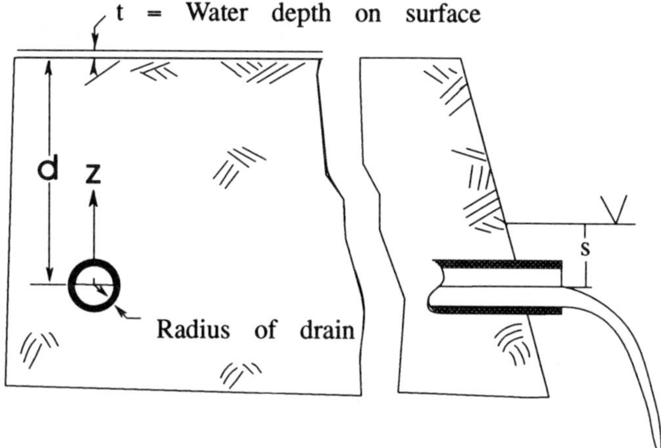

Figure 10.5. The configuration for a single buried subsurface drain in a homogeneous isotropic soil. The drain outlets into a ditch in which the free water surface is a distance, s, above the centerline of the drain. The centerline is the datum for the potential.

$$h = d + t, \text{ when } \quad z = d$$
$$h = s, \quad \text{when } x^2 + z^2 = r^2 \qquad 10.37$$

Solution

A general solution to equation 10.37 for the boundary conditions given in 10.35 is known and very little conceptual value would be gained by its derivation. A detailed derivation can be found in other texts such as Hagen et al. (1967) and will not be repeated here. Fitting the boundary conditions is done as follows.

The general solution is:

$$h = 2C\left(\ln\left[\frac{\sqrt{x^2 + z^2}}{r}\right] - \ln\left[\frac{\sqrt{x^2 + (2d - z)^2}}{r}\right]\right) + B \qquad 10.38$$

which has two undetermined constants, B and C. The two boundary conditions given in equation 10.37 can be used to determine the particular solution to the situation in figure 10.5. To obtain the particular solution, note that the first boundary condition is independent of x. Therefore, we assume x = 0. Under this condition, by applying the first boundary condition to equation 10.38:

$$d + t = 2C\left[\ln\left(\frac{\sqrt{d^2}}{r}\right) - \ln\left(\frac{\sqrt{d^2}}{r}\right) + B\right]$$

Reduces to:

$$B = d + t \qquad 10.39$$

The second boundary condition is more difficult to apply because either z = +r or z = −r can be used. To simplify the second boundary condition, we will assume that it is being applied at x = 0. For the first logarithmic term no problem is created since either value of z gives the argument of the logarithm as unity. However, to obtain a reasonable value for the argument of the second logarithm it is convenient to approximate the boundary condition by taking the average value of z using both ±r at x = 0. This operation yields for the argument of the second logarithmic term:

$$\frac{\frac{(2d - r) + (2d + r)}{2}}{r} = \frac{2d}{r}$$

So that the second boundary condition applied to the second logarithmic term yields:

$$s = 2C\left[-\ln\left(\frac{\sqrt{(2d)^2}}{r}\right) + d + t\right]$$

After some manipulation, application to equation 10.38 yields:

$$C = \frac{1}{2} \frac{d + t - s}{\ln\left(\frac{2d}{r}\right)} \tag{10.40}$$

The particular solution can be written by substituting equation 10.39 and equation 10.40 into equation 10.38 and rearranging to eliminate r from the denominator of the logarithmic terms. It is convenient to eliminate the radical from the argument of the logarithmic terms whereupon the solution becomes:

$$h = \left(\frac{d + t - s}{\ln\left(\frac{2d}{r}\right)}\right) \frac{1}{2}\left\{\ln\left[x^2 + z^2\right] - \ln\left[x^2 + (2d - z)^2\right]\right\} + d + t \tag{10.41}$$

Equation 10.41 is the solution of Laplace's equation in two dimensions particularized for the situation depicted in figure 10.5.

It is possible to model the rate of flow into the single drain installed in a homogeneous isotropic soil. We can apply the first-order phenomenological form that we assumed in the derivation of Laplace's equation, i.e.:

$$q_z = -K_z \frac{\partial h}{\partial z}$$

The solution can be obtained by noting that at the soil surface z = d, all flow takes place vertically and, therefore, there is no variation of the potential h with respect to the x-coordinate direction. Thus:

$$\frac{\partial h}{\partial x} = 0$$

The conservation of mass and the absence of any sources or sinks in the saturated soil mass requires that all water that eventually enters the drain must pass through the surface.

The flow rate through the surface then must equal the flow rate into the drain and we need only to determine the flow into the surface to know what is being drained from the soil. To find the flow through the surface, we must determine the potential gradient at the soil surface be differentiating equation 10.41 with respect to z. This yields:

$$-\frac{\partial h}{\partial z} = \frac{1}{2}\left[\frac{d + t - s}{\ln\left(\frac{2d}{r}\right)}\right] \times \left[\frac{2z}{(x^2 + z^2)} + \frac{2(2d - z)}{x^2 + (2d - z)^2}\right] \tag{10.42}$$

where the gradient is negative downward because the z-axis was taken positive upward. Since we are determining the gradient of the potential at the soil surface, we evaluate $\partial h/\partial z$ at z = d to obtain:

$$-\frac{\partial h}{\partial z} = \frac{2(d + t - s)}{\ln\left(\frac{2d}{r}\right)} \times \frac{d}{x^2 + d^2} \tag{10.43}$$

The flux of water through the surface can now be determined by substituting equation 10.43 into equation 10.41. The result is:

Table 10.1. Flux into an unlined drain as a function of distance from the centerline of the drain for a drain buried 1 m deep

Distance (m)	0	1	2	3	4	5
Relative flux	1	0.50	0.20	0.10	0.059	0.03 8

$$q_z = K_z \left(-\frac{\partial h}{\partial z}\right)\Bigg]_{z \infty d} = \frac{2K_z(d+t-s)}{\ln \frac{2d}{r}} \times \frac{d}{x^2 + d^2}$$

$$= \frac{2K_z(d+t-s)}{\ln \frac{2d}{r}} \times f(x) \qquad 10.44$$

The flux defined by equation 10.44 has the form of a constant multiplied by a function of the horizontal distance x from the drain, f (x). Figure 10.6 is a graph of f(x). It is readily seen that most of the value of f(x) comes from close to the drain since at x >> 0 the value of the function rapidly approaches zero. For example, for a drain buried 1 m deep, i.e, d = 1 m, immediately over the drain at x = 0, the function has the value unity. Therefore, the flux into the drain immediately over the centerline of the drain is equal to:

$$q_z = \frac{2K_z(1 + t - s)}{\ln\left(\frac{2}{r}\right)}$$

However, at a distance of 2 m laterally away from the drain on either side, the integrand is only 1/5 as large as it is immediately over the drain:

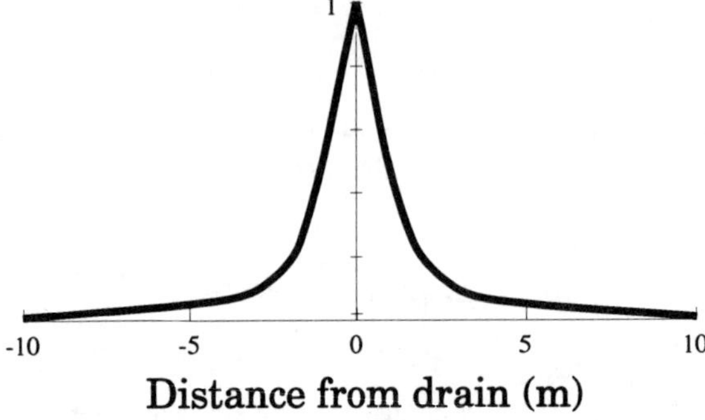

Figure 10.6. The integrand magnitude of the flow solution to a single, unlined drain buried 1 m deep in a homogeneous, isotropic medium. The flow decreases to almost nothing with increasing distance away from the drain.

$$q_z = \frac{2K_z(1 + t - s)}{\ln\left(\frac{2}{r}\right)} \times \frac{1}{5}$$

Thus the flux of water through the surface is reduced to 0.2 of the value which exists over the drain. The magnitude of the flux continues to decrease as the inverse square of the distance away from the drain, as shown in table 10.1.

The total flow rate into the drain for a unit length (say 1 m) of drain can be obtained by integrating the flux over the surface from $-\infty < x < \infty$. Equation 10.44 can be used to obtain the integrand:

$$Q = \int_{-\infty}^{\infty} q dx$$

$$= \frac{2K_z(d + t - s)}{\ln\left(\frac{2}{r}\right)} \int_{-\infty}^{\infty} \frac{d}{x^2 + d^2} dx$$

Since the function under the integral is an even function, it can be multiplied by 2 and integrated from the zero to infinity

$$Q = \frac{4K_z(d + t - s)}{\ln\left(\frac{2d}{r}\right)} \int_{0}^{\infty} \frac{d}{x^2 + d^2} dx \qquad 10.45$$

Equation 10.45 can be readily evaluated using a table of integrals:

Example 10.3. Calculation of flow into an infinite, homogeneous, isotropic medium.
Determine the flow into a 0.1 m diameter drain in a homogeneous, isotropic soil by evaluating the integral of equation 10.44. Assuming the drain is 1 m deep, the soil is saturated to the surface but no ponding is occurring, and the water level at the outlet is level with the top of the drain. Assume the hydraulic conductivity is 0.02 m/h.

Solution
By reference to a table of integrals, the integral in equation 10.44 is evaluated as:

$$\int_{0}^{\infty} \frac{d}{\eta^2 + d^2} d\eta = \frac{\pi}{2}$$

If there is no ponding at the soil surface, $t = 0$. Further, if the depth of water at the outlet is level with the top of the drain, $s = r$. Therefore:

$$Q = \frac{2 \times \pi \times 0.02 \frac{m}{h} \times (1\ m + 0 - 0.05\ m)}{\ln\left(\frac{2 \cdot 1\ m}{0.05\ m}\right)}$$

$$= 0.0323622 \frac{m^3}{m \cdot h} = 0.0033 \frac{m^3}{m \cdot h}$$

Unsaturated Flow from a Water Table

We now will investigate equation 10.12 to see how it is applied to obtain a model for water movement in an unsaturated soil. Let us begin by looking at the parameters associated with equation 10.12 as they are influenced by unsaturated soil conditions.

The water content of a saturated soil is a function of only the amount of pore space available to be filled with water. This is not true for an unsaturated soil. In unsaturated soils the water in the soil forms a film over individual soil particles and particles contact each other necks of water are formed. The result is that the water content is a function of the thickness of the water film over the particles and of the effective radius of the curved air-water interfaces formed by the water at the points of particle contact.

As water is supplied to, or removed from, an unsaturated soil, the thickness of water film and radius of the air-water interface change, thereby changing the volumetric water content of the soil. The result is that the volumetric water content, θ, becomes a time-varying parameter. The term $\partial\theta/\partial\tau$ must remain in equation 10.12 when it is to be used to obtain a model for flow in an unsaturated soil.

Whether Q, the source or sink term, need be included must be determined by the conditions peculiar to the problem being considered. Certainly, depending on the physiological needs of the plant, a soil mass with plant roots present may have water removed from the mass as a result of the action of root hairs. Thus, a sink term may be necessary for proper application of equation 10.12. Problems arise in this case because of the difficulty of locating the exact positions of the roots in the soil mass. For purposes of the equation development, we will consider a soil with no sources or sinks. These assumptions will permit development of the concepts associated with unsaturated flow in soils.

Hydraulic Conductivity of Unsaturated Soil

The hydraulic conductivity of an unsaturated soil differs from that of a saturated soil for several reasons. In a saturated soil, movement is from pore to pore and the hydraulic conductivity is, in part, determined by the size and arrangement of the pore spaces. The movement of water through unsaturated soils, however, takes place along the water films surrounding soil particles, through the water at the points of contact of soil particles, and in the form of water vapor. Since the water content of an unsaturated soil is related to the thickness of water films, and since the amount of water that can be moved along a film depends on the thickness of the film, the hydraulic conductivity of an unsaturated soil depends on the water content of the soil.

The discussion during the development of the water potential function has shown that water content and water potential are related. From what has been stated, we find a complex situation in which the hydraulic conductivity is related to the water potential distribution in the soil. For this reason, the hydraulic conductivity, even for a homogeneous, isotropic unsaturated soil, is not a constant, but is a function of position and therefore cannot be removed from the partial differentiation.

Therefore, equation 10.12 for an unsaturated soil that is assumed to be homogeneous and isotropic becomes:

$$\frac{\partial}{\partial x}\left(K\frac{\partial\psi}{\partial x}\right) + \frac{\partial}{\partial y}\left(K\frac{\partial\psi}{\partial y}\right) + \frac{\partial}{\partial z}\left(K\frac{\partial\psi}{\partial z}\right) = \frac{\partial\theta}{\partial\tau} \qquad 10.46$$

Let us determine the proper form of the water potential function that applies for an initially saturated but draining homogeneous isotropic soil. If the soil mass contains no semipermeable barriers, then the osmotic component, ψ_s, of the water potential does not exist. Also, within a saturated soil the only pressure term present normally is hydrostatic pressure and, therefore, ψ_π pressure due to confinement is zero.

As a soil looses water and changes from the saturated to the unsaturated phase, the capillary component of the matric potential, ψ_τ, appears as curved air-water interfaces occur in the soil pores.

As a soil loses additional water, the matric potential becomes more and more predominant. A state of water content is soon reached for which the contribution to the total water potential due to matric potential, ψ_m, exceeds the contribution due to gravitational forces, ψ_g. When this occurs, with regard to equation 10.45, any gravitational contribution is experienced by the water only for flow in a vertical direction. Purely horizontal flow within the soil mass is a function only of ψ_m the matrix potential. Therefore, equation 10.46 can be written:

$$\frac{\partial}{\partial x}\left(K\frac{\partial\psi_m}{\partial x}\right) + \frac{\partial}{\partial y}\left(K\frac{\partial\psi_m}{\partial y}\right) + \frac{\partial}{\partial z}\left(K\frac{\partial\psi_m + \psi_g}{\partial z}\right) = \frac{\partial\theta}{\partial\tau} \qquad 10.47$$

where the z-coordinate is oriented vertically in the direction of gravitational attraction. The resulting equation is in a form from which we may begin to examine properties of unsaturated flow in homogeneous, isotropic soils.

We note in passing that, although the same basic equation must apply to water flow in plants, the presence of cells whose walls contain semipermeable membranes requires including ψ_s and ψ_π as components of the total water potential. For green plants, therefore, a significantly more complex form of the equation must be solved. No solutions of the general equation are known for flow in plants at the present time.

Example 10.4. The capillary component of matrix potential
One end of a column of a loamy sand is placed in a container of water. After sufficient time has elapsed to reach a state of relative equilibrium, moisture is found to have risen to 0.45 m above the level of the water. (a) What is the water potential just inside the wetting front in the column? (b) What is it at 0.40 m above the level of the water? (c) What is it at a point 0.50 m above the level of the water? Assume the datum for ψ_g is the free water surface.

Solution
(a) At a point inside the wetting front the water potential must be composed of only the matrix potential component, and, therefore, must be equal to the height column of water being supported by the curved air-water interfaces in the soil matrix:

$$\psi_{45} = \psi_t = -0.45\ m_{H_2O}$$

$$= -\frac{0.45\ m}{\frac{10.33\ m}{100\ kPa}}$$

$$= -4.4\ kPa$$

(b) At a point 0.40 m above the water level, ψ_τ, must support 0.05 m less water, so:

$$\psi_{40} = -0.45\ m_{H_2O} + 0.05\ m_{H_2O} = -0.40\ m_{H_2O}$$

$$= \frac{-0.40\ m}{\frac{10.33\ m}{100\ kPa}}$$

$$= -3.9\ kPa$$

(c) At a point 50 cm above the water level the water potential will be a function of the water content at that point. This information is not given so no determination can be made.

Solution for Unsaturated Flow from a Water Table

To develop an appreciation of the problems associated with mathematical modeling of the unsaturated flow of water through soils, we will examine a situation familiar to many researchers. The techniques used by those who have been successful in obtaining a solution for the unsaturated flow case will be useful to illustrate specific points important for an understanding of the unsaturated flow process.

The physical situation that can be modeled using the approach presented here is one in which water is moving upward through the soil mass at some constant rate. Such a situation would occur where a water table was present below the surface at some depth, d, and water moved to the surface where evaporation was occurring at some constant rate, q, per unit area of the surface. A similar situation involves movement from a water table at depth d upward to the lower boundary of the root zone of a crop, under the assumption that the root zone absorbs water as a plane sink at some constant rate of per unit area. The solution could be used to determine the rate at which water might be withdrawn by a crop. Figures 10.7a and b illustrate the nature of the assumptions for which the model is derived.

For the situation described, equation 10.46 can be modified as follows. Since unsaturated flow occurs in the vertical direction only, no variation of the flux occurs along either the x- or y-axes. Therefore:

$$\frac{\partial}{\partial z}\left(\frac{K(\psi_m + \psi_g)}{\partial z}\right) = \frac{\partial \theta}{\partial \tau}$$

Although the water content of the soil above the water table will vary with position, for the situation under examination a steady-state case where no variation in water content occurs with time is assumed to exist. Thus:

$$\frac{\partial \theta}{\partial \tau} = 0$$

is appropriate.

Referring to figure 10.5, note that the coordinate system chosen is such that the x-y plane is at the water table and the z-axis is positive upward. The gravitational component of the water potential must be negative. We will use this information in a later stage of the derivation and solve the differential equation. The simplifications made thus far enable equation 10.47 to be written:

$$\frac{\partial}{\partial z}\left(K\frac{\partial(\psi_m + \psi_g)}{\partial z}\right) = 0 \qquad 10.48$$

To obtain the boundary conditions necessary to solve equation 10.47, note that it would be very useful to determine the maximum flux, which could be maintained from the location of the free water surface in the soil where $\psi = 0$ to the plane of the roots, where ψ is at some very large negative value. The boundary conditions can be written:

$$\begin{aligned} q &= q_{max}; \quad 0 \leq d \\ \psi &= \infty; \quad \text{all } \tau \end{aligned} \qquad 10.49$$

Equation 10.48 can be integrated once to obtain:

$$K\frac{\partial(\psi_m + \psi_g)}{\partial z} = q_{max} \qquad 10.50$$

where the constant of integration is the upward flux of water, since the left side of equation 10.49 is the first-order phenomenological form (Darcy's Law) that governs the flux upward from the water table.

To facilitate the remainder of the solution, it is useful to modify equation 10.50. As was done earlier, the water potentials ψ_m and ψ_g can be written as a head of water rather than a pressure.

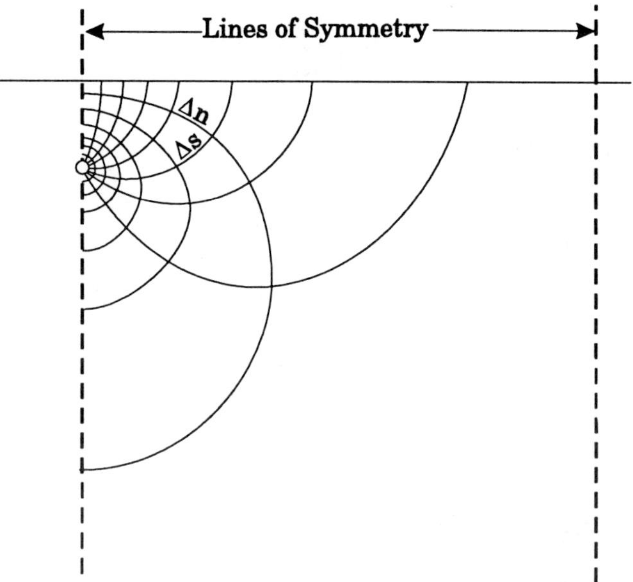

Figure 10.7. Equipotential lines and flow lines for a drain in an infinitely deep, homogeneous, isotropic soil. The flow and equipotential lines, though curved, meet at right angles and form "curvilinear" squares. Only one-half of the flow net is shown because of symmetry about the vertical axis.

After making this change, and noting that the z-axis has been chosen positive upward while the force of gravity is directed downward, we can express equation 10.50 as:

$$K\left(\frac{\partial(\psi_m)}{\partial z} - 1\right) = q_{max} \qquad 10.51$$

This result follows because the gravitational contribution to the water potential gradient, $\partial h/\partial z$, is unity since the dimensions of h are the same as those of the z-coordinate direction.

Because the soil is not completely saturated at values of water potential less than zero, the paths that water must follow in moving through the soil need to be altered. First, since the pores are not completely filled with water, water can only move by films that cover the soil particles. The films are obviously thinner and offer less cross-sectional area for flow to occur than the pores in a saturated soil. The result is that the hydraulic conductivity is less than it would be for a saturated situation. Second, no longer is it possible for a molecule of water to move directly from pore to pore. Rather, the molecule must travel over the surface of each particle of soil, a more tortuous path than it followed when the soil was saturated. The result is that the hydraulic conductivity is reduced still more.

It is obvious that the hydraulic conductivity of unsaturated soil must depend on the amount of water in the soil as does the water potential in the unsaturated condition. The result is that the hydraulic conductivity must be expressed as some function of the water potential. A convenient expression is one proposed by Gardner:

$$K(\psi) = \frac{c}{[(-\psi)^n + b]} \qquad 10.52$$

Equation 10.51 can be simplified. Rearrange the equation to read:

$$\frac{\partial \psi}{\partial z} = \frac{q_{max}}{K(\psi)} + 1$$

Then multiply both sides by dz and note:

$$d\psi = \frac{\partial \psi}{\partial z} = dz$$

Equation 10.51 becomes:

$$dz = \frac{d\psi}{\frac{q}{K(\psi)}} + 1$$

After using the expression for $K(\psi)$ given in equation 10.51:

$$dz = \frac{d\psi}{1 + \frac{b \times q_{max}}{c} + (-\psi)^n \frac{q_{max}}{c}} \qquad 10.53$$

where b, c, and n are constants for a given soil, and ψ, the soil water potential, is expressed in dimensions of length (pressure head). Note that since ψ is zero for a saturated soil, equation 10.53 makes it apparent that the hydraulic conductivity of a soil decreases with decreasing soil water content and matrix potential. The value of b in equation 10.53 is very small, while n ranges from 2 for fine-textured soils to 3 or 4 for coarser soils. For the present development, we will assume equation 10.53 governs, but leave c, b, and n undetermined to obtain the most general form for the solution. To obtain a solution, equation 10.53 must be integrated from $z = 0$ where $\psi = 0$ to $z = d$ and according to the boundary conditions assumed, $\psi = -\infty$:

$$\int_0^d dz = \int_0^\infty \frac{d\psi}{1 + \frac{b \times q_{max}}{c} + \frac{(\psi)^n \times q_{max}}{c}} \qquad 10.54$$

Integration of equation 10.54 is not straightforward and is beyond the scope of this text. However, it has been solved. Cisler (1969) found with some rearrangement and the use of an expanded table of integrals that equation 10.54 could be integrated to yield:

$$d = \left[\frac{b^{\frac{1}{n}}}{n} \frac{\pi c}{bq_{max}} \left(\frac{c}{b \cdot q_{max}} + 1\right)^{\left(\frac{1+n}{n}\right)} \frac{1}{\sin\left(\frac{\pi}{n}\right)}\right] \qquad 10.55$$

When ψ is very near zero, b and ψ may be approximately the same magnitude. However, as the soil has less and less water, i.e., as ψ becomes very small and approaches zero, the absolute value of $(-\psi)^n$ becomes large. In accordance with the rules to determine significant digits when numbers are summed, the significance of b becomes negligible. In this case:

$$\left(\frac{c}{b \times q_{max}} + 1\right) \infty \frac{c}{b \times q_{max}}$$

The solution, equation 10.55 can be written approximately as:

$$d = \left(\frac{c}{q_{max}}\right)^{\frac{1}{n}} \times \frac{\pi}{n \times \sin\left(\frac{\pi}{n}\right)}$$

From which the maximum flux can be obtained. This yields:

$$q_{max} = c\left[\frac{\pi}{n \times d \times \sin\frac{\pi}{n}}\right]^{n} \qquad 10.56$$

Equation 10.55 was obtained by assuming that the matrix potential in the soil water at a distance d from the water table was infinite. In practice, with ψ sufficiently small but still much greater than negative infinity, the value of the integrand in equation 10.54 is very small. Thus, the limiting case is approached rapidly at values of soil water tension much greater than negative infinity. Depending on the values of n and c for the soil being considered, the integrand approaches zero within the range of soil water potentials present in the atmosphere above the soil surface. Therefore, equation 10.56 can serve as a model to determine the maximum flux of water that can be furnished to supply evaporation from a soil surface for a water table at a depth d below the surface.

Another problem for which the model can be used is to determine at what distance below the bottom of the plant root system a water table should be maintained to insure a minimum flux of water upward to the plant to supply evapotranspiration needs. This is demonstrated through the solution of a pair of example problems.

Example 10.5. Unsaturated hydraulic conductivity computation
Assume equation 10.51 is valid. Compute the hydraulic conductivity in m/d of a soil for which n = 2 for ψ of –0.5 atm, –5 atm, and –15 atm. Assume b = 0.04 m while c = 4×10^{-2} m^4/d.

Solution
Convert the water potentials to m of water:

$$\psi_{-0.5\ atm} = \frac{(-0.5\ atm)}{1} \times \frac{10.33\ m}{atm} = -5.2\ m \qquad 10.57\ (a)$$

Similarly,

$$\psi_{-5\ atm} = 5 \times 10.33 = 52\ m \qquad 10.57\ (b)$$

and:

$$\psi_{-15\ atm} = -15 \times 10.33 = -155\ m \qquad 10.57\ (c)$$

Then:

$$K_{-0.5\text{ atm}} = \frac{4 \times 10^{-4}\,\frac{m^3}{d}}{(|-5.2\text{ m}|)^2 + 0.04\text{ m}^2} = 1.5 \times 10^{-5}\,\frac{m}{d} \qquad 10.57\text{ (d)}$$

$$K_{-5\text{ atm}} = \frac{4 \times 10^{-4}}{(|-52|)^2 + 0.04} = 1.5 \times 10^{-7}\,\frac{m}{d} \qquad 10.57\text{ (e)}$$

$$K_{-15\text{ atm}} = \frac{4 \times 10^{-4}}{(|-155|)^2 + 0.04} = 1.7 \times 10^{-8}\,\frac{m}{d} \qquad 10.57\text{ (f)}$$

Example 10.6. Application to subirrigation
A crop under certain conditions of temperature and relative humidity has an evapotranspiration demand of 5 mm of water per day. If this occurs over a 10-h period and the water is supplied by a constant water table. At what depth below the surface should the water table level be maintained? Assume the following constants:

(a) for a sandy loam; $n = 3$; $c = 3.2 \times 10^{-3}\,\frac{m^4}{d}$

(b) for a clay loam; $n = 2$; $c = 1.1 \times 10^{-3}\,\frac{m^3}{d}$

Solution
A 5 mm/day rate of water removal is equivalent to a flux of:

$$5 \times 10^{-3}\,\frac{m^3}{m^2 \cdot d}$$

Using the solution to equation 10.55 for d:
(a)

$$d = \left(\frac{3.2 \times 10^{-2}\frac{m^4}{d}}{5 \times 10^{-3}\frac{m}{d}}\right)^{\frac{1}{3}} \times \left(\frac{\pi}{3 \sin\left(\frac{\pi}{3}\right)}\right)$$

$$= 2.2\text{ m}$$

(b)

$$d = \left(\frac{\frac{1.1 \times 10^{-3} m^3}{d}}{\frac{5 \times 10^{-3}\, m^3}{m^3 \cdot d}}\right)^{\frac{1}{2}} \times \left(\frac{\pi}{3 \sin\left(\frac{\pi}{2}\right)}\right)$$

$$= 0.49\text{ m}$$

We recognize that an uneven distribution and penetration of roots toward the water table will modify this requirement, since a single root deep in the soil could withdraw a significant amount of water. Also, since a clay loam will hold more water than a sandy loam, the plant growing in a clay loam could withdraw considerable water from the soil surrounding the roots, thus not requiring as much water from the water table. Over a long period, however, the plant in a clay loam would exhaust all the available water and then would have to be supplied only by upward flux from the water table.

Unsaturated Flow Toward Plant Roots

Development of the Differential Equation

Water movement from the soil to the plant root is not well understood because of the complexity of the root system and the difficulty of measuring changes in the parameters of the soil-plant continuum over small distances such as are involved in the rooting region. The differential equation derived previously, however, can be applied to a simplified situation to gain information and illustrate some of the concepts involved. The knowledge thus gained will help the engineer appreciate the behavior of a plant under conditions that may be created in the field as a result of modifying the plant environment through engineering practices.

The plant root is the absorbing surface for water intake in the root system. The exact mechanism of absorption and the resistances involved are largely unknown but, regardless of the mechanism, the source of water must be the soil matric.

Within the soil, the plant root can be thought of as an elongated cylinder whose axis may change direction many times in its total length. Over significant portions, however, the path followed by the root may be straight. Therefore the physical situation closely resembles an infinitely long absorbing cylinder. Certainly in these regions flow toward the root is two-dimensional and largely radial. The model to be examined, therefore, involves two-dimensional flow to an absorbing cylinder which has a water potential at its surface lower than that in the soil at some distance away. Due to the geometry of the system, it is logical to use cylindrical coordinates for the analysis. Equation 10.12 can be expressed in cylindrical coordinates in two dimensions as:

$$\frac{1}{r}\frac{\partial}{\partial r}\left[r\left(K_r\frac{\partial \psi}{\partial r}\right)\right] = \frac{\partial \theta}{\partial \tau} \qquad 10.58$$

where it is assumed that only radial flow toward the root takes place.

Solution of the Equation

To obtain a particular solution to equation 10.58, we assume that the following situation exists. Initially all the soil surrounding the root is at some uniform potential, ψ_m. A steady rate of flow is then demanded of the soil reservoir by the absorbing surface of the root. The magnitude of this flow is q_r, and after a time the potential distribution near the root will be modified so as to maintain the flow rate. Then the boundary conditions will become:

$$\psi = \psi_o \text{ for all r}$$

$$2\pi AK_r\frac{\partial \psi}{\partial r} = q_r \text{ for all } \tau > 0 \text{ at } r = A$$

$$2\pi RK_r\frac{\partial \psi}{\partial r} = q_r \text{ for } R > A \text{ and } \tau > 0$$

where

A = radius of the root

K_r = hydraulic conductivity in the radial direction

R = radius of the zone from which water is being absorbed by the root

ψ_o = matric potential of the soil when root begins absorbing water
q_r = flux of water being absorbed by the root [volume/(time, length of root)]
τ = time

We use dimensions volume rate per unit length of root rather than per unit area for convenience because it does not require knowing the radius of the root. Since, after a sufficiently long time, a steady-state situation will become established near the root in some region A. In this case equation 10.58 reverts to the homogeneous equation:

$$\frac{\partial}{\partial r}\left(K\frac{\partial \psi}{\partial r}\right) = 0$$

Which can immediately be integrated to obtain:

$$r \times K\frac{\partial \mu}{\partial r} = C \qquad 10.59$$

From the boundary conditions:

$$A \times K\frac{\partial \psi}{\partial r} = C = \frac{q_r}{2\pi} \qquad 10.60$$

For unsaturated conditions, when the soil water potential is significantly less than zero, the term "b" in equation 10.5 is sufficiently small that the hydraulic conductivity can be approximated as the following function of the soil water potential:

$$K(\psi) = \frac{c}{|-\psi|^n} \qquad 10.61$$

Equation 10.58 can be multiplied by dr and rearranged to obtain:

$$K\frac{\partial \psi}{\partial r}dr = \frac{q_r}{2\pi}\frac{dr}{r}$$

or:

$$Kd\psi = \frac{q_r}{2\pi}\frac{dr}{r} \qquad 10.62$$

By making use of the model in equation 10.60, we can write:

$$\frac{c}{|-\psi|^n}d\psi = \frac{q_r}{2\pi}\frac{dr}{r}$$

Which can be integrated between the limits A ± r ± R:

Table 10.2. Hydraulic conductivity and parameters for equation 10.61 for K_r

Soil Type	Hydraulic Conductivity	c for K (m/day)	c for K (m/h)	n
Chino clay	$[c/(\lvert\psi\rvert^n)]$	$1.1 \times 10^{-1}\ m^3/d$	4.58×10^{-3}	2
Yolo light clay	$[c/(\lvert\psi\rvert^n)]$	$4.0 \times 10^{-4}\ m^3/d$	1.67×10^{-5}	2
Pachappa sandy loam	$[c/(\lvert\psi\rvert^n)]$	$3.2 \times 10^{-3}\ m^3/d$	1.33×10^{-4}	3

$$c\int_{\psi}^{\psi_o} |\psi|^{-n} = \frac{q_r}{2\pi}\int_A^R \frac{d\eta}{\eta}$$

to obtain:

$$\frac{c}{(1-n)}\left(|-\psi_o|^{(1-n)} - |-\psi_A|^{(1-n)}\right) = \frac{q_r}{2\pi}\ln\left(\frac{R}{A}\right) \qquad 10.63$$

Table 10.2 gives some values for c and n for use in equation 10.61. We make use of this information to illustrate equation 10.63 for a clay loam soil where n = 2. In this case the equation becomes:

$$\frac{1}{\psi_A} - \frac{1}{\psi_o} = \frac{q_r}{2\pi c}\ln\left(\frac{R}{A}\right) \qquad 10.64$$

The term $1/\psi_A$ of 10.64 can be expanded in a Taylor series about ψ_o and the result substituted into equation 10.63 to obtain:

$$-\frac{1}{\psi_o} + \left[-\frac{1}{\psi_o} - \frac{1}{\psi_o^2} = \frac{(\psi_A - \psi_o)^2}{2 \cdot \psi_o^3} + \cdots + R_{err}\right]$$

If ψ_o is sufficiently less than its value at field capacity, say –5 atm, the terms of order greater than unity can be disregarded to obtain:

$$-(\psi_A - \psi_o) = \frac{q_r \times \psi_o^2}{2\pi \times c}\ln\left(\frac{R}{A}\right) \qquad 10.65$$

The significance of the solution can be observed by noting that the difference in potential between the root surface, ψ_A, and the soil at distance, R, ψ_o is directly related to the flux, and to the square of the initial soil water potential. Thus, in a dry soil the water potential at the root surface will have to be much lower to maintain a given q than in an initially moist soil where ψ_o has a larger (less negative) value.

If the plant physiological processes are affected by the water stress imposed on it by the soil, the dryer soil will exert more effect on the plant. Note that the ratio of radius of influence, R, to diameter of root, A, enters into the equation only as a logarithmic term, and therefore root radius has appropriately less effect upon the potential difference than does the flux, q, or the initial potential, ψ_o.

A different treatment of equation 10.63 yields information on the maximum rate a plant root can withdraw water from the soil. The maximum steady-state rate for any given radius of influence R will occur when the potential difference between the soil and the root surface has its largest value. The limiting case occurs as the magnitude of ψ_o becomes large. We investigate the limit by assuming $\psi_{A\rightarrow}\infty$. As this occurs, from equation 10.59:

$$q = \frac{2\pi c}{-\psi_o \ln\left(\frac{R}{A}\right)} \qquad 10.66$$

The withdrawal rate predicted by equation 10.66 is inversely related to the magnitude of the initial soil water potential, ψ_o exists at some distance, R, from the center of the root. It is also directly related to the coefficient c in the model for the unsaturated hydraulic conductivity.

Equation 10.66 indicates that during periods of high demand by the plant for water the soil may not be able to supply sufficient water to the root. Under these conditions, transpiration may withdraw water from the plant and cause temporary wilting. This kind of wilting is often observed during the hotter portions of the day, especially in soils that have been depleted of water and thus have a low initial soil water

Example 10.7. Potentials near an absorbing root
A bean plant occupies about 0.09 m^2 of the surface area of a Yolo light clay soil and transpires 2 mm water/day. Assume transpiration occurs over a 12-h period every day and the root radius is 0.01 mm while the radius of influence is 10 mm. If the active water absorbing length of the root system is 10 m, what potential exists between the soil and the root to maintain a steady-state flux equivalent to the transpiration rate under each of the following conditions:

(a) If the initial soil water potential is –200 kPa?

(b) If the initial soil water potential is –500 kPa?

(c) If the radius of influence is increased to 50 mm, what is the answer to part (a)?

Solution
First compute the flux, q, to the absorbing surface of the root:

$$q = \frac{2\ \text{mm}}{12\ \text{h}} \times 0.09\ \text{m}^2 \times 10^6\ \frac{\text{mm}^2}{\text{m}^2}$$

$$= 1.5 \times 10^4\ \frac{\text{mm}^3}{\text{h}}$$

The flow rate per meter of absorbing root surface is:

$$\text{rate} = \frac{1.5 \times 10^4\ \frac{\text{mm}^3}{\text{h}}}{0.01\ \text{mm} \times \pi \times 10\ \text{m} \times \left(10^3\ \frac{\text{mm}}{\text{m}}\right)^2}$$

$$= 4.77 \times 10^{-4}\ \frac{\text{mm}^3}{\text{mm}^2 \times \text{h}}$$

We can apply equation 10.60 after first converting the water potentials into units of length to be consistent with the dimensions in the hydraulic conductivity model (eq. 10.51):

$$\psi_o = -200 \text{ kPa} \times \frac{10.33 m_{H_2O}}{100 \text{ kPa}}$$

$$= -20.7 \; m_{H_2O}$$

So that the potential difference for part (a) is:

$$\psi_A - \psi_o = \frac{4.8 \times 10^{-4} \dfrac{mm^3}{mm^2 \times h} \times (-20.7 \text{ m})^2}{2 \times \pi \times 0.003 \dfrac{m^2}{h}} \times \ln\left(\frac{10 \text{ mm}}{1000 \text{ mm}}\right)$$

$$= 50.2 \text{ m} \times \frac{1 \text{ atm}}{10.33 \; m_{H_2O}}$$

$$= -4.9 \text{ atm}$$

For part (b) the soil potential is initially –500 kPa, or:

$$\psi_A - \psi_o = \frac{500 \times 10.33}{100} = 51.7 \text{ m}$$

so:

$$\psi_A - \psi_o = \frac{4.8 \times 10^{-4} \times (-51.7)^2}{2 \times \pi \times 0.003} \times \ln\left(\frac{10}{1000}\right)$$

$$= \frac{313}{10.33} \text{ atm}$$

$$= -30.3 \text{ atm}$$

If R = 50 mm, then:

$$\ln\left(\frac{R}{A}\right) = \ln\left(\frac{50}{1000}\right)$$

Thus, (a) becomes:

$$\psi_A - \psi_o = \frac{4.8 \times 10^{-4} \times (-20.7)^2}{2 \times \pi \times 0.0013} - 2.9957$$

$$= -32.7 \text{ m}$$

$$= -3.2 \text{ atm}$$

Note that the radius of influence of the root has little effect on the answer. Decreasing the initial soil water potential results in a much larger potential difference to maintain the same flux to the root. Little is known about the total effective length of root in a growing plant. It is possible that other portions of the root would become active in response to a decreased initial soil water potential, thus lowering the flux per unit length of root.

Exercises

10.1. Derive the equivalent of equation 10.8 assuming the volume element in figure 10.1 is situated in a region of soil where the flow is downward only.

10.2. Why is the boundary of a submerged inlet or outlet a line of constant potential?

10.3. The figure below depicts a situation in which a stream is diked to prevent flooding. The dike is built on a layer of sand over a layer of clay. The drainage ditch on the field side of the dike is pumped to prevent flooding. For the geometry shown, what height would you expect water to rise in tubes driven into the sandy layer at the locations shown? Assume the datum is the bottom of the drainage ditch.

10.4. If the sand has a hydraulic conductivity of 0.29 m/h and the dike borders the field for a distance of 0.3 km, at what rate (in m^3/h)?

10.5. At what rate must the ditch be pumped to maintain the water at the level shown? Assume the only source of water into the drainage ditch is the stream.

10.6. Show that equation 10.30 is a general solution to the two dimensional Laplace's equation (10.28). (Hint: First eliminate r by performing the division indicated by the subtraction of logarithms, then eliminate the square root by multiplying the logarithm by the exponent to attain a form on which the partial differentiation may be quickly performed.)

10.7. Evaluate equation 10.37 to determine the model for flow into a single tile buried in homogeneous soil per unit of length

10.8. Compute the flow per day into 1 m of tile in figure 10.6, if the hydraulic conductivity is 2×10 Sm/h, the drain is buried 1 m deep, t = 0, while r = s = 50 mm.

10.9. A county drain commissioner in Michigan is asked to provide a random drain that is to be 1.3 km long (4265 ft). The design calls for a drain that will just flow full if the entire length of drain is contributing to the flow. If the velocity of flow in the drain is described by:

$$v = \frac{1}{0.015} \times \left(\frac{r}{2}\right)^{0.67} \times \sqrt{s}$$

where s = 0.001 m/m and r is the radius of the drain. The soil has an average hydraulic conductivity, K = 0.01 m/h and the drain is 1 m deep. What radius drain be used? (Determine this based on the drain size at the outlet.)

10.10. Based on the information given in example 10.6 and table 10.2, compute the maximum upward flux of water for a Yolo light clay soil when the water table is 0.5 m below the evaporating surface? What is the flux if the water table is 2 m below the evaporating surface?

10.11. Derive equation 10.60 for a sandy soil where n = 3.

10.12. Show that if one starts with n = 2 in equation 10.54, and assumes that b is very small, it is possible to obtain the same result as is derived from equation 10.51 or 10.52 by substituting n = 2 in these equations. Note that the integration for this special case can be done without use of an expanded table of integrals.

10.13. Assume the water flux (per unit length) into a plant root is 6×10^{-6} m^2/d. If the root radius is A = 0.01 mm, the plant is in a Yolo clay soil, and ψ_o is –8 atm, compute if:

(a) R = 10 m

(b) R = 5 mm

10.14. What is the maximum water flux (per unit length of root per hour) for the in example 9?

10.15. If the conditions are those used in example 10 and each plant occupied an average of 0.09 m^2 of soil surface, what length of root system would be needed if the plant transpired 0.4 mm/h?

Bibliography

Chapter 3

Bosen, J. F. 1960. A formula for approximation of the saturation vapor pressure over water. *Monthly Weather Rev.* 88(8):275-276.

Hyland, R. W. and A. Wexler. 1983. Formulations for the thermodynamic properties of the saturated phases of H_2O from 173.15 to 473.15 K. *ASHRAE Trans.* 89(2A):520-535. (From 1989 ASHRAE Handbook, Fundamentals. Am. Soc. of Heating, Refrigeration, and Air Conditioning Engr. 1791 Tullie Circle, N.E., Atlanta, GA 30329.)

Maxwell, J. C. 1965. Will there be enough water? *Am. Scientist* 35:97-103.

Threlkeld, J. L. 1970. *Thermal Environmental Engineering,* 2nd Ed. Englewood Cliffs, N.J.: Prentice-Hall.

Washburn, E. W., ed. 1928. International Critical Tables 3:210-212. New York: McGraw Hill.

Chapter 5

Bonner, J. and A. Galston. 1952. *Principles of Plant Physiology*. San Francisco: W. H. Freeman Co.

Clark, W. C. 1989. Managing planet earth. *Sci. Am.* 261(3):47-69

Cutter, E. G. 1969. *Plant Anatomy.* Reading, Mass.: Addison Wesley.

Dittmer, H. J. 1972. *Modern Plant Biology.* New York: Van Nostrand Reinhold Co.

Dowling, R. N. 1928. *Sugar Beet and Beet Sugar.* London: Ernest Benn.

Eames, A. J. and L. H. McDaniels. 1947. *An Introduction to Plant Physiology*, 2nd Ed. New York: McGraw-Hill Book Co.

Esau, K. 1965. *Plant Anatomy.* New York: John Wiley & Sons.

Faries, V. C. 1970. *Thermodynamics.* New York: MacMillan Co.

Fuller, H. J. and O. Tippo. 1954. *College Botany.* New York: Henry Holt Co.

Galston, A. W. 1961. *The Life of the Green Plant.* Englewood Cliffs, N.J.: Prentice-Hall.

Heck, W. W. 1966. The use of plants as indicators of air pollution. *Intern. J. Air Water Pollution* 10:99-111.

Hill, C. A. 1971. Vegetation: A sink for atmospheric pollutants. Paper presented at 64th meeting of the Air Pollution Control Assoc., Pittsburgh, Pa.

Kozlowski, T. T. 1972. *Water Deficits and Plant Growth,* Vol. 3. New York: Academic Press.

Meyer, B. S., D. B. Anderson and R. Bohning. 1960. *Elementary Plant Physiology.* Princeton, N.J.: D. Van Nostrand Co.

Narayan, C. V. and B. A. Stout. 1972. Mechanical checking of navy beans. *Transactions of the ASAE* 15(1):191-194.

Pearson, R. W. 1966. Soil environment and root development. In *Plant Environment and Efficient Water Use.* Madison, Wis.: Am. Soc. Agron. and Soil Sci. Soc. of Am.

Ray, P. M. 1972. *The Living Plant*, 2nd Ed. New York: Holt, Reinhart and Winston.

Reeve, P. A. 1964. Mechanical thinning pays. *Sugar Beet J.* 27(3):2.

Scott, R. R. 1977. *Plant Root Systems.* New York: McGraw-Hill Book Co.

Slayter R. O. 1967. *Plant-Water Relationships.* New York: Academic Press.

Tortora G. J., D. R. Cicero and H. L. Parish. 1970. *Plant Form and Function.* New York: Macmillan Co.

Wiersum, L. K. 1957. The relationship of the size and structural rigidity of pores to their penetration by roots. *Pl. Soil* 9:75-85.

Chapter 6

Bange, G. G. J. 1953. On the quantitative explanation of stomatal transpiration. *Acta. Botan. Neerl.* 2:255-297.

Carnes, A. 1934. Soil crusts. *Agr. Eng.* 15:167-169, 171.

Corbett, J. L. 1969. The nutritional value of grassland herbage. In *Assessment of and Factors Affecting Requirements of Farm Livestock*, ed. D. P. Cuthbertson. New York: Pergamon Press.

Criddle, R. S. 1969. Structural proteins of chloroplasts and mitochondria. *Ann. Rev. Plant Physiol.* 20:239.

Dixon, H. H. 1924. *The Transpiration Stream.* London: Univ. of London Press.

Flatt, W. P. 1969. Methods of calorimetry (B) indirect. In *The Science of Nutrition of Farm Livestock*, ed. D. P. Cuthbertson. New York: Pergamon Press.

Hanks, R. J. and F. C. Thorp. 1956. Seedling emergence of wheat as related to soil moisture content, bulk density, oxygen diffusion rate and crust strength. *Soil Sci. Soc. Am. Proc.* 20:307-310.

Hillel, D. 1972. Soil moisture and seed germination. In *Water Deficits and Plant Growth*, Vol. 3. ed. T. T. Kozlowski. New York: Academic Press.

Kaufmann, M. R. 1972. Water deficits and reproductive growth. In *Water Deficits and Plant Growth*, ed. T. T. Kozlowski. New York: Academic Press

Kozlowski, T. T. 1972. Shrinking and swelling of plant tissues. In *Water Deficits and Plant Growth*, Vol. 3. ed. T. T. Kozlowski. New York: Academic Press.

Meyer, B. S., D. B. Anderson and R. H. Bohning. 1960. *Introduction to Plant Physiology.* Princeton, N.J.: D. Van Nostrand Co.

Slayter, R. O. 1967. *Plant-Water Relationships.* New York: Academic Press.

Weatherley, P. F. 1965. Discussion on terminology. In *Water Stress in Plants*, ed. B. Slavik. Proc. of Symp. in Prague, Czechoslovakia, 30 Sept.-4 Oct. 1963. The Hague, Netherlands: Dr. W. Lunk Publishers.

Wiebe, H. H. et al. 1971. Measurement of plant and soil water status. Utah Agr. Exp. Sta. Bull. 484.

Chapter 7

Bear, F. E. 1965. *Soils in Relation to Crop Growth.* New York: Van Nostrand Reinhold Co.

Black, C. A. 1968. *Soil Plant Relationships.* New York: John Wiley & Sons.

Buckman, H. O. and N. C. Brady. 1969. *The Nature and Properties of Soils*, 7th Ed. New York: Macmillan Co.

Donahue, R. L. 1970. *Our Soils and Their Management*, 3rd Ed. Danville, Ill.: Interstate Printers and Publishers

Humble, G. D. and K. Rascke. 1971. Stomatal opening quantitatively related to potassium transport. *Plant Physiol.* 48:447-453.

Kossovich, P. S. 1916. *A Short Course of General Soil Science*, 2nd Ed. Springfield, Va.: U.S. Dept. of Commerce Clearinghouse for Federal and Tech. Inform.

Millar, C. E., L. M. Turk and H. D. Foth. 1969. *Fundamentals of Soil Science*, 4th Ed. New York: John Wiley & Sons.

Chapter 8

Barrow, G. M. 1966. *Physical Chemistry.* New York: McGraw Hill Book Co.

Bates, M. 1960. *The Forest and the Sea.* New York: Random House.

Geiger, R. 1965. *The Climate Near the Ground.* Cambridge, Mass.: Harvard University Press.

Goody, R-M. 1964. *Atmospheric Radiation.* London: Clarendon Press.

Halliday, D. and R. Resnick. 1962. *Physics*, Part II. New York: John Wiley & Sons.

Kondrat'yev, K. Y. 1969. *Radiation in the Atmosphere.* New York: Academic Press.

Lemon, E. R., A. H. Glaser and L. E. Satterwhite. 1957. Some aspects of soil plant and meteorological factors to evapotranspiration. *Soil Sci. Soc. Proc.* 21:464-465.

Ritchie, J. T. 1971. Dryland evaporative flux in a subhumid climate. I. Micrometeorological influences. *Agron. J.* 63:51-62.

Robinson, N. 1966. *Solar Radiation-American.* New York: Elsevier Publishing Co.

Van Wijk, W. R. 1965. *Physics of Plant Environment.* New York: John Wiley & Son.

Chapter 9

Beskow, G. 1947. *Soil Freezing and frost heaving: with special reference to highways and railroads.* Evanston, Ill.: Northwestern Univ.

Carslaw, H. S. and J. C. Jaeger. 1959. *Conduction of Heat in Solids*, 2nd Ed. New York: Oxford Univ. Press.

Chang, J. H. 1958. *Ground temperature*, Vol. 1. Res. Rept. Quartermaster Res. and Develop. Command, U.S. Army Contract No. DA 19-129-QM-348.

Crawford, C. B. 1952. Soil temperatures, a review of published records. In *Frost action in soils,* Special Rept. 2, Proc. of Symp. 30th Ann. Meeting of Transportation Res. Board. 1951. Washington, D.C.: Natl. Acad. Sci-Natl. Res. Council.

David, R. K. 1940. Studies on soil temperatures in relation to other factors controlling the disposal of solar radiation. *Indian J. Agr. Sci.* 10: 352-357.

Geiger, R. 1950. *The Climate Near the Ground,* lst Ed. Cambridge, Mass.: Harvard Univ. Press.

————. 1965. *The Climate Near the Ground,* 2nd Ed. Cambridge, Mass.: Harvard Univ. Press.

Glinski, J. and J. Lipiec. 1990. *Soil Physical Conditions and Plant Roots,* Boca Raton, Fla.: CRC Press.

Ingersoll, L. R., O. J. Zobel and A. C. Ingersoll. 1954. *Heat Conduction.* Madison, Wisc.: Univ. of Wisconsin Press.

Johnston, C. N. 1937. Heat conductivity of soil governs heat losses from heated oil lines. *Petrol. Engr.* 9(1):41-44, 46-45.

Kersten, M. S. 1949. Thermal properties of soils. Univ. Minn., Inst. Tech. Bull. 28

Lowry, W. P. 1967. Weather and Life. Corvallis, Oreg.: Oregon State Univ.

Marshall, T. J. and J. W. Holmes. *Soil Physics.* London: Cambridge Univ. Press.

Mayer, A. M. and A. Pojakoff-Mayber. 1963. *The Germination of Seeds.* New York: Macmillan Co.

Natl. Oceanic and Atmospheric Admin. Climatological Data. U.S. Dept. of Commerce. Washington, D.C.: GPO.

Pollock, B. M. 1972. Effects of environment after sowing on viability. In *Viability of Seeds*, ed. E. H. Roberts. London: Chapman & Hall.

Smith, W. O. 1939. Thermal conductivities in moist soil. *Soil Sci. Soc. Am. Proc.* 4:32-40.

Steishnoff, C. and L. F. Hough 1965. Response of blueberry seed germination to soil temperature, light, potassium nitrate and coumarin. *Proc. Am. Hort. Soc.* 93:260-266.

Sutton, O. G. 1953. *Micrometeorology.* New York: McGraw-Hill Book Co.

Van Wijk, W. R. 1963. *Physics of Plant Environment.* New York: John Wiley & Sons.

Washburn, E. W. 1929. Thermal conductivity of nonmetallic solids. In *International Critical Tables*, Vol. 5. New York: McGraw-Hill Book Co.

Chapter 10

Cisler, J. 1969. The solution for maximum velocity of isothermal steady state flow of water upward from water table to soil surface. *Soil Sci.* 108:148

Gardner, W. R. 1958. Some steady-state solutions of the unsaturated moisture flow equation with application to evaporation from a water table. *Soil Sci.* 85:228-232.

Hagen, R. E., H. R. Hoise and T. W. Edminster. 1967. *Irrigation of Agricultural Lands.* Madison, Wis.: Am. Soc. of Agron.

Harr, M. E. 1962. *Groundwater and Seepage.* New York: McGraw-Hill Book Co.

Luthin, J. N. 1967. *Drainage of Agricultural Lands.* Madison, Wisc.: Am. Soc. of Agron.

Rose, C. W. 1966. *Agricultural Physics.* New York: Pergamon Press.

Slayter, R. O. 1967. *Plant-Water Relationships.* New York: Academic Press.

Index